菲迪克（FIDIC）文献译丛

施工合同条件
（原书 2017 年版）

Conditions of
Contract for Construction

国际咨询工程师联合会 编

唐　萍　张瑞杰　等译

（正式使用发生争执时，以英文原版为准）

机械工业出版社

本《施工合同条件》(红皮书)2017年版,是1999年版红皮书的升级版。它继承了原有合同条件的优点,延续了菲迪克(FIDIC)平衡风险分担的基本原则,并在萃取过去近20年使用中获得和总结的丰富经验的基础上编写。与1999年版合同通用条件相比,本版的合同通用条件的篇幅大幅度增加。此外,本书还具有以下特色:

1)通知和其他通信交流的要求更加详细和明确;
2)平等对待雇主和承包商的索赔,并将其与争端分别处理的规定;
3)争端避免的机制;
4)质量管理和承包商合同合规验证的详细规定。

本书内容包括施工合同条件的通用条件和专用条件编写指南,附有争端避免/裁决委员会协议书一般条件,以及各担保函格式以及投标函、合同协议书和争端避免/裁决委员会协议书等格式。

本书推荐用于由雇主或其代表工程师设计的建筑或工程项目。这种合同的通常情况是,由承包商按照雇主提供的设计进行工程施工。但这些工程可包括承包商设计的土木、机械、电气和/或构筑物的某些部分。

读者对象:工程咨询(设计)单位,从事投资、金融和工程项目管理的部门和组织,各类项目业主,建筑施工监理企业,工程承包企业,环保企业,会计师/律师事务所,保险公司以及有关高等院校等单位和人员。

版权所有。未经出版者事先书面许可,对本出版物的任何部分不得以任何方式或途径复制或传播,包括但不限于复印、录制、录音,或通过任何数据库、信息或可检索的系统。
本书封面贴有机械工业出版社和国际咨询工程师联合会(FIDIC,菲迪克)双方的防伪标签,无标签或标签不全者不得使用和销售。

北京市版权局著作权合同登记 图字:01-2020-1315

图书在版编目(CIP)数据

施工合同条件:原书2017年版/国际咨询工程师联合会编;唐萍等译. —北京:机械工业出版社,2021.3(2023.6重印)
(菲迪克(FIDIC)文献译丛)
书名原文:Conditions of Contract for Construction for Building and Engineering Works Designed by the Employer
ISBN 978-7-111-67587-7

Ⅰ.①施… Ⅱ.①国… ②唐… Ⅲ.①建筑施工-经济合同-研究 Ⅳ.①TU723.1

中国版本图书馆CIP数据核字(2021)第031985号

机械工业出版社(北京市百万庄大街22号 邮政编码100037)
策划编辑:何文军　责任编辑:何文军　时　颂
责任校对:梁　倩　封面设计:张　静
责任印制:常天培
北京铭成印刷有限公司印刷
2023年6月第1版第2次印刷
210mm×297mm・27印张・1273千字
标准书号:ISBN 978-7-111-67587-7
定价:299.00元

电话服务　　　　　　网络服务
客服电话:010-88361066　机 工 官 网:www.cmpbook.com
　　　　　010-88379833　机 工 官 博:weibo.com/cmp1952
　　　　　010-68326294　金 书 网:www.golden-book.com
封底无防伪标均为盗版　机工教育服务网:www.cmpedu.com

译者的话

本书是由国际咨询工程师联合会（FIDIC，菲迪克）编写，于 2017 年出版的第 2 版《施工合同条件》。2017 年版《施工合同条件》继承了 1999 年第 1 版菲迪克《施工合同条件》的优点，根据多年来在实践中取得的经验以及专家、学者和相关各方的意见和建议，针对 1999 年版在应用过程中产生的问题进行了修订。2017 年版《施工合同条件》的格式与 1999 年版大体保持一致，但文本篇幅增加了 80%，1999 年版保留下来的内容未做重大修改或修订。2017 年版《施工合同条件》的条款更加清晰、透明和明确，以减少合同双方争端的发生，使项目更加成功。2017 年版《施工合同条件》中，通知和其他通信交流的要求更加具体和明确，处理雇主和承包商索赔的规定更加公平，增加了争端避免的机制，详细规定了有关质量管理以及承包商合同合规验证的内容。

希望此译本的出版，对我国广大从事工程咨询（设计）、投资、金融和项目管理的部门和组织，各类项目业主，建筑施工监理企业，工程承包企业，环保企业，会计 / 律师事务所，保险公司以及有关高等院校等人员在学习和运用菲迪克合同条件，有效地解决在国际、国内工程咨询和工程承包活动中的合同管理问题，更好地开拓国内外工程咨询和工程承包市场，提高工程建设的投资效益和社会效益，建立和完善工程项目总承包制度，促进我国工程建设管理体制与国际惯例接轨，推动我国工程咨询事业和工程建设管理模式及体制的全面深化改革会有所帮助。

翻译过程中，我们虽然尽力想使译文准确通顺、能完整地传达原文的内容，汉语表达规范易懂，但限于专业知识与语言水平，译文中可能出现不妥乃至错误之处，敬请读者指正。本书以中英文对照方式编排，以便用户核对中译文，从而更准确地理解本书。

本书由唐萍、张瑞杰、贾志成、史骏、邓冰茹、张辰旭、郭文涛、李莉萍、莫伟平、郑海燕、秦春燕、曾家平、张荣芹等翻译，唐萍、张瑞杰、贾志成校译，唐萍、张瑞杰、邓冰茹、张辰旭审校。

译者

FIDIC is the international federation of national Member Associations of consulting engineers.

FIDIC was founded in 1913 by three national associations of consulting engineers within Europe. The objectives of forming the federation were to promote in common the professional interests of the Member Associations and to disseminate information of interest to their members. Today, FIDIC membership covers some 90 countries from all parts of the globe and encompassing most of the private practice consulting engineers.

FIDIC is charged with promoting and implementing the consulting engineer industry's strategic goals on behalf of Member Associations. Its strategic objectives are to: represent world-wide the majority of firms providing technology-based intellectual services for the built and natural environment; assist members with issues relating to business practice; define and actively promote conformance to a code of ethics; enhance the image of consulting engineers as leaders and wealth creators in society; promote the commitment to environmental sustainability; support and promote young professionals as future leaders.

FIDIC arranges seminars, conferences and other events in the furtherance of its goals: maintenance of high ethical and professional standards; exchange of views and information; discussion of problems of mutual concern among Member Associations and representatives of the international financial institutions; and development of the consulting engineering industry in developing countries.

FIDIC members endorse FIDIC's statutes and policy statements and comply with FIDIC's Code of Ethics which calls for professional competence, impartial advice and open and fair competition.

FIDIC, in the furtherance of its goals, publishes international standard forms of contracts for works (Short Form, Construction, Plant and Design Build, EPC/Turnkey, Design, Build and Operate) and agreements (for clients, consultants, sub-consultants, joint ventures, and representatives), together with related materials such as standard pre-qualification forms.

FIDIC also publishes business practice documents such as policy statements, position papers, guides, guidelines, training manuals, and training resource kits in the areas of management systems (quality management, risk management, integrity management, sustainability management) and business processes (consultant selection, quality based selection, tendering, procurement procedure, insurance, liability, technology transfer, capacity building, definition of services).

FIDIC organises an extensive programme of seminars, conferences, capacity building workshops, and training courses.

FIDIC aims to maintain high ethical and professional standards throughout the consulting engineering industry through the exchange of views and information, with discussion of problems of mutual concern among Member Associations and representatives of the multilateral development banks and other international financial institutions.

FIDIC publications and details about training courses and conferences are available from the Secretariat in Geneva, Switzerland, Specific activities are detailed in an annual business plan and the FIDIC website, www.FIDIC.org, gives extensive background information.

菲迪克（FIDIC）是咨询工程师国家（地区）成员协会国际联合会。

菲迪克（FIDIC）是由欧洲三个国家的咨询工程师协会于 1913 年成立的。组建联合会的目的是共同促进成员协会的职业利益，以及向其会员传播有益信息。今天，菲迪克（FIDIC）已有来自于全球各地 90 个国家（地区）的会员，包括大多数私人执业的咨询工程师。

菲迪克（FIDIC）代表成员协会负责促进和实施咨询工程师行业的战略目标。其战略目标是：代表全世界为建设和自然环境提供以技术为基础的智力服务的大多数公司；协助会员处理与业务实践相关的问题；制定并积极促进遵守职业道德规范；提升咨询工程师作为社会领导者和财富创造者的形象；促进对环境可持续性的承诺；支持和促进青年咨询工程师成为未来的领导者。

菲迪克（FIDIC）举办各类研讨会、会议及其他活动，以促进其目标：维护高水平的道德和职业标准；交流观点和信息；讨论成员协会和国际金融机构代表共同关心的问题；以及发展中国家工程咨询业的发展。

菲迪克（FIDIC）会员认可菲迪克（FIDIC）章程和政策声明，并遵守其职业道德规范要求的专业技能、公正的建议和公开公平的竞争。

菲迪克（FIDIC）为了实现其目标，发布了国际标准格式的工程合同（简明格式、施工、生产设备和设计－施工、EPC/ 交钥匙、设计、施工和运营）和协议书（针对客户、咨询工程师、分包咨询工程师、联营体和代表），以及资格预审标准格式等相关资料。

菲迪克（FIDIC）还出版比如政策声明、行动报告、指南、指导方针、培训手册和管理体系领域的培训资料包（质量管理、风险管理、廉洁管理、可持续管理）以及业务流程［咨询工程师 / 单位的选择、根据质量选择（咨询服务）、招标、采购程序、保险、责任、技术转让、实力建设、服务定义］的业务实践文件。

菲迪克（FIDIC）组织研讨会、会议、实力建设研讨会和培训课程等各类活动。

菲迪克（FIDIC）致力于通过交换观点和信息，在成员协会和多边开发银行及其他国际金融机构代表之间讨论共同关心的问题，在整个工程咨询行业维护高水平的道德和职业标准。

菲迪克（FIDIC）出版物以及培训课程和会议的详细信息，可以从设在瑞士日内瓦的菲迪克（FIDIC）秘书处得到。具体活动详见年度业务计划和菲迪克（FIDIC）网站 www.FIDIC.org，该网站提供了大量的背景信息。

COPYRIGHT

© FIDIC 2021 All rights reserved.

The Copyright owner of this document is the International Federation of Consulting Engineers - FIDIC.

Translation from English to Chinese has been performed by China Machine Press with FIDIC's permission.

The lawful purchaser of this document has the right to make a single copy of the duly purchased document for his or her personal and private use. No part of this publication may be shared reproduced, distributed, translated, adapted, stored in a retrieval system, or communicated, in any form or by any means, mechanical, electronic, magnetic, photocopying, recording or otherwise, without prior written permission from FIDIC. To request such permission, please contact FIDIC, Case 311, CH-1215 Geneva 15, Switzerland; fax +41 22 799 49 01, e-mail: fidic@fidic.org. Electronic copies can be obtained from FIDIC at www.fidic.org/bookshop.

FIDIC considers the official and authentic text to be the version in the English language and assumes no liability whatsoever for the completeness, correctness, adequacy or otherwise of the translation into Chinese for any use to which this document may be put.

Disclaimer

While FIDIC aims to ensure that its publications represent the best in business practice, the Federation accepts or assumes no liability or responsibility for any events or the consequences thereof that derive from the use of its publications, including their translations. FIDIC publications are provided "as is" without warranty of any kind, either express or implied, including, without limitation, warranties of merchantability, fitness for a particular purpose and non-infringement. FIDIC publications are not exhaustive and are only intended to provide general guidance. They should not be relied upon in a specific situation or issue. Expert legal advice should be obtained whenever appropriate, and particularly before entering or terminating a contract.

版权

©FIDIC 2021 版权所有。

本文件的版权所有人为国际咨询工程师联合会——菲迪克（FIDIC）。

经菲迪克（FIDIC）许可，中国机械工业出版社完成了英文版的中文翻译。

本文件的合法购买者有权将正式购买的文件制作副本，供其个人和私人目的使用。未经菲迪克(FIDIC)事先书面许可，不得将本出版物的任何部分分享复制、分发、翻译、改编、存储在检索系统中，或以任何格式或通过任何方式以机械、电子、磁性、影印、记录或其他方式传送。如有意获得此类许可，请联系菲迪克（FIDIC），地址：Case 311, CH-1215 Geneva 15, Switzerland；传真：+41 22 799 49 01，电子邮件：fidic@fidic.org。电子版本可从菲迪克（FIDIC）获取，网址为www.fidic.org/bookshop。

菲迪克（FIDIC）认为正式的和权威性的文本为英语版本，并对本文件的任何用途的中文翻译文本的完整性、正确性、充分性或其他方面不承担任何责任。

免责声明

尽管菲迪克（FIDIC）的目标是确保其出版物代表最佳业务实践，但对于使用其出版物及其翻译文本而引起的任何事件或后果，联合会（菲迪克，FIDIC）不承担任何责任或义务。菲迪克（FIDIC）出版物按"原样"提供，没有任何明示或暗示的保证，包括对可销售性、特定用途的适用性和非侵权性的无限保证。菲迪克（FIDIC）出版物并非详尽无遗，仅提供一般性指导。在特定的情况或问题上不应依赖它们。除此之外，适当时，尤其在签订或终止合同前，应获得专家法律建议。

CONTENTS

Acknowledgements

Notes

General Conditions

    Contents 2
    Clauses 1 to 21 12
    Appendix: General Conditions of Dispute Avoidance/Adjudication Agreement... 214
    Annex: DAAB Procedural Rules 232
    Index of Sub-Clauses 244

Guidance for the Preparation of Particular Conditions

    Contents 256
    Introductory Guidance Notes 258
    Particular Conditions Part A – Contract Data 260
    Particular Conditions Part B – Special Provisions 270
        Notes on the Preparation of Tender Documents 274
        Notes on the Preparation of Special Provisions 278
        Clauses 280
    Advisory Notes to Users of FIDIC Contracts Where the Project is to Include Building Information Modelling Systems 362
Annexes: Forms of Securities 368

Forms of Letter of Tender, Letter of Acceptance, Contract Agreement and Dispute Adjudication/Avoidance Agreement

    Letter of Tender 392
    Letter of Acceptance 394
    Contract Agreement 396
    Dispute Avoidance/Adjudication Agreement 398

目录

致谢

说明

通用条件

  目录 ·· 3
  条款 1 至 21 条 ··· 13
  附录 争端避免/裁决委员会协议书一般条件 ··· 215
  附件 争端避免/裁决委员会程序规则 ··· 233
  条款索引 ·· 245

专用条件编写指南

  目录 ·· 257
  介绍性指南说明 ·· 259
  专用条件 A 部分——合同数据 ·· 261
  专用条件 B 部分——特别规定 ·· 271
    编写招标文件注意事项 ·· 275
    编写特别规定注意事项 ·· 279
    条款 ·· 281
  项目使用建筑信息模型系统的 FIDIC 合同用户的建议说明 ····················· 363

附件 担保函格式 ·· 369

投标函、中标函、合同协议书和争端避免/裁决委员会协议书格式

  投标函 ·· 393
  中标函 ·· 395
  合同协议书 ·· 397
  争端避免/裁决委员会协议书 ·· 399

ACKNOWLEDGEMENTS

Fédération Internationale des Ingénieurs-Conseils (FIDIC) extends special thanks to the following persons who prepared, assisted and contributed in the preparation of the Second Edition of FIDIC's three standard forms of contract (Conditions of Contract for Construction, Conditions of Contract for Plant and Design-Build, and Conditions of Contract for EPC/Turnkey Projects):

FIDIC Contracts Committee's Updates Special Group:
Zoltán Záhonyi, Z&Partners Consulting Engineers, Hungary; Siobhan Fahey, Consulting Engineer, Ireland (Principal Drafter); Christoph Theune, GKW Consult GmbH, Germany; and William Howard, CDM Smith, USA (Executive Committee liaison).

Initial Update Task Group:
Svend Poulsen, Atkins/COWI, Denmark (Group Leader); Aisha Nadar, Advokatfirman Runeland AB, Sweden (task group principal drafter); Robin Schonfeld, SMEC, Australia (task group principal drafter); Darko Plamenac, Consulting Engineer, Serbia; Jan Ziepke, Consulting Engineer, Germany; and Zoltán Záhonyi, Z&Partners Consulting Engineers, Hungary (Contracts Committee liaison).

Second Stage Update Task Group:
Simon Worley, EIA Ltd., UK (Group Leader); John Greenhalgh, Greenhalgh Associates, UK; Leo Grutters, C2S Global, Germany; Aisha Nadar, AdvokatfirmanRuneland AB, Sweden; William Godwin, Matrix Seminars, UK; Siobhan Fahey, Consulting Engineer, Ireland (Contracts Committee liaison); and Shelley Adams, EIA Ltd., UK (secretary to the task group).

The preparation was carried out under the general direction of the FIDIC Contracts Committee:
Philip Jenkinson, Atkins, UK (past Chairman); Zoltán Záhonyi, Z&Partners Consulting Engineers, Hungary (Chairman); Vincent Leloup, Exequatur, France; KajMöller, SWECO, Sweden; Siobhan Fahey, Consulting Engineer, Ireland; Mike Roberts, Mott MacDonald, UK; Des Barry, Consulting Engineer, Ireland; Christoph Theune, GKW Consult GmbH, Germany; Enrico Vink, FIDIC Managing Director; and Christophe Sisto, FIDIC Design & Edition Manager.

together with liaisons to the FIDIC Contracts Committee:
Geoff French, Scott Wilson, UK (past EC liaison); Kaj Möller, SWECO, Sweden (past EC liaison); José Amorim Faria, SOPSEC, Portugal (past EFCA liaison); William Howard, CDM Smith, USA (EC primary liaison); Aisha Nadar, Advokatfirman Runeland AB, Sweden (EC secondary liaison; and Pawel Zejer, AECOM, Poland (EFCA liaison).

Advisory Notes to Users of FIDIC Contracts Where the Project is Using Building Information Modelling Systems were provided by:
Anthony Barry, Aurecon, Australia (FIDIC Executive Committee), Andrew Read, Pedersen Read Consulting, New Zealand (chairman FIDIC Business Practice Committee) and Stephen Jenkins, Aurecon, New Zealand (chairman FIDIC Risk, Liability and Quality Committee).

Special Advisers to the Contracts Committee provided invaluable and continued support in the various drafting and revision stages:
Christopher Seppälä, White & Case LLP, France (legal adviser, assisted by Dimitar Kondev, White & Case LLP, France); Nael G Bunni, Ireland (risk and insurance adviser); Axel-V. Jaeger, Germany; Michael Mortimer Hawkins, UK/Sweden; Christopher Wade, UK; Nicholas Gould and Jeremy Glover, Fenwick Elliott, UK.

致谢

国际咨询工程师联合会（FIDIC，菲迪克）向编写、协助和参与编写其三本标准合同格式（施工合同条件、生产设备和设计-施工合同条件、设计采购施工（EPC）/交钥匙工程合同条件）第 2 版的以下成员特致谢意。

菲迪克（FIDIC）合同委员会新版特别工作组：
匈牙利 Z&Partners 咨询工程师公司的 Zoltán Záhonyi；爱尔兰咨询工程师 Siobhan Fahey（主要起草人）；德国 GKW Consult GmbH 公司的 Christoph Theune；以及美国 CDM Smith 公司的 William Howard（执行委员会联系人）。

初始阶段修订工作组：
丹麦 Atkins/COWI 公司的 Svend Poulsen（组长）；瑞典 Advokatfirman Runeland AB 公司的 Aisha Nadar（工作组主要起草人）；澳大利亚 SMEC 的 Robin Schonfeld（工作组主要起草人）；塞尔维亚咨询工程师 Darko Plamenac；德国咨询工程师 Jan Ziepke；以及匈牙利 Z&Partners 咨询工程师公司的 Zoltán Záhonyi（合同委员会联系人）。

第二阶段修订工作组：
英国 EIA 有限公司的 Simon Worley（组长）；英国 Greenhalgh Associates 公司的 John Greenhalgh；德国 C2S Global 公司的 Leo Grutters；瑞典 Advokatfirman Runeland AB 公司的 Aisha Nadar；英国 Matrix Seminars 公司的 William Godwin；爱尔兰咨询工程师 Siobhan Fahey（合同委员会联系人）；以及英国 EIA 有限公司的 Shelley Adams（工作组秘书）。

本书是在 FIDIC 合同委员会指导下编写的：
英国 Atkins 公司的 Philip Jenkinson（前董事长）；匈牙利 Z&Partners 咨询工程师公司的 Zoltán Záhonyi（董事长）；法国 Exequatur 公司的 Vincent Leloup；瑞典 SWECO 公司的 KajMöller；爱尔兰咨询工程师 Siobhan Fahey；英国 Mott MacDonald 公司的 Mike Roberts；爱尔兰咨询工程师 Des Barry；德国 GKW Consult GmbH 公司的 Christoph Theune；FIDIC 秘书长 Enrico Vink；以及 FIDIC 设计和编辑经理 Christophe Sisto。

还有与 FIDIC 合同委员会联络的联系人：
英国 Scott Wilson 公司的 Geoff French（执行委员会前联系人）；瑞典 SWECO 公司的 Kaj Möller（执行委员会前联系人）；葡萄牙 SOPSEC 公司的 José Amorim Faria（EFCA 前联系人）；美国 CDM Smith 公司的 William Howard（执行委员会主要联系人）；瑞典 Advokatfirman Runeland AB 公司的 Aisha Nadar（执行委员会第二联系人）；以及波兰 AECOM 公司的 Pawel Zejer（EFCA 联系人）。

项目使用建筑信息模型系统的 FIDIC 合同用户的建议说明由以下人员提供：
澳大利亚 Aurecon 公司的 Anthony Barry（FIDIC 执行委员会），新西兰 Pedersen Read Consulting 公司的 Andrew Read（FIDIC 业务实践委员会主席）；以及新西兰 Aurecon 公司的 Stephen Jenkins（FIDIC 风险、责任和质量委员会主席）。

下列合同委员会特别顾问在各个起草和修订阶段提供了宝贵和持续的支持：
法国 White & Case LLP 公司的 Christopher Seppälä（法律顾问，并由其公司的 Dimitar Kondev 协助）；爱尔兰的 Nael G Bunni（风险和保险顾问）；德国的 Axel-V. Jaeger；英国/瑞典的 Michael Mortimer Hawkins；英国的 Christopher Wade；英国 Fenwick Elliott 公司的 Nicholas Gould 和 Jeremy Glover。

Drafts were reviewed by many persons and organisations, including those listed below. Their comments were duly studied by the FIDIC Contracts Committee's Updates Special Group and Update Task Groups and, where considered appropriate, have influenced the wording of the clauses.

Mahmoud Abu Hussein, Dolphin Energy, United Arab Emirates; Ihab Abu-Zahra, CRC Hassan Dorra, Egypt; Mushtaq Ahmad Smore, Engineer, Pakistan; Richard Appuhn, Engineer, Italy; Ulrik Bang Olsen, Bang Olsen & Partners Law Firm P/S, Denmark; Hartmut Bruehl, Engineer, Germany; Donald Charrett, MTECC, Australia; Edward Corbett, Corbett & Co International Construction Lawyers Ltd, UK; Cremona Cotovelea, SCPA Tecuci-Paltineanu, Romania; Mark Etheridge, UWP Consulting Pty (Ltd), UK; European International Contractors, Berlin, Germany; Ciaran Fahy, Engineer, Ireland; Stephane Giraud, Egis, France; Karen Gough, 39 Essex Chambers, UK; Sarwono Hardjomuljadi, Special Adviser to the Minister of Public Works, Indonesia; Sebastian Hoek, Kanzlei Dr.Hök Stieglmeier & Kollegen Berlin, Germany; Tomohide Ichiguchi, JICA, Japan; Reza Ikani, Tehran Berkeley Group of Companies, Iran; Levent Irmak, MC2 Modern, Turkey; Gordon Jaynes, Lawyer, UK; Nabeel Khokhar, Driver Group, UK; Humphrey Lloyd, Queen Mary University London, UK; Liu Luobing, Shanghai SICC Planning & Architectural Design, People's Republic of China; Husni Madi, Shura Construction Management, Jordan; Malith Mendis, ACESL, Sri Lanka; Benjamin Mellors, Holman Fenwick Willan LLP, UK; Christopher Miers, Probyn Miers, UK; Henry Musonda, Kiran & Musonda Associates, Zambia; Kjeld B Nielsen, Sweco, Denmark; Patrizia Palmitessa-Savric, Ginder Palmitessa Pty Ltd, Botswana; James Perry, PS Consulting, France; Mikko Pulkkinen, Wärtsilä Corporation, Finland; John Ritchie, Consultant, Canada; Munther Sakhet, Allied Planning & Engineering Co, Jordan; Michael Sergeant, HFW, UK; Christian Siemer, Fichtner Consulting Engineers, Germany; Evgeny Smirnov, EBRD, UK; Jakob B. Sorensen, M Holst, Advokater, Denmark; Benjamin Valloire, Syntec Ingénierie, France; Kitty Villani, Council of Europe Development Bank, France; Ahmed Faty Waly, WALY Arbitration, Egypt.

Acknowledgement of all reviewers above does not mean that such persons or organisations approve the wording of all clauses.

FIDIC very much appreciates the time and effort devoted by all the above persons.

The ultimate decision on the form and content of the document rests with FIDIC.

书稿曾经下列许多人员和组织审阅，他们的意见已由菲迪克（FIDIC）合同委员会新版特别工作组和修订工作组充分研究，认为适宜的意见已反映在条款措辞中。这些人员和组织包括：

阿拉伯联合酋长国 Dolphin Energy 公司的 Mahmoud Abu Hussein；埃及 CRC Hassan Dorra 公司的 Ihab Abu-Zahra；巴基斯坦工程师 Mushtaq Ahmad Smore；意大利工程师 Richard Appuhn；丹麦 Bang Olsen & Partners Law Firm P/S 公司的 Ulrik Bang Olsen；德国工程师 Hartmut Bruehl；澳大利亚 MTECC 公司的 Donald Charrett；英国 Corbett & Co International Construction Lawyers 有限公司的 Edward Corbett；罗马尼亚 SCPA Tecuci-Paltineanu 公司的 Cremona Cotovelea；英国 UWP Consulting Pty 公司的 Mark Etheridge；德国柏林的欧洲国际承包商组织；爱尔兰工程师 Ciaran Fahy；法国 Egis 公司的 Stephane Giraud；英国 39 Essex Chambers 的 Karen Gough；印度尼西亚公共工程部长特别顾问 Sarwono Hardjomuljadi；德国柏林 Kanzlei Dr.Hök Stieglmeier & Kollegen 公司的 Sebastian Hoek；日本 JICA 公司的 Tomohide Ichiguchi；伊朗德黑兰伯克利公司集团的 Reza Ikani；土耳其 MC2 Modern 公司的 Levent Irmak；英国律师 Gordon Jaynes；英国 Driver 集团的 Nabeel Khokhar；英国伦敦玛丽女王大学的 Humphrey Lloyd；中国上海投资咨询集团规划与建筑设计公司的 Liu Luobing；约旦 Shura 施工管理公司的 Husni Madi；斯里兰卡 ACESL 公司的 Malith Mendis；英国 Holman Fenwick Willan LLP 公司的 Benjamin Mellors；英国 Probyn Miers 公司的 Christopher Miers；赞比亚 Kiran & Musonda Associate 公司的 Henry Musonda；丹麦 Sweco 公司的 Kjeld B Nielsen；博茨瓦纳 Ginder Palmitessa Pty 有限公司的 Patrizia Palmitessa-Savric；法国 PS Consulting 公司的 James Perry；芬兰 Wärtsilä Corporation 公司的 Mikko Pulkkinen；加拿大咨询师 John Ritchie；约旦 Allied Planning & Engineering 公司的 Munther Sakhet；英国 HFW 公司的 Michael Sergeant；德国 Fichtner 咨询工程公司的 Christian Siemer；英国 EBRD 公司的 Evgeny Smirnov；丹麦 M Holst, Advokater 公司的 Jakob B. Sorensen；法国 Syntec Ingénierie 公司的 Benjamin Valloire；法国欧洲开发银行委员会的 Kitty Villani；埃及 WALY 仲裁委员会的 Ahmed Faty Waly。

对上述审稿人的致谢，并不表示审稿人和审稿组织对所有条款措辞的赞同。

菲迪克（FIDIC）对所有上述人员付出的时间和精力表示非常感谢。

本文件格式和内容的最终确定由菲迪克（FIDIC）负责。

NOTES

This Second Edition of the Conditions of Contract for Construction has been published by the Fédération Internationale des Ingénieurs-Conseils (FIDIC) as an update of the FIDIC 1999 Conditions of Contract for Construction (Red Book), First Edition.

Along with the FIDIC 1999 Yellow Book (the Conditions of Contract for Plant and Design-Build) and the FIDIC 1999 Silver Book (the Conditions of Contract for EPC/Turnkey Projects), the FIDIC 1999 Red Book has been in widespread use for nearly two decades. In particular, it has been recognised for, among other things, its principles of balanced risk sharing between the Employer and the Contractor in projects where the Contractor constructs the works in accordance with a design provided by the Employer. However, the works may include some elements of Contractor-designed civil, mechanical, electrical and/or construction works.

This Second Edition of the FIDIC Red Book continues FIDIC's fundamental principles of balanced risk sharing while seeking to build on the substantial experience gained from its use over the past 18 years. For example, this edition provides:

1) greater detail and clarity on the requirements for notices and other communications;

2) provisions to address Employers' and Contractors' claims treated equally and separated from disputes;

3) mechanisms for dispute avoidance and

4) detailed provisions for quality management, and verification of Contractor's contractual compliance.

These Conditions of Contract for Construction include conditions, which are likely to apply to the majority of such contracts. Essential items of information which are particular to each individual contract are to be included in the Particular Conditions Part A – Contract Data.

In addition, it is recognised that many Employers, especially governmental agencies, may require special conditions of contract, or particular procedures, which differ from those included in the General Conditions. These should be included in Part B – Special Provisions.

It should be noted, that the General Conditions and the Particular Conditions (Part A – Contract Data and Part B – Special Provisions) are all part of the Conditions of Contract.

To assist Employers in preparing tender documents and in drafting Particular Conditions of Contract for specific contracts, this publication includes Notes on the Preparation of Tender Documents and Notes on the Preparation of Special Provisions, which provide important advice to drafters of contract documents, in particular the Specifications and Special Provisions. In drafting Special Provisions, if clauses in the General Conditions are to be replaced or supplemented and before incorporating any example wording, Employers are urged to seek legal and engineering advice in an effort to avoid ambiguity and to ensure completeness and consistency with the other provisions of the contract.

This publication begins with a series of comprehensive flow charts which typically show, in visual form, the sequences of activities which characterise the FIDIC Construction form of contract. The charts are illustrative, however, and must not be taken into consideration in the interpretation of the Conditions of Contract.

说明

第 2 版《施工合同条件》已由国际咨询工程师联合会（FIDIC，菲迪克）出版，是菲迪克（FIDIC）1999 年第 1 版《施工合同条件》（红皮书）的修订版。

菲迪克（FIDIC）1999 年版红皮书《施工合同条件》连同菲迪克（FIDIC）1999 年版黄皮书《生产设备和设计 - 施工合同条件》和菲迪克（FIDIC）1999 年版银皮书《设计采购施工（EPC）/ 交钥匙工程合同条件》已经广泛使用了近 20 年。特别是，其中**承包商**按照雇主提供的设计在其工程施工的项目中，**雇主**和**承包商**之间平衡风险分担的原则已得到认可。然而，该工程可能包括由**承包商**设计的土木、机械、电气和 / 或构筑物的任意组合。

第 2 版菲迪克（FIDIC）红皮书延续了菲迪克（FIDIC）平衡风险分担的基本原则，同时寻求在过去 18 年使用中获得的丰富经验的基础上编写。例如，本版本提供：

1） 通知和其他通信交流的要求更加详细和明确；

2） 平等对待**雇主**和**承包商**的索赔，并将其与争端分别处理的规定；

3） 争端避免的机制；（以及）

4） 质量管理和**承包商**合同合规验证的详细规定。

这些**施工合同条件**包含了可能适用于大多数此类合同的条件。每份单独合同所特有的基本信息项应包含在**专用条件 A 部分——合同数据**中。

此外，人们认识到许多**雇主**，特别是政府机构，可能需要特别的合同条件或特定程序，这些条件或程序与**通用条件**中所含的不同。这些不同的条件或程序应包括在 **B 部分——特别规定**中。

应注意的是，**通用条件**和**专用条件**（**A 部分——合同数据**和 **B 部分——特别规定**）全都是**合同条件**的一部分。

为协助**雇主**编写投标文件和起草特定合同的**合同专用条件**，本出版物包括**编写投标文件注意事项**和**编写特别规定注意事项**，为合同文件的起草者，尤其是起草规范要求和**特别规定**时，提供了重要的建议。在起草**特别规定**时，如果要替换或补充**通用条件**中的条款，在采用任何范例措辞之前，应敦促**雇主**寻求法律和工程方面的建议，以避免含糊不清，并确保与合同其他规定保持完整和一致。

本出版物以一系列综合流程图开始，这些流程图以可视化形式展示了菲迪克（FIDIC）合同施工形式的活动顺序。然而，这些图只是说明性的，不应作为**合同条件**的解释。

This publication also includes a number of sample forms to help both Parties to develop a common understanding of what is required by third parties such as providers of securities and guarantees.

Drafters of contract documents are reminded that the General Conditions of all FIDIC contracts are protected by copyright and trademark and may not be changed without specific written consent, usually in the form of a licence to amend, from FIDIC. If drafters wish to amend the provisions found in the General Conditions, the place for doing this is in the Particular Conditions Part B – Special Provisions, as mentioned above, and not by making changes in the General Conditions as published.

FIDIC considers the official and authentic texts to be the versions in the English language.

本出版物还包括一些样本格式，以帮助**双方**就第三方如担保和保函的提供方的要求达成共识。

合同文件起草者应注意，所有**菲迪克（FIDIC）**合同的**通用条件**受版权和商标保护，未经**菲迪克（FIDIC）**通常以修改许可证的形式授予的特别书面同意，不得修改。如果起草者希望修改**通用条件**中的规定，则应在**专用条件 B 部分——特别规定**（如上所述）中修改，而不是修改已发布的**通用条件**。

菲迪克（FIDIC）认为，正式的、权威性的文本应为英文版本。

Typical Sequence of Principal Events During Contracts for Construction

1. The Time for Completion is to be stated (in the Contract Data) as a number of days, to which is added any extensions of time under Sub-Clause 8.5.
2. In order to indicate the sequence of events, the diagram below is based upon the example of the Contractor failing to comply with Sub-Clause 8.2.
3. The Defects Notification Period is to be stated (in the Contract Data) as a number of days, to which is added any extensions under Sub-Clause 11.3.
4. Depending on the type of Works, Tests after Completion may also be required.

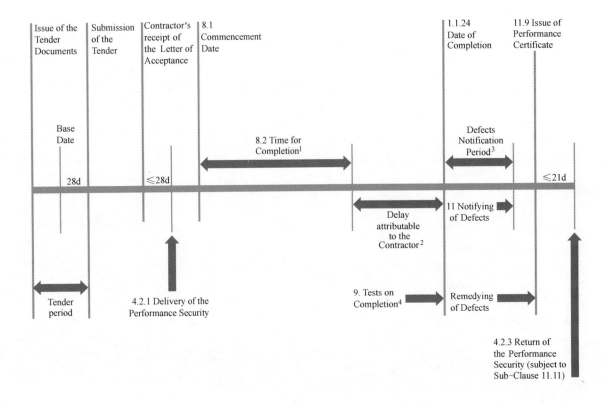

施工合同中主要事项的典型顺序

1. **竣工时间**（在**合同数据**中）用天数表示，加上根据第 8.5 款规定的任何延长期。

2. 为了表示事项的顺序，下图以**承包商**未能遵守**第 8.2 款**的规定为例。

3. **缺陷通知期限**（在**合同数据**中）用天数表示，加上根据第 11.3 款规定的任何延长期。

4. 根据工程的类型，可能还需要**竣工后试验**。

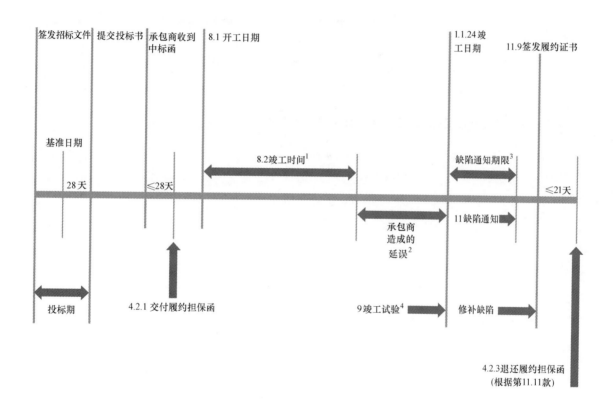

Typical Sequence of Payment Events Envisaged in Clause 14

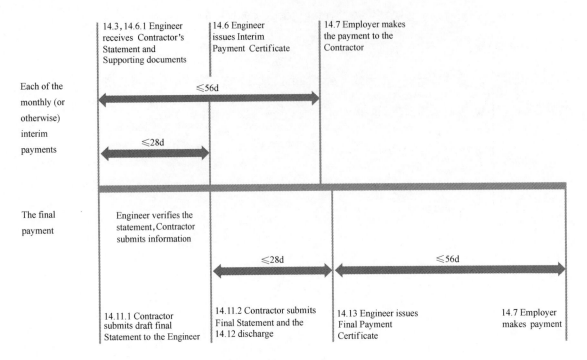

Typical Sequence of Dispute Events Envisaged in Clause 21

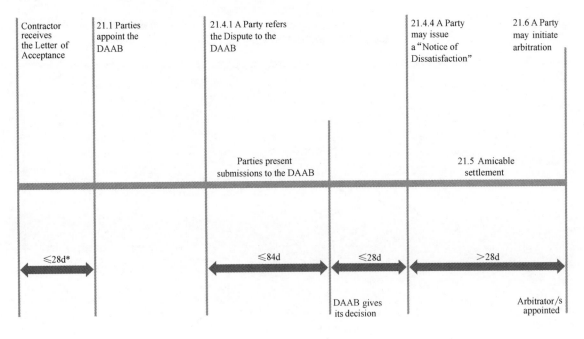

* If not stated otherwise in the Contract Data (Sub-Clause 21.1)

第 14 条中设想的付款事项的典型顺序

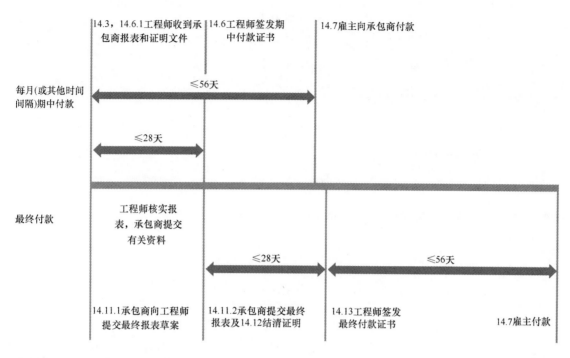

第 21 条中设想的争端事项的典型顺序

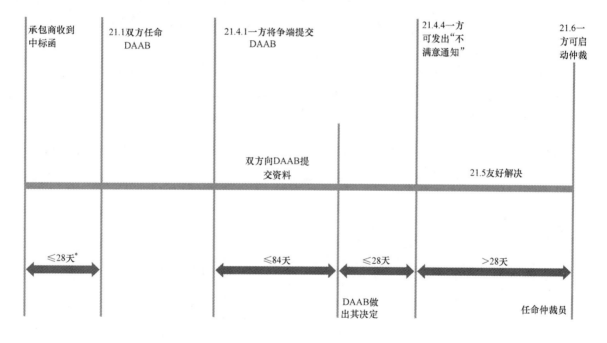

* 如果合同数据（第 21.1 款）中未另行规定。

Typical Sequence of Events in Agreement or Determination under Sub-Clause 3.7

Scenario 1[1]

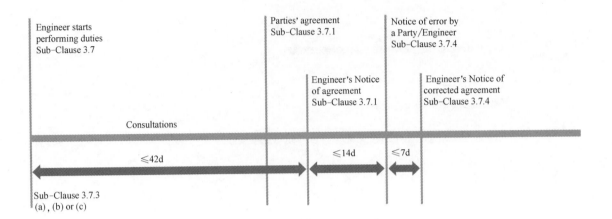

Scenario 2[2]

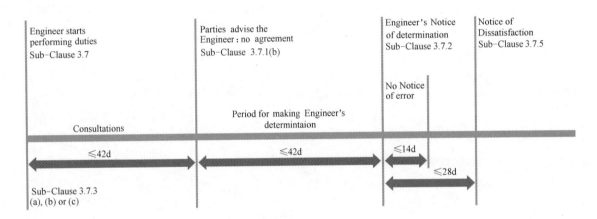

Scenario 3[3]

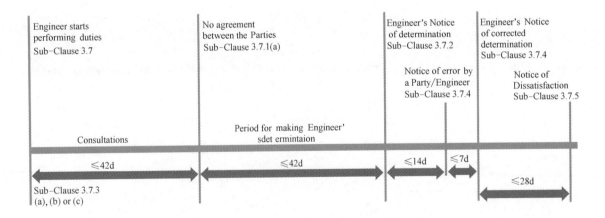

1. Agreement is reached within 42 days, error found in the Engineer's Notice of agreement and corrected.

2. The Parties' early advice that agreement cannot be reached and so Engineer's determination is necessary, no error in Engineer's determination.

3. No agreement within 42 days, Engineer determines within 42 days, error found in the Engineer's determination and corrected.

第3.7款中商定或确定事项的典型顺序

情景 1[1]

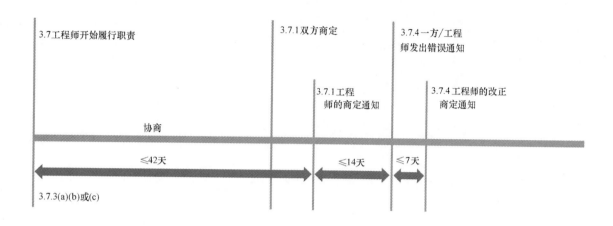

情景 2[2]

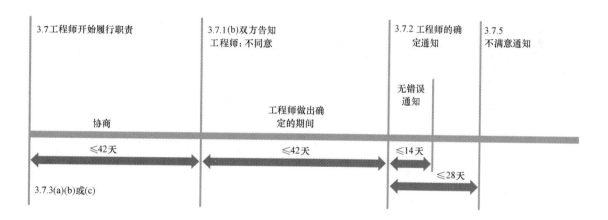

情景 3[3]

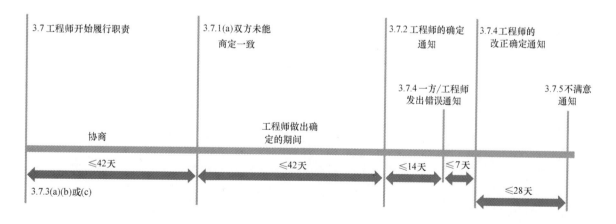

1. 42天内达成商定，在**工程师**的商定**通知**中发现错误并改正。

2. 双方早期的建议通过商定未能达成，所以**工程师**的确定是必要的，**工程师**的确定没有错误。

3. 42天内未达成商定，**工程师**在42天内做出确定，并在其确定中发现错误并改正。

施工合同条件
Conditions of Contract for
CONSTRUCTION

用于由雇主设计的建筑和工程
FOR BUILDING AND ENGINEERING WORKS
DESIGNED BY THE EMPLOYER

通用条件
General Conditions

通用条件
GENERAL CONDITIONS

专用条件编写指南和附件：
担保函格式
GUIDANCE FOR
THE PREPARATION OF
PARTICULAR CONDITIONS
AND ANNEXES: FORMS
OF SECURITIES

投标函、中标函、合同协议书
和争端避免/裁决委员会协议书格式
FORMS OF LETTER OF
TENDER, LETTER OF
ACCEPTANCE, CONTRACT
AGREEMENT AND DAAB
AGREEMENT

General Conditions

CONTENTS

1 GENERAL PROVISIONS 12

- 1.1 Definitions
- 1.2 Interpretation
- 1.3 Notices and Other Communications
- 1.4 Law and Language
- 1.5 Priority of Documents
- 1.6 Contract Agreement
- 1.7 Assignment
- 1.8 Care and Supply of Documents
- 1.9 Delayed Drawings or Instructions
- 1.10 Employer's Use of Contractor's Documents
- 1.11 Contractor's Use of Employer's Documents
- 1.12 Confidentiality
- 1.13 Compliance with Laws
- 1.14 Joint and Several Liability
- 1.15 Limitation of Liability
- 1.16 Contract Termination

2 THE EMPLOYER 38

- 2.1 Right of Access to the Site
- 2.2 Assistance
- 2.3 Employer's Personnel and Other Contractors
- 2.4 Employer's Financial Arrangements
- 2.5 Site Data and Items of Reference
- 2.6 Employer-Supplied Materials and Employer's Equipment

3 THE ENGINEER 42

- 3.1 The Engineer
- 3.2 Engineer's Duties and Authority
- 3.3 The Engineer's Representative
- 3.4 Delegation by the Engineer
- 3.5 Engineer's Instructions
- 3.6 Replacement of the Engineer
- 3.7 Agreement or Determination
- 3.8 Meetings

通用条件

目 录

1　一般规定 ·· 13

 1.1　定义
 1.2　解释
 1.3　通知和其他通信交流
 1.4　法律和语言
 1.5　文件优先次序
 1.6　合同协议书
 1.7　权益转让
 1.8　文件的照管和提供
 1.9　延误的图纸或指示
 1.10　雇主使用承包商文件
 1.11　承包商使用雇主文件
 1.12　保密
 1.13　遵守法律
 1.14　共同的和各自的责任
 1.15　责任限度
 1.16　合同终止

2　雇主 ·· 39

 2.1　现场进入权
 2.2　协助
 2.3　雇主人员和其他承包商
 2.4　雇主的资金安排
 2.5　现场数据和参考事项
 2.6　雇主提供的材料和雇主设备

3　工程师 ·· 43

 3.1　工程师
 3.2　工程师的任务和权利
 3.3　工程师代表
 3.4　由工程师付托
 3.5　工程师的指示
 3.6　工程师的替代
 3.7　商定或确定
 3.8　会议

4	**THE CONTRACTOR**	**52**
4.1	Contractor's General Obligations	
4.2	Performance Security	
4.3	Contractor's Representative	
4.4	Contractor's Documents	
4.5	Training	
4.6	Co-operation	
4.7	Setting Out	
4.8	Health and Safety Obligations	
4.9	Quality Management and Compliance Verification Systems	
4.10	Use of Site Data	
4.11	Sufficiency of the Accepted Contract Amount	
4.12	Unforeseeable Physical Conditions	
4.13	Rights of Way and Facilities	
4.14	Avoidance of Interference	
4.15	Access Route	
4.16	Transport of Goods	
4.17	Contractor's Equipment	
4.18	Protection of the Environment	
4.19	Temporary Utilities	
4.20	Progress Reports	
4.21	Security of the Site	
4.22	Contractor's Operations on Site	
4.23	Archaeological and Geological Findings	
5	**SUBCONTRACTING**	**82**
5.1	Subcontractors	
5.2	Nominated Subcontractors	
6	**STAFF AND LABOUR**	**86**
6.1	Engagement of Staff and Labour	
6.2	Rates of Wages and Conditions of Labour	
6.3	Recruitment of Persons	
6.4	Labour Laws	
6.5	Working Hours	
6.6	Facilities for Staff and Labour	
6.7	Health and Safety of Personnel	
6.8	Contractor's Superintendence	
6.9	Contractor's Personnel	
6.10	Contractor's Records	
6.11	Disorderly Conduct	
6.12	Key Personnel	
7	**PLANT, MATERIALS AND WORKMANSHIP**	**92**
7.1	Manner of Execution	
7.2	Samples	
7.3	Inspection	
7.4	Testing by the Contractor	
7.5	Defects and Rejection	
7.6	Remedial Work	

| 4 | 承包商 | 53 |

- 4.1 承包商的一般义务
- 4.2 履约担保
- 4.3 承包商代表
- 4.4 承包商文件
- 4.5 培训
- 4.6 合作
- 4.7 放线
- 4.8 健康和安全义务
- 4.9 质量管理和合规验证体系
- 4.10 现场数据的使用
- 4.11 中标合同金额的充分性
- 4.12 不可预见的物质条件
- 4.13 道路通行权和设施
- 4.14 避免干扰
- 4.15 进场通路
- 4.16 货物运输
- 4.17 承包商设备
- 4.18 环境保护
- 4.19 临时公用设施
- 4.20 进度报告
- 4.21 现场安保
- 4.22 承包商的现场作业
- 4.23 考古和地质发现

| 5 | 分包 | 83 |

- 5.1 分包商
- 5.2 指定分包商

| 6 | 员工 | 87 |

- 6.1 员工的雇用
- 6.2 工资标准和劳动条件
- 6.3 招聘人员
- 6.4 劳动法
- 6.5 工作时间
- 6.6 为员工提供设施
- 6.7 人员的健康和安全
- 6.8 承包商的监督
- 6.9 承包商人员
- 6.10 承包商的记录
- 6.11 无序行为
- 6.12 关键人员

| 7 | 生产设备、材料和工艺 | 93 |

- 7.1 实施方法
- 7.2 样品
- 7.3 检验
- 7.4 由承包商试验
- 7.5 缺陷和拒收
- 7.6 修补工作

5

| 7.7 | Ownership of Plant and Materials |
| 7.8 | Royalties |

| 8 | **COMMENCEMENT, DELAYS AND SUSPENSION** .... 98 |

8.1	Commencement of Works
8.2	Time for Completion
8.3	Programme
8.4	Advance Warning
8.5	Extension of Time for Completion
8.6	Delays Caused by Authorities
8.7	Rate of Progress
8.8	Delay Damages
8.9	Employer's Suspension
8.10	Consequences of Employer's Suspension
8.11	Payment for Plant and Materials after Employer's Suspension
8.12	Prolonged Suspension
8.13	Resumption of Work

| 9 | **TESTS ON COMPLETION** .... 110 |

9.1	Contractor's Obligations
9.2	Delayed Tests
9.3	Retesting
9.4	Failure to Pass Tests on Completion

| 10 | **EMPLOYER'S TAKING OVER** .... 114 |

10.1	Taking Over the Works and Sections
10.2	Taking Over Parts
10.3	Interference with Tests on Completion
10.4	Surfaces Requiring Reinstatement

| 11 | **DEFECTS AFTER TAKING OVER** .... 118 |

11.1	Completion of Outstanding Work and Remedying Defects
11.2	Cost of Remedying Defects
11.3	Extension of Defects Notification Period
11.4	Failure to Remedy Defects
11.5	Remedying of Defective Work off Site
11.6	Further Tests after Remedying Defects
11.7	Right of Access after Taking Over
11.8	Contractor to Search
11.9	Performance Certificate
11.10	Unfulfilled Obligations
11.11	Clearance of Site

| 12 | **MEASUREMENT AND VALUATION** .... 128 |

12.1	Works to be Measured
12.2	Method of Measurement
12.3	Valuation of the Works
12.4	Omissions

7.7	生产设备和材料的所有权
7.8	土地（矿区）使用费

8　开工、延误和暂停 ······ 99

8.1	工程的开工
8.2	竣工时间
8.3	进度计划
8.4	预先警示
8.5	竣工时间的延长
8.6	部门造成的延误
8.7	工程进度
8.8	误期损害赔偿费
8.9	雇主暂停
8.10	雇主暂停的后果
8.11	雇主暂停后对生产设备和材料的付款
8.12	拖长的暂停
8.13	复工

9　竣工试验 ······ 111

9.1	承包商的义务
9.2	延误的试验
9.3	重新试验
9.4	未能通过竣工试验

10　雇主的接收 ······ 115

10.1	工程和分项工程的接收
10.2	部分工程的接收
10.3	对竣工试验的干扰
10.4	需要复原的地面

11　接收后的缺陷 ······ 119

11.1	完成扫尾工作和修补缺陷
11.2	修补缺陷的费用
11.3	缺陷通知期限的延长
11.4	未能修补缺陷
11.5	现场外缺陷工程的修补
11.6	修补缺陷后进一步试验
11.7	接收后的进入权
11.8	承包商调查
11.9	履约证书
11.10	未履行的义务
11.11	现场清理

12　测量和估价 ······ 129

12.1	需测量的工程
12.2	测量方法
12.3	工程的估价
12.4	删减

7

| 13 | VARIATIONS AND ADJUSTMENTS | 132 |

- 13.1 Right to Vary
- 13.2 Value Engineering
- 13.3 Variation Procedure
- 13.4 Provisional Sums
- 13.5 Daywork
- 13.6 Adjustments for Changes in Laws
- 13.7 Adjustments for Changes in Cost

| 14 | CONTRACT PRICE AND PAYMENT | 144 |

- 14.1 The Contract Price
- 14.2 Advance Payment
- 14.3 Application for Interim Payment
- 14.4 Schedule of Payments
- 14.5 Plant and Materials intended for the Works
- 14.6 Issue of IPC
- 14.7 Payment
- 14.8 Delayed Payment
- 14.9 Release of Retention Money
- 14.10 Statement at Completion
- 14.11 Final Statement
- 14.12 Discharge
- 14.13 Issue of FPC
- 14.14 Cessation of Employer's Liability
- 14.15 Currencies of Payment

| 15 | TERMINATION BY EMPLOYER | 164 |

- 15.1 Notice to Correct
- 15.2 Termination for Contractor's Default
- 15.3 Valuation after Termination for Contractor's Default
- 15.4 Payment after Termination for Contractor's Default
- 15.5 Termination for Employer's Convenience
- 15.6 Valuation after Termination for Employer's Convenience
- 15.7 Payment after Termination for Employer's Convenience

| 16 | SUSPENSION AND TERMINATION BY CONTRACTOR | 172 |

- 16.1 Suspension by Contractor
- 16.2 Termination by Contractor
- 16.3 Contractor's Obligations After Termination
- 16.4 Payment after Termination by Contractor

| 17 | CARE OF THE WORKS AND INDEMNITIES | 176 |

- 17.1 Responsibility for Care of the Works
- 17.2 Liability for Care of the Works
- 17.3 Intellectual and Industrial Property Rights
- 17.4 Indemnities by Contractor
- 17.5 Indemnities by Employer
- 17.6 Shared Indemnities

| 13 | 变更和调整 | 133 |

13.1 变更权
13.2 价值工程
13.3 变更程序
13.4 暂列金额
13.5 计日工作
13.6 因法律改变的调整
13.7 因成本改变的调整

| 14 | 合同价格和付款 | 145 |

14.1 合同价格
14.2 预付款
14.3 期中付款的申请
14.4 付款计划表
14.5 拟用于工程的生产设备和材料
14.6 期中付款证书的签发
14.7 付款
14.8 延误的付款
14.9 保留金的发放
14.10 竣工报表
14.11 最终报表
14.12 结清证明
14.13 最终付款证书的签发
14.14 雇主责任的中止
14.15 支付的货币

| 15 | 由雇主终止 | 165 |

15.1 通知改正
15.2 因承包商违约的终止
15.3 因承包商违约终止后的估价
15.4 因承包商违约终止后的付款
15.5 为雇主便利的终止
15.6 为雇主便利终止后的估价
15.7 为雇主便利终止后的付款

| 16 | 由承包商暂停和终止 | 173 |

16.1 由承包商暂停
16.2 由承包商终止
16.3 终止后承包商的义务
16.4 由承包商终止后的付款

| 17 | 工程照管和保障 | 177 |

17.1 工程照管的职责
17.2 工程照管的责任
17.3 知识产权和工业产权
17.4 由承包商保障
17.5 由雇主保障
17.6 保障分担

9

18	**EXCEPTIONAL EVENTS**	**182**

- 18.1 Exceptional Events
- 18.2 Notice of an Exceptional Event
- 18.3 Duty to Minimise Delay
- 18.4 Consequences of an Exceptional Event
- 18.5 Optional Termination
- 18.6 Release from Performance under the Law

19	**INSURANCE**	**188**

- 19.1 General Requirements
- 19.2 Insurance to be provided by the Contractor

20	**EMPLOYER'S AND CONTRACTOR'S CLAIMS**	**194**

- 20.1 Claims
- 20.2 Claims For Payment and/or EOT

21	**DISPUTES AND ARBITRATION**	**202**

- 21.1 Constitution of the DAAB
- 21.2 Failure to Appoint DAAB Member(s)
- 21.3 Avoidance of Disputes
- 21.4 Obtaining DAAB's Decision
- 21.5 Amicable Settlement
- 21.6 Arbitration
- 21.7 Failure to Comply with DAAB's Decision
- 21.8 No DAAB In Place

APPENDIX	**214**

General Conditions of Dispute Avoidance/Adjudication agreement

INDEX OF SUB-CLAUSES	**244**

18	**例外事件**	183
18.1	例外事件	
18.2	例外事件的通知	
18.3	将延误减至最小的义务	
18.4	例外事件的后果	
18.5	自主选择终止	
18.6	依法解除履约	
19	**保险**	189
19.1	一般要求	
19.2	由承包商提供的保险	
20	**雇主和承包商的索赔**	195
20.1	索赔	
20.2	付款和/或竣工时间延长的索赔	
21	**争端和仲裁**	203
21.1	争端避免/裁决委员会的组成	
21.2	未能任命争端避免/裁决委员会成员	
21.3	争端避免	
21.4	取得争端避免/裁决委员会的决定	
21.5	友好解决	
21.6	仲裁	
21.7	未能遵守争端避免/裁决委员会的决定	
21.8	未设立争端避免/裁决委员会	

附录	215
争端避免/裁决委员会协议书一般条件	

条款索引	245

General Conditions

1 General Provisions

1.1 Definitions

In the Contract the following words and expressions shall have the meanings stated, except where the context requires otherwise

1.1.1 "**Accepted Contract Amount**" means the amount accepted in the Letter of Acceptance for the execution of the Works in accordance with the Contract.

1.1.2 "**Advance Payment Certificate**" means a Payment Certificate issued by the Engineer for advance payment under Sub-Clause 14.2.2 [*Advance Payment Certificate*].

1.1.3 "**Advance Payment Guarantee**" means the guarantee under Sub-Clause 14.2.1 [*Advance Payment Guarantee*].

1.1.4 "**Base Date**" means the date 28 days before the latest date for submission of the Tender.

1.1.5 "**Bill of Quantities**" means the document entitled bill of quantities (if any) included in the Schedules.

1.1.6 "**Claim**" means a request or assertion by one Party to the other Party for an entitlement or relief under any Clause of these Conditions or otherwise in connection with, or arising out of, the Contract or the execution of the Works.

1.1.7 "**Commencement Date**" means the date as stated in the Engineer's Notice issued under Sub-Clause 8.1 [*Commencement of Works*].

1.1.8 "**Compliance Verification System**" means the compliance verification system to be prepared and implemented by the Contractor for the Works in accordance with Sub-Clause 4.9.2 [*Compliance Verification System*].

1.1.9 "**Conditions of Contract**" or "**these Conditions**" means these General Conditions as amended by the Particular Conditions.

1.1.10 "**Contract**" means the Contract Agreement, the Letter of Acceptance, the Letter of Tender, any addenda referred to in the Contract Agreement, these Conditions, the Specification, the Drawings, the Schedules, the Contractor's Proposal, the JV Undertaking (if applicable) and the further documents (if any) which are listed in the Contract Agreement or in the Letter of Acceptance.

1.1.11 "**Contract Agreement**" means the agreement entered into by both Parties in accordance with Sub-Clause 1.6 [*Contract Agreement*].

1.1.12 "**Contract Data**" means the pages, entitled contract data which constitute Part A of the Particular Conditions.

通用条件

1 一般规定

1.1 定义

除上下文另有要求外，下列词语和措辞应具有以下所述的含义。

1.1.1 "**中标合同金额**"系指在**中标函**中按照**合同**规定所认可的**工程**施工所需的费用。

1.1.2 "**预付款证书**"系指**工程师**根据第 **14.2.2** 项［*预付款证书*］的规定签发的证书。

1.1.3 "**预付款保函**"系指根据第 **14.2.1** 项［*预付款保函*］所规定的担保。

1.1.4 "**基准日期**"系指递交**投标书**截止日期前 28 天的日期。

1.1.5 "**工程量表（工程量清单）**"系指包含在**资料表**中如此标名的文件（如果有）。

1.1.6 "**索赔**"系指一方向另一方提出的要求或主张，要求根据本**条件**的任何**条款**或与**合同**或**工程**施工有关的由其引起的权利或救济。

1.1.7 "**开工日期**"系指根据第 **8.1** 款［*工程的开工*］发出的**工程师**通知中规定的日期。

1.1.8 "**合规验证体系**"系指**承包商**根据第 **4.9.2** 项［*合规验证体系*］的规定，为**工程**编写和实施的合规验证体系。

1.1.9 "**合同条件**"或"**本条件**"系指由**专用条件**修订的**通用条件**。

1.1.10 "**合同**"系指**合同协议书**、**中标函**、**投标函**，**合同协议书**中提到的任何附录、**本条件**、**规范要求**、**图纸**、**资料表**、**联营体承诺书**（如适用）以及**合同协议书**或**中标函**中列出的添加文件（如果有）。[一]

1.1.11 "**合同协议书**"系指双方根据第 **1.6** 款［*合同协议书*］所述签订的合同协议书。

1.1.12 "**合同数据**"系指构成**专用条件** A 部分——**合同数据**的文本。

[一] 此处中文按照勘误表改正后的英文翻译。——译者注

1.1.13 "**Contract Price**" means the price defined in Sub-Clause 14.1 [*The Contract Price*].

1.1.14 "**Contractor**" means the person(s) named as contractor in the Letter of Tender accepted by the Employer and the legal successors in title of such person(s).

1.1.15 "**Contractor's Documents**" means the documents prepared by the Contractor as described in Sub-Clause 4.4 [*Contractor's Documents*], including calculations, digital files, computer programs and other software, drawings, manuals, models, specifications and other documents of a technical nature.

1.1.16 "**Contractor's Equipment**" means all apparatus, equipment, machinery, construction plant, vehicles and other items required by the Contractor for the execution of the Works. Contractor's Equipment excludes Temporary Works, Plant, Materials and any other things intended to form or forming part of the Permanent Works.

1.1.17 "**Contractor's Personnel**" means the Contractor's Representative and all personnel whom the Contractor utilises on Site or other places where the Works are being carried out, including the staff, labour and other employees of the Contractor and of each Subcontractor; and any other personnel assisting the Contractor in the execution of the Works.

1.1.18 "**Contractor's Representative**" means the natural person named by the Contractor in the Contract or appointed by the Contractor under Sub-Clause 4.3 [*Contractor's Representative*], who acts on behalf of the Contractor.

1.1.19 "**Cost**" means all expenditure reasonably incurred (or to be incurred) by the Contractor in performing the Contract, whether on or off the Site, including taxes, overheads and similar charges, but does not include profit. Where the Contractor is entitled under a Sub-Clause of these Conditions to payment of Cost, it shall be added to the Contract Price.

1.1.20 "**Cost Plus Profit**" means Cost plus the applicable percentage for profit stated in the Contract Data (if not stated, five percent (5%)). Such percentage shall only be added to Cost, and Cost Plus Profit shall only be added to the Contract Price, where the Contractor is entitled under a Sub-Clause of these Conditions to payment of Cost Plus Profit.

1.1.21 "**Country**" means the country in which the Site (or most of it) is located, where the Permanent Works are to be executed.

1.1.22 "**DAAB**" or "**Dispute Avoidance/Adjudication Board**" means the sole member or three members (as the case may be) so named in the Contract, or appointed under Sub-Clause 21.1 [*Constitution of the DAAB*] or Sub-Clause 21.2 [*Failure to Appoint DAAB Member(s)*].

1.1.23 "**DAAB Agreement**" means the agreement signed or deemed to have been signed by both Parties and the sole member or each of the three members (as the case may be) of the DAAB in accordance with Sub-Clause 21.1 [*Constitution of the DAAB*] or Sub-Clause 21.2 [*Failure to Appoint DAAB Member(s)*], incorporating by reference the General Conditions of Dispute Avoidance/Adjudication Agreement contained in the Appendix to these General Conditions with such amendments as are agreed.

1.1.13 "合同价格"系指第 14.1 款 [合同价格] 规定的价格。

1.1.14 "承包商"系指在雇主接受的投标函中称为承包商的当事人，及其财产所有权的合法继承人。

1.1.15 "承包商文件"系指第 4.4 款 [承包商文件] 中所述的，由承包商根据合同编写的文件，包括所有计算、数字文件、计算机程序和其他软件、图纸、手册、模型、规范要求和其他技术性文件。

1.1.16 "承包商设备"系指承包商为实施工程所需的所有仪器、设备、机械、施工设备、车辆和其他物品。承包商设备不包括临时工程、生产设备、材料以及拟构成或正构成永久工程一部分的任何其他物品。

1.1.17 "承包商人员"系指承包商代表和承包商在现场或其他工程正在进行的地方聘用的所有人员，包括承包商和每个分包商的职员、工人和其他雇员；以及所有其他帮助承包商实施工程的人员。

1.1.18 "承包商代表"系指由承包商在合同中指定的自然人，或根据第 4.3 款 [承包商代表] 的规定，由承包商任命为其代表的人员。

1.1.19 "成本（费用）"系指承包商在履行合同中在现场内外发生的（或将要发生的）所有合理开支，包括税费、管理费和类似支出，但不包括利润。如果承包商根据本条件的任一条款有权获得成本（费用），应将其加到合同价格中。

1.1.20 "成本加利润"系指合同数据中规定的成本加上适当比例的利润（如未规定，应加 5%）。只有在承包商根据本条件中的条款有权获得成本加利润时，该百分比才能加到成本中，成本加利润才能加进合同价格。

1.1.21 "工程所在国"系指实施永久工程的现场（或其大部分）所在的国家。

1.1.22 "DAAB（争端避免/裁决委员会）"系指合同中唯一的一名或三名人员（视情况而定），或根据第 21.1 款 [争端避免/裁决委员会的组成] 或第 21.2 款 [未能任命争端避免/裁决委员会成员] 的规定任命的人员。

1.1.23 "DAAB 协议书"系指由双方和 DAAB 的唯一成员或三名成员（视情况而定）中的每位成员，根据第 21.1 款 [争端避免/裁决委员会的组成] 或第 21.2 款 [未能任命争端避免/裁决委员会成员] 的规定签署或视为已签署的协议书，包括本通用条件的附件所含的争端避免/裁决委员会协议书一般条件及其商定的修改。

1.1.24 "**Date of Completion**" means the date stated in the Taking-Over Certificate issued by the Engineer; or, if the last paragraph of Sub-Clause 10.1 [*Taking Over the Works and Sections*] applies, the date on which the Works or Section are deemed to have been completed in accordance with the Contract; or, if Sub-Clause 10.2 [*Taking Over Parts*] or Sub-Clause 10.3. [*Interference with Tests on Completion*] applies, the date on which the Works or Section or Part are deemed to have been taken over by the Employer.

1.1.25 "**day**" means a calendar day.

1.1.26 "**Daywork Schedule**" means the document entitled daywork schedule (if any) included in the Contract, showing the amounts and manner of payments to be made to the Contractor for labour, materials and equipment used for daywork under Sub-Clause 13.5 [*Daywork*].

1.1.27 "**Defects Notification Period**" or "**DNP**" means the period for notifying defects and/or damage in the Works or a Section or a Part (as the case may be) under Sub-Clause 11.1 [*Completion of Outstanding Work and Remedying Defects*], as stated in the Contract Data (if not stated, one year), and as may be extended under Sub-Clause 11.3 [*Extension of Defects Notification Period*]. This period is calculated from the Date of Completion of the Works or Section or Part.

1.1.28 "**Delay Damages**" means the damages for which the Contractor shall be liable under Sub-Clause 8.8 [*Delay Damages*] for failure to comply with Sub-Clause 8.2 [*Time for Completion*].

1.1.29 "**Dispute**" means any situation where:

(a) one Party makes a claim against the other Party (which may be a Claim, as defined in these Conditions, or a matter to be determined by the Engineer under these Conditions, or otherwise);
(b) the other Party (or the Engineer under Sub-Clause 3.7.2 [*Engineer's Determination*]) rejects the claim in whole or in part; and
(c) the first Party does not acquiesce (by giving a NOD under Sub-Clause 3.7.5 [*Dissatisfaction with Engineer's determination*] or otherwise),

provided however that a failure by the other Party (or the Engineer) to oppose or respond to the claim, in whole or in part, may constitute a rejection if, in the circumstances, the DAAB or the arbitrator(s), as the case may be, deem it reasonable for it to do so.

1.1.30 "**Drawings**" means the drawings of the Works included in the Contract, and any additional and modified drawings issued by (or on behalf of) the Employer in accordance with the Contract.

1.1.31 "**Employer**" means the person named as the employer in the Contract Data and the legal successors in title to this person.

1.1.32 "**Employer's Equipment**" means the apparatus, equipment, machinery, construction plant and/or vehicles (if any) to be made available by the Employer for the use of the Contractor under Sub-Clause 2.6 [*Employer-Supplied Materials and Employer's Equipment*]; but does not include Plant which has not been taken over under Clause 10 [*Employer's Taking Over*].

1.1.24 "竣工日期"系指工程师签发的接收证书中规定的日期；或如果第 10.1 款［工程和分项工程的接收］的最后一段适用，工程和分项工程按照合同被视为已完成的日期；或如果根据第 10.2 款［部分工程的接收］或第 10.3 款［对竣工试验的干扰］适用，竣工日期视为雇主已接收工程、分项工程或部分工程的日期。

1.1.25 "天（日）"系指一个日历日。

1.1.26 "计日工作计划表"系指合同中题为计日工作计划的文件（如果有），根据第 13.5 款［计日工作］的规定，向承包商支付计日工作使用的人工、材料和设备的金额和付款方式。

1.1.27 "缺陷通知期限"或"DNP"系指根据第 11.1 款［完成扫尾工作和修补缺陷］的规定，通知工程或某分项工程或某部分工程（视情况而定）存在缺陷和/或损害的期限，如合同数据中所述（如未规定，则为一年），也可根据第 11.3 款［缺陷通知期限的延长］的规定提出延长期。该期限从工程或分项工程或部分工程的竣工日开始计算。

1.1.28 "误期损害赔偿费"系指承包商应根据第 8.8 款［误期损害赔偿费］的规定，对其未能遵守第 8.2 款［竣工时间］的要求而承担的损害赔偿。

1.1.29 "争端"系指下列情况：

（a） 一方向另一方提出索赔（可能是本条件所定义的索赔，或是工程师根据本条件决定的事项，或其他事项）；

（b） 另一方（或工程师根据第 3.7.2 项［工程师的确定］的规定）拒绝全部或部分索赔；（以及）

（c） 第一方不认可（根据第 3.7.5 项［对工程师的确定不满意］发出不满意通知或其他方式），

但另一方（或工程师）未能全部或部分反对或回复索赔，在这种情况下，如果 DAAB 或仲裁员视情况认为拒绝是合理的，则可以构成拒绝。

1.1.30 "图纸"系指包含在合同中的工程图纸，以及由雇主（或其代表）按照合同发出的任何补充和修改的图纸。

1.1.31 "雇主"系指在合同数据中称为雇主的当事人，及其财产所有权的合法继承人。

1.1.32 "雇主设备"系指雇主根据第 2.6 款［雇主提供的材料和雇主设备］的规定，由雇主向承包商提供使用的仪器、设备、机械、施工设备和/或车辆（如果有）；但不包括根据第 10 条［雇主的接收］的规定，尚未经雇主接收的生产设备。

1.1.33 "**Employer's Personnel**" means the Engineer, the Engineer's Representative (if appointed), the assistants described in Sub-Clause 3.4 [*Delegation by the Engineer*] and all other staff, labour and other employees of the Engineer and of the Employer engaged in fulfilling the Employer's obligations under the Contract; and any other personnel identified as Employer's Personnel, by a Notice from the Employer or the Engineer to the Contractor.

1.1.34 "**Employer-Supplied Materials**" means the materials (if any) to be supplied by the Employer to the Contractor under Sub-Clause 2.6 [*Employer-Supplied Materials and Employer's Equipment*].

1.1.35 "**Engineer**" means the person named in the Contract Data appointed by the Employer to act as the Engineer for the purposes of the Contract, or any replacement appointed under Sub-Clause 3.6 [*Replacement of the Engineer*].

1.1.36 "**Engineer's Representative**" means the natural person who may be appointed by the Engineer under Sub-Clause 3.3 [*Engineer's Representative*].

1.1.37 "**Exceptional Event**" means an event or circumstance as defined in Sub-Clause 18.1 [*Exceptional Events*].

1.1.38 "**Extension of Time**" or "**EOT**" means an extension of the Time for Completion under Sub-Clause 8.5 [*Extension of Time for Completion*].

1.1.39 "**FIDIC**" means the Fédération Internationale des Ingénieurs-Conseils, the International Federation of Consulting Engineers.

1.1.40 "**Final Payment Certificate**" or "**FPC**" means the payment certificate issued by the Engineer under Sub-Clause 14.13 [*Issue of FPC*].

1.1.41 "**Final Statement**" means the Statement defined in Sub-Clause 14.11.2 [*Agreed Final Statement*].

1.1.42 "**Foreign Currency**" means a currency in which part (or all) of the Contract Price is payable, but not the Local Currency.

1.1.43 "**General Conditions**" means this document entitled "Conditions of Contract for Construction for Building and Engineering Works designed by the Employer", as published by FIDIC.

1.1.44 "**Goods**" means Contractor's Equipment, Materials, Plant and Temporary Works, or any of them as appropriate.

1.1.45 "**Interim Payment Certificate**" or "**IPC**" means a Payment Certificate issued by the Engineer for an interim payment under Sub-Clause 14.6 [*Issue of IPC*].

1.1.46 "**Joint Venture**" or "**JV**" means a joint venture, association, consortium or other unincorporated grouping of two or more persons, whether in the form of a partnership or otherwise.

1.1.47 "**JV Undertaking**" means the letter provided to the Employer as part of the Tender setting out the legal undertaking between the two or more persons constituting the Contractor as a JV. This letter shall be signed by all the persons who are members of the JV, shall be addressed to the Employer and shall include:

1.1.33 "雇主人员"系指工程师、工程师代表（如任命）、第3.4款[*由工程师付托*]中规定的助手，以及工程师和雇主根据合同履行雇主义务的所有其他职员、工人和其他雇员；以及由雇主或工程师通知承包商作为雇主人员的任何其他人员。

1.1.34 "雇主提供的材料"系指雇主根据第2.6款[*雇主提供的材料和雇主设备*]的规定，向承包商提供的材料（如果有）。

1.1.35 "工程师"系指由雇主任命并在合同数据中指名，为实施合同担任工程师的人员，或根据第3.6款[*工程师的替代*]的规定，由雇主任命的任何其他人员。

1.1.36 "工程师代表"系指根据第3.3款[*工程师代表*]的规定，可由工程师任命的自然人。

1.1.37 "例外事件"系指第18.1款[*例外事件*]中定义的事件或情况。

1.1.38 "延长时间"或"竣工时间的延长"系根据第8.5款[*竣工时间的延长*]的规定延长竣工时间。

1.1.39 "菲迪克（FIDIC）"系指国际咨询工程师联合会。

1.1.40 "最终付款证书"或"FPC"系指工程师根据第14.13款[*最终付款证书的签发*]的规定签发的付款证书。

1.1.41 "最终报表"系指第14.11.2项[*商定的最终报表*]中规定的报表。

1.1.42 "外币"系指可用于支付合同价格中部分（或全部）款项的当地货币以外的某种货币。

1.1.43 "通用条件"系指由菲迪克（FIDIC）出版的，名为"由雇主设计的建筑和工程施工合同条件"的文件。

1.1.44 "货物"系指承包商设备、材料、生产设备和临时工程，或视情况指其中任何一种。

1.1.45 "期中付款证书"或"IPC"系指工程师根据第14.6款[*期中付款证书的签发*]的规定，为期中付款签发的付款证书。

1.1.46 "联营体"或"JV"系指由两人或两人以上组成的联营企业、联盟、财团或其他非法人团体，其形式是合伙或其他形式。

1.1.47 "联营体承诺书"系指作为投标书的一部分提供给雇主的信函，说明承包商组建联营体的两个或多个人员之间的法律承诺。该信函应由联营体所有成员签字，并应寄给雇主，内容应包括：

(a) each such member's undertaking to be jointly and severally liable to the Employer for the performance of the Contractor's obligations under the Contract;
(b) identification and authorisation of the leader of the JV; and
(c) identification of the separate scope or part of the Works (if any) to be carried out by each member of the JV.

1.1.48 "**Key Personnel**" means the positions (if any) of the Contractor's Personnel, other than the Contractor's Representative, that are stated in the Specification.

1.1.49 "**Laws**" means all national (or state or provincial) legislation, statutes, acts, decrees, rules, ordinances, orders, treaties, international law and other laws, and regulations and by-laws of any legally constituted public authority.

1.1.50 "**Letter of Acceptance**" means the letter of formal acceptance, signed by the Employer, of the Letter of Tender, including any annexed memoranda comprising agreements between and signed by both Parties. If there is no such letter of acceptance, the expression "Letter of Acceptance" means the Contract Agreement and the date of issuing or receiving the Letter of Acceptance means the date of signing the Contract Agreement.

1.1.51 "**Letter of Tender**" means the letter of tender, signed by the Contractor, stating the Contractor's offer to the Employer for the execution of the Works.

1.1.52 "**Local Currency**" means the currency of the Country.

1.1.53 "**Materials**" means things of all kinds (other than Plant), whether on the Site or otherwise allocated to the Contract and intended to form or forming part of the Permanent Works, including the supply-only materials (if any) to be supplied by the Contractor under the Contract.

1.1.54 "**month**" is a calendar month (according to the Gregorian calendar).

1.1.55 "**No-objection**" means that the Engineer has no objection to the Contractor's Documents, or other documents submitted by the Contractor under these Conditions, and such Contractor's Documents or other documents may be used for the Works.

1.1.56 "**Notice**" means a written communication identified as a Notice and issued in accordance with Sub-Clause 1.3 [*Notices and Other Communications*].

1.1.57 "**Notice of Dissatisfaction**" or "**NOD**" means the Notice one Party may give to the other Party if it is dissatisfied, either with an Engineer's determination under Sub-Clause 3.7 [*Agreement or Determination*] or with a DAAB's decision under Sub-Clause 21.4 [*Obtaining DAAB's Decision*].

1.1.58 "**Part**" means a part of the Works or part of a Section (as the case may be) which is used by the Employer and deemed to have been taken over under Sub-Clause 10.2 [*Taking Over Parts*].

1.1.59 "**Particular Conditions**" means the document entitled particular conditions of contract included in the Contract, which consists of Part A - Contract Data and Part B – Special Provisions.

1.1.60 "**Party**" means the Employer or the Contractor, as the context requires. "**Parties**" means both the Employer and the Contractor.

（a） 各成员就履行**合同**规定的**承包商**义务，承诺对**雇主**承担共同的和各自的责任；

（b） **联营体**负责人的确定及授权；（以及）

（c） 确定**联营体**每个成员分别从事的范围或部分**工程**（如果有）。

1.1.48 "**关键人员**"系指**规范要求**中规定的除**承包商代表**外的**承包商人员**的职位（如果有）。

1.1.49 "**法律**"系指所有全国性（或州或省）的法律、条例、法令、法案、规则、条令、命令、条约、国际法和其他法律，以及任何合法成立的公共部门制定的规则和细则等。

1.1.50 "**中标函**"系指**雇主**签署的正式接受**投标函**的信函，包括其所附的由双方间签署的协议的任何备忘录。如无此类中标函，则"**中标函**"系指**合同协议书**，签发或收到**中标函**的日期系指签署**合同协议书**的日期。

1.1.51 "**投标函**"系指由**承包商**签署的投标函，说明**承包商**向**雇主**提出的实施**工程**的报价。

1.1.52 "**当地货币**"系指**工程**所在国的货币。

1.1.53 "**材料**"系指拟构成或正构成永久**工程**一部分的各类物品（**生产设备**除外），无论在**现场**或以其他方式按**合同**分配，包括根据**合同**要由**承包商**供应的只供材料（如果有）。

1.1.54 "**月**"系指一个日历月（按公历）。

1.1.55 "**不反对**"系指**工程师**不反对**承包商文件**或**承包商**根据本**条件**提交的其他文件，而此类**承包商文件**或其他文件可用于**工程**。

1.1.56 "**通知**"系指按照第 1.3 款［*通知和其他通信交流*］发出的书面**通知**。

1.1.57 "**不满意通知**"或"**NOD**"系指一方对**工程师**根据第 3.7 款［*商定或确定*］的规定做出的确定，或 **DAAB** 根据第 21.4 款［*取得争端避免/裁决委员会的决定*］的规定做出的确定不满意时，可向另一方发出的**通知**。

1.1.58 "**部分工程**"系指**雇主**使用并根据第 10.2 款［*部分工程的接收*］的规定，被视为已接收的部分**工程**或部分**分项工程**（视情况而定）。

1.1.59 "**专用条件**"系指包含在**合同**中标题为合同专用条件的文件，由 **A** 部分——合同数据和 **B** 部分——**特别规定**组成。

1.1.60 "**当事方（一方）**"系指**雇主**或**承包商**，根据上下文的需要。"**双方（所有当事方）**"系指**雇主**和**承包商**。

1.1.61 "**Payment Certificate**" means a payment certificate issued by the Engineer under Clause 14 [*Contract Price and Payment*].

1.1.62 "**Performance Certificate**" means the certificate issued by the Engineer (or deemed to be issued) under Sub-Clause 11.9 [*Performance Certificate*].

1.1.63 "**Performance Security**" means the security under Sub-Clause 4.2 [*Performance Security*].

1.1.64 "**Permanent Works**" means the works of a permanent nature which are to be executed by the Contractor under the Contract.

1.1.65 "**Plant**" means the apparatus, equipment, machinery and vehicles (including any components) whether on the Site or otherwise allocated to the Contract and intended to form or forming part of the Permanent Works.

1.1.66 "**Programme**" means a detailed time programme prepared and submitted by the Contractor to which the Engineer has given (or is deemed to have given) a Notice of No-objection under Sub-Clause 8.3 [*Programme*].

1.1.67 "**Provisional Sum**" means a sum (if any) which is specified in the Contract by the Employer as a provisional sum, for the execution of any part of the Works or for the supply of Plant, Materials or services under Sub-Clause 13.4 [*Provisional Sums*].

1.1.68 "**QM System**" means the Contractor's quality management system (as may be updated and/or revised from time to time) in accordance with Sub-Clause 4.9.1 [*Quality Management System*].

1.1.69 "**Retention Money**" means the accumulated retention moneys which the Employer retains under Sub-Clause 14.3 [*Application for Interim Payment*] and pays under Sub-Clause 14.9 [*Release of Retention Money*].

1.1.70 "**Review**" means examination and consideration by the Engineer of a Contractor's submission in order to assess whether (and to what extent) it complies with the Contract and/or with the Contractor's obligations under or in connection with the Contract.

1.1.71 "**Schedules**" means the document(s) entitled schedules prepared by the Employer and completed by the Contractor, as attached to the Letter of Tender and included in the Contract. Such document(s) may include data, lists and schedules of payments and/or rates and prices, and guarantees.

1.1.72 "**Schedule of Payments**" means the document(s) entitled schedule of payments (if any) in the Schedules showing the amounts and manner of payments to be made to the Contractor.

1.1.73 "**Section**" means a part of the Works specified in the Contract Data as a Section (if any).

1.1.74 "**Site**" means the places where the Permanent Works are to be executed and to which Plant and Materials are to be delivered, and any other places specified in the Contract as forming part of the Site.

1.1.75 "**Special Provisions**" means the document (if any), entitled special provisions which constitutes Part B of the Particular Conditions.

1.1.76 "**Specification**" means the document entitled specification included in the Contract, and any additions and modifications to the specification in accordance with the Contract. Such document specifies the Works.

1.1.61 "付款证书"系指根据第 14 条[合同价格和付款]的规定签发的付款证书。

1.1.62 "履约证书"系指根据第 11.9 款[履约证书]的规定签发的(或视为已签发的)证书。

1.1.63 "履约担保"系指根据第 4.2 款[履约担保]规定的担保。

1.1.64 "永久工程"系指按照合同规定要由承包商实施的永久性工程。

1.1.65 "生产设备"系指在现场或以其他方式按合同分配,并构成或拟构成永久工程一部分的装备、设备、机械和车辆(包括任何部件)。

1.1.66 "进度计划"系指由承包商编制和提交的详细时间进度计划,工程师已根据第 8.3 款[进度计划]的规定,向承包商发出(或视为已发出)不反对通知。

1.1.67 "暂列金额"系指雇主在合同中规定为暂列金额的一笔款项(如果有),根据第 13.4 款[暂列金额]的规定,用于实施工程的任何部分,或用于提供生产设备、材料或服务。

1.1.68 "质量管理体系"系指承包商根据第 4.9.1 项[质量管理体系]的规定,制定的质量管理体系(可不时更新和/或修订)。

1.1.69 "保留金"系指雇主根据第 14.3 款[期中付款的申请]的规定扣留的累计保留金,根据第 14.9 款[保留金的发放]的规定进行支付。

1.1.70 "审核"系指工程师对承包商提交的材料进行审查和审议,以评估其是否(以及在多大程度上)符合本合同和/或本合同规定的或与本合同有关的承包商义务。

1.1.71 "资料表"系指合同中名为各种表的文件,由雇主编制、由承包商填写并随投标函一起提交。此类文件可包括数据、表册、付款计划表、费率和/或价格表以及保函。

1.1.72 "付款计划表"系指资料表中名为付款计划表的文件(如果有),其中说明了应向承包商付款的金额和方式。

1.1.73 "分项工程"系指在合同数据中确定为分项工程(如果有)的工程组成部分。

1.1.74 "现场"系将实施永久工程和运送生产设备与材料到达的地点,以及合同中指定为现场组成部分的任何其他场所。

1.1.75 "特别规定"系指名为特别规定的文件(如果有),其构成专用条件 B 部分。

1.1.76 "规范要求"系指包含在合同中名为规范要求的文件,以及按照合同对规范要求所做的任何补充和修改。此类文件规定对工程的要求。

1.1.77 "**Statement**" means a statement submitted by the Contractor as part of an application for a Payment Certificate under Sub-Clause 14.3 [*Application for Interim Payment*], Sub-Clause 14.10 [*Statement at Completion*] or Sub-Clause 14.11 [*Final Statement*].

1.1.78 "**Subcontractor**" means any person named in the Contract as a subcontractor, or any person appointed by the Contractor as a subcontractor or designer, for a part of the Works; and the legal successors in title to each of these persons.

1.1.79 "**Taking-Over Certificate**" means a certificate issued (or deemed to be issued) by the Engineer in accordance with Clause 10 [*Employer's Taking Over*].

1.1.80 "**Temporary Works**" means all temporary works of every kind (other than Contractor's Equipment) required on Site for the execution of the Works.

1.1.81 "**Tender**" means the Letter of Tender, the Contractor's Proposal, the JV Undertaking (if applicable), and all other documents which the Contractor submitted with the Letter of Tender, as included in the Contract.

1.1.82 "**Tests after Completion**" means the tests (if any) which are stated in the Specification and which are carried out in accordance with the Special Provisions after the Works or a Section (as the case may be) are taken over under Clause 10 [*Employer's Taking Over*].

1.1.83 "**Tests on Completion**" means the tests which are specified in the Contract or agreed by both Parties or instructed as a Variation, and which are carried out under Clause 9 [*Tests on Completion*] before the Works or a Section (as the case may be) are taken over under Clause 10 [*Employer's Taking Over*].

1.1.84 "**Time for Completion**" means the time for completing the Works or a Section (as the case may be) under Sub-Clause 8.2 [*Time for Completion*], as stated in the Contract Data as may be extended under Sub-Clause 8.5 [*Extension of Time for Completion*], calculated from the Commencement Date.

1.1.85 "**Unforeseeable**" means not reasonably foreseeable by an experienced contractor by the Base Date.

1.1.86 "**Variation**" means any change to the Works, which is instructed as a variation under Clause 13 [*Variations and Adjustments*].

1.1.87 "**Works**" mean the Permanent Works and the Temporary Works, or either of them as appropriate.

1.1.88 "**year**" means 365 days.

1.2 Interpretation

In the Contract, except where the context requires otherwise:

(a) words indicating one gender include all genders; and "he", "his" and "himself" shall be read as "he/she", "his/her" and "himself/herself" respectively;
(b) words indicating the singular also include the plural and words indicating the plural also include the singular;
(c) provisions including the word "agree", "agreed" or "agreement" require the agreement to be recorded in writing;

1.1.77 "报表"系指承包商根据第14.2.1项[预付款担保函](如果适用)、第14.3款[期中付款的申请]、第14.10款[竣工报表]或第14.11款[最终报表]的规定,提交的作为付款证书申请组成部分的报表。⊖

1.1.78 "分包商"系指在合同中为分包商的任何人,或承包商为部分工程任命为分包商或设计师的任何人员;以及这些人员各自财产所有权的合法继承人。

1.1.79 "接收证书"系指工程师根据第10条[雇主的接收]的规定签发(或视为已签发)的证书。

1.1.80 "临时工程"系指为实施工程,在现场所需的所有各类临时工程(承包商设备除外)。

1.1.81 "投标书"系指投标函和合同中包括的由承包商随投标函一起提交的联营体承诺书(如适用)和所有其他文件。⊖

1.1.82 "竣工后试验"系指在规范要求中规定的,在工程或某分项工程(视情况而定),按照第10条[雇主的接收]的规定接收后,根据特别规定的要求进行的试验(如果有)。

1.1.83 "竣工试验"系指在合同中规定的,或双方商定的,或按指示作为一项变更的,在工程或某分项工程(视情况而定)根据第10条[雇主的接收]的规定接收前,按照第9条[竣工试验]的要求进行的试验。

1.1.84 "竣工时间"系指合同数据中规定的,自开工日期算起至工程或某分项工程(视情况而定)根据第8.2款[竣工时间]规定的要求竣工,连同根据第8.5款[竣工时间的延长]的规定提出的延长期的全部时间。

1.1.85 "不可预见的"系指一个有经验的承包商在提交基准日期前不能合理预见。

1.1.86 "变更"系指根据第13条[变更和调整]的规定,经指示作为变更的、对工程所做的任何更改。

1.1.87 "工程"系指永久工程和临时工程,或其中任何一项(视情况而定)。

1.1.88 "年"系指365天。

1.2 解释

在合同中,除上下文中另有要求外:

(a) 表示一性别的词,包括所有性别;"他""他的"和"他自己"应分别理解为"他/她""他/她的"和"他/她自己";

(b) 单数形式的词,也包括复数含义,反之亦然;

(c) 包括"同意(商定)""达成(取得)一致",或"协议"等词的各项规定都要求用书面记载;

⊖ 此处中文按照勘误表改正后的英文翻译。——译者注

(d) "written" or "in writing" means hand-written, type-written, printed or electronically made, and resulting in a permanent record;

(e) "may" means that the Party or person referred to has the choice of whether to act or not in the matter referred to;

(f) "shall" means that the Party or person referred to has an obligation under the Contract to perform the duty referred to;

(g) "consent" means that the Employer, the Contractor or the Engineer (as the case may be) agrees to or gives permission for, the requested matter;

(h) "including", "include" and "includes" shall be interpreted as not being limited to, or qualified by, the stated items that follow;

(i) words indicating persons or parties shall be interpreted as referring to natural and legal persons (including corporations and other legal entities); and

(j) "execute the Works" or "execution of the Works" means the construction and completion of the Works and the remedying of any defects (and shall be deemed to include design to the extent, if any, specified in the Contract)

In any list in these Conditions, where the second-last item of the list is followed by "and" or "or" or "and/or" then all of the list items going before this item shall also be read as if they are followed by "and" or "or" or "and/or" (as the case may be).

The marginal words and other headings shall not be taken into consideration in the interpretation of these Conditions.

1.3 Notices and Other Communications

Wherever these Conditions provide for the giving of a Notice (including a Notice of Dissatisfaction) or the issuing, providing, sending, submitting or transmitting of another type of communication (including acceptance, acknowledgement, advising, agreement, approval, certificate, Claim, consent, decision, determination, discharge, instruction, No-objection, record(s) of meeting, permission, proposal, record, reply, report, request, Review, Statement, statement, submission or any other similar type of communication), the Notice or other communication shall be in writing and:

(a) shall be:

(i) a paper-original signed by the Contractor's Representative, the Engineer, or the authorised representative of the Employer (as the case may be); or

(ii) an electronic original generated from any of the systems of electronic transmission stated in the Contract Data (if not stated, system(s) acceptable to the Engineer), where the electronic original is transmitted by the electronic address uniquely assigned to each of such authorised representatives,

or both, as stated in these Conditions; and

(b) if it is a Notice, it shall be identified as a Notice. If it is another form of communication, it shall be identified as such and include reference to the provision(s) of the Contract under which it is issued where appropriate;

(c) delivered by hand (against receipt), or sent by mail or courier (against receipt), or transmitted using any of the systems of electronic transmission under sub-paragraph (a)(ii) above; and

(d) delivered, sent or transmitted to the address for the recipient's communications as stated in the Contract Data. However, if the recipient gives a Notice of another address, all Notices and other communications shall be delivered accordingly after the sender receives such Notice.

(d) "书面"或"以书面形式"系指手写、打字、印刷或电子制作，并形成永久性记录；

(e) "可以"系指一方（当事人）或个人有权选择是否就所述事项采取行动；

(f) "应当"系指一方（当事人）或个人根据**合同**规定有义务履行规定的职责；

(g) "同意"系指**雇主**、**承包商**或**工程师**（视情况而定）同意或准许所要求的事项；

(h) "包括……在内""包括"和"包含"应解释为不限于或不受后述事项的限制；

(i) 表示人或当事方的词语，应解释为自然人和法人（包括公司和其他法律实体）；（以及）

(j) "**实施工程**"或"**工程实施**"系指**工程**的施工和竣工以及任何缺陷的修补（应视为包括在**合同**规定范围内的设计，如果有）。

在本**条件**的任何排列中，如果倒数第 2 项后为"和／以及"，或"或"，或"和／或"，则该项之前的所有列项也应理解为"和／以及"，或"或"，或"和／或"（视情况而定）。

旁注和其他标题在本**条件**的解释中不应考虑。

1.3
通知和其他通信交流

本**条件**不论在何种场合规定发出**通知**（包括**不满意通知**）或签发、提供、发送、提交，或另一种通信方式传送（包括接受、确认、通知、建议、协议、批准、证明书、索赔、同意、决定、确定、解除、指示、**不反对**、会议记录、许可、提议、记录、答复、报告、请求、**审核**、**报表**、说明，提交或任何其他类似的通信交流），**通知**或其他通信交流都应以书面形式，并且：

(a) 应为：

(i) 由**承包商代表**、**工程师**或**雇主**授权代表（视情况而定）签署的纸质原件；（或）

(ii) 由**合同数据**规定的任何电子传输方式生成的电子原件（如未规定，由**工程师**可接受的方式），电子原件通过唯一分配给每个此类授权代表的电子邮箱发送；

或按照本**条件**所述的两种方式发送；（以及）

(b) 如果是**通知**，则应注明为**通知**。如果是另一种通信方式，则应同样注明，并包括在适当情况下按照**合同**规定签发的参考条款；

(c) 由人面交（取得对方收据），或通过邮寄或信差传送（取得对方收据），或用上述（a）(ii) 目中提出的任何的电子传输方式发送；（以及）

(d) 交付、传送或传输至**合同数据**中规定的接收人的地址。但，如接收人**通知**了另外地址，寄件人收到**通知**后，所有**通知**和其他通信交流均应按新地址发送。

Where these Conditions state that a Notice or NOD or other communication is to be delivered, given, issued, provided, sent, submitted or transmitted, it shall have effect when it is received (or deemed to have been received) at the recipient's current address under sub-paragraph (d) above. An electronically transmitted Notice or other communication is deemed to have been received on the day after transmission, provided no non-delivery notification was received by the sender.

All Notices, and all other types of communication as referred to above, shall not be unreasonably withheld or delayed.

When a Notice or NOD or certificate is issued by a Party or the Engineer, the paper and/or electronic original shall be sent to the intended recipient and a copy shall be sent to the Engineer or the other Party, as the case may be. All other communications shall be copied to the Parties and/or the Engineer as stated under these Conditions or elsewhere in the Contract.

1.4 Law and Language

The Contract shall be governed by the law of the country (or other jurisdiction) stated in the Contract Data (if not stated, the law of the Country), excluding any conflict of law rules.

The ruling language of the Contract shall be that stated in the Contract Data (if not stated, the language of these Conditions). If there are versions of any part of the Contract which are written in more than one language, the version which is in the ruling language shall prevail.

The language for communications shall be that stated in the Contract Data. If no language is stated there, the language for communications shall be the ruling language of the Contract.

1.5 Priority of Documents

The documents forming the Contract are to be taken as mutually explanatory of one another. If there is any conflict, ambiguity or discrepancy, the priority of the documents shall be in accordance with the following sequence:

(a) the Contract Agreement;
(b) the Letter of Acceptance;
(c) the Letter of Tender;
(d) the Particular Conditions Part A – Contract Data;
(e) the Particular Conditions Part B – Special Provisions;
(f) these General Conditions;
(g) the Specification;
(h) the Drawings;
(i) the Schedules;
(j) the JV Undertaking (if the Contractor is a JV); and
(k) any other documents forming part of the Contract.

If a Party finds an ambiguity or discrepancy in the documents, that Party shall promptly give a Notice to the Engineer, describing the ambiguity or discrepancy. After receiving such Notice, or if the Engineer finds an ambiguity or discrepancy in the documents, the Engineer shall issue the necessary clarification or instruction.

1.6 Contract Agreement

The Parties shall sign a Contract Agreement within 35 days after the Contractor receives the Letter of Acceptance, unless they agree otherwise. The Contract Agreement shall be based on the form annexed to the Particular Conditions. The costs of stamp duties and similar charges (if any) imposed by law in connection with entry into the Contract Agreement shall be borne by the Employer.

本**条件**规定**通知**或**不满意通知**或其他通信交流要交付、发出、签发、提供、发送、提交或传送的，在接收人根据上述（d）段规定的现时地址收到（或视为已收到）时生效。如果寄件人没有收到未送达通知，电子发送的**通知**或其他通信交流在发送后的第二天视为已收到。

上述所有**通知**和其他种类的通信交流，不得无故被扣压或拖延。

一方或**工程师**签发**通知**或**不满意通知**或证书时，纸质和/或电子原件应发送给预定接收人，并且亦应根据情况抄送**工程师**或另一方。所有其他通信交流均应按本**条件**或合同其他规定抄送给双方和/或**工程师**。

1.4 法律和语言

合同应受**合同数据**中所述国家（或其他司法管辖区）的法律管辖［如未规定，则由工程**所在国**的法律管辖］，不包括任何法律规则的冲突。

合同的主导语言应为**合同数据**中规定的语言（如未规定，则为本**条件**的语言）。如果**合同**任何部分的文本采用一种以上语言编写，则应以**合同数据**规定的主导语言文本为准。

通信交流应使用**合同数据**中规定的语言。如未规定，应使用**合同**的主导语言。

1.5 文件优先次序

构成合同的文件应能够相互说明。如有任何冲突、歧义或不一致，文件的优先次序如下：

（a）合同协议书；
（b）中标函；
（c）投标函；
（d）专用条件 A 部分——合同数据；
（e）专用条件 B 部分——特别规定；
（f）本通用条件；
（g）规范要求；
（h）图纸；
（i）资料表；
（j）联营体承诺书（如果承包商是联营体）；（以及）
（k）构成合同组成部分的任何其他文件。

如一方发现文件有歧义或不一致，应立即向**工程师**发出**通知**，说明歧义或不一致之处。收到此类**通知**后，或如果**工程师**发现文件中的歧义或不一致之处，**工程师**应发出必要的澄清或指示。

1.6 合同协议书

除非另有协议，双方应在**承包商**收到**中标函**后35天内签订合同协议书。合同协议书应以**专用条件**所附格式为依据。与签订合同协议书有关的、依法征收的印花税和类似费用（如果有）应由**雇主**承担。

If the Contractor comprises a JV, the authorised representative of each member of the JV shall sign the Contract Agreement.

1.7 Assignment

Neither Party shall assign the whole or any part of the Contract or any benefit or interest in or under the Contract. However, either Party:

(a) may assign the whole or any part of the Contract with the prior agreement of the other Party, at the sole discretion of such other Party; and

(b) may, as security in favour of a bank or financial institution, assign the Party's right to any moneys due, or to become due, under the Contract without the prior agreement of the other Party.

1.8 Care and Supply of Documents

The Specification and Drawings shall be in the custody and care of the Employer. Unless otherwise stated in the Contract, two copies of the Contract and of each subsequent Drawing shall be supplied to the Contractor, who may make or request further copies at the cost of the Contractor.

Each of the Contractor's Documents shall be in the custody and care of the Contractor, unless and until submitted to the Engineer. The Contractor shall supply to the Engineer one paper-original, one electronic copy (in the form as stated in the Specification or, if not stated, a form acceptable to the Engineer) and additional paper copies (if any) as stated in the Contract Data of each of the Contractor's Documents.

The Contractor shall keep at all times, on the Site, a copy of:

(a) the Contract;
(b) the records under Sub-Clause 6.10 [*Contractor's Records*] and Sub-Clause 20.2.3 [*Contemporary records*];
(c) the publications (if any) named in the Specification;
(d) the Contractor's Documents;
(e) the Drawings; and
(f) Variations, Notices and other communications given under the Contract.

The Employer's Personnel shall have right of access to all these documents during all normal working hours, or as otherwise agreed with the Contractor.

If a Party (or the Engineer) becomes aware of an error or defect (whether of a technical nature or otherwise) in a document which was prepared for use in the execution of the Works, the Party (or the Engineer) shall promptly give a Notice of such error or defect to the other Party (or to the Parties).

1.9 Delayed Drawings or Instructions

The Contractor shall give a Notice to the Engineer whenever the Works are likely to be delayed or disrupted if any necessary drawing or instruction is not issued to the Contractor within a particular time, which shall be reasonable. The Notice shall include details of the necessary drawing or instruction, details of why and by when it should be issued, and details of the nature and amount of the delay or disruption likely to be suffered if it is late.

If the Contractor suffers delay and/or incurs Cost as a result of a failure of the Engineer to issue the notified drawing or instruction within a time which is reasonable and is specified in the Notice with supporting details, the Contractor shall be entitled subject to Sub-Clause 20.2 [*Claims For Payment and/or EOT*] to EOT and/or payment of such Cost Plus Profit.

如果**承包商**由**联营体**组成，**联营体**各成员的授权代表应签署合同协议书。

1.7 权益转让

任一方都不应将**合同**的全部或任何部分，或在**合同**中或由**合同**规定的任何利益或权益转让他人。但任一方：

(a) 在另一方完全自主决定的情况下，事先征得其同意后，可以将合同的全部或部分转让；（以及）
(b) 可作为有利于银行或金融机构的担保，在未经另一方事先同意的情况下，转让其根据**合同**规定的任何到期或将到期应得款项的权利。

1.8 文件的照管和提供

规范要求和**图纸**应由**雇主**保存和照管。除非**合同**中另有规定，应给**承包商**一式两份合同文本和后续**图纸**，**承包商**可以自费复制或要求提供更多份数。

每份**承包商文件**都应由**承包商**保存和照管，除非并直到提交**工程师**为止，**承包商**应向**工程师**提供一份纸质原件、一份电子文件（按**规范要求**规定的格式，如未规定，按**工程师**接受的格式），以及按**合同数据**规定的**承包商文件**的附加纸质副本（如果有）。

承包商应随时在**现场**保存一份：

(a) **合同**；
(b) 第 6.10 款［*承包商的记录*］和第 20.2.3 项［*同期记录*］规定的记录；
(c) **规范要求**中指名的出版物（如果有）；
(d) **承包商文件**；
(e) **图纸**；（以及）
(f) **合同**规定的**变更**、**通知**和其他通信交流。

雇主人员有权在所有正常工作时间，或与**承包商**商定的其他时间，使用所有这些文件。

如果一方（或**工程师**）发现为实施**工程**编制的文件中有错误或缺陷（无论是技术性的或是其他的），该方（或**工程师**）应迅速将该错误或缺陷**通知**另一方（或双方）。

1.9 延误的图纸或指示

如果任何必需的图纸或指示未能在合理的特定时间内发至**承包商**，致使**工程**可能拖延或中断时，**承包商**应**通知工程师**。**通知**应包括必需的图纸或指示的细节，为何和何时前必须发出的详细理由，以及如果晚发出可能遭受的延误或中断的性质和程度的详情。

如果由于**工程师**未能在合理的并在**承包商**附有支持细节的**通知**中规定的时间内发出图纸或指示，使**承包商**遭受延误和／或招致增加**费用**，**承包商**有权根据第 20.2 款［*付款和／或竣工时间延长的索赔*］的规定，获得**竣工时间的延长**和／或此类**成本加利润**的支付。

However, if and to the extent that the Engineer's failure was caused by any error or delay by the Contractor, including an error in, or delay in the submission of, any of the Contractor's Documents, the Contractor shall not be entitled to such EOT and/or Cost Plus Profit.

1.10 Employer's Use of Contractor's Documents

As between the Parties, the Contractor shall retain the copyright and other intellectual property rights in the Contractor's Documents (and other design documents, if any, made by (or on behalf of) the Contractor).

The Contractor shall be deemed (by signing the Contract Agreement) to give to the Employer a non-terminable transferable non-exclusive royalty-free licence to copy, use and communicate the Contractor's Documents (and such other design documents, if any), including making and using modifications of them. This licence shall:

(a) apply throughout the actual or intended operational life (whichever is longer) of the relevant parts of the Works;
(b) entitle any person in proper possession of the relevant part of the Works to copy, use and communicate the Contractor's Documents (and such other design documents, if any) for the purposes of completing, operating, maintaining, altering, adjusting, repairing and demolishing the Works;
(c) in the case of Contractor's Documents (and such other design documents, if any) which are in the form of electronic or digital files, computer programs and other software, permit their use on any computer on the Site and/or at the locations of the Employer and the Engineer and/or at other places as envisaged by the Contract; and
(d) in the event of termination of the Contract:

   (i) under Sub-Clause 15.2 [*Termination for Contractor's Default*], entitle the Employer to copy, use and communicate the Contractor's Documents (and other design documents made by or for the Contractor, if any); or
   (ii) under Sub-Clause 15.5 [*Termination for Employer's Convenience*], Sub-Clause 16.2 [*Termination by Contractor*] or Sub-Clause 18.5 [*Optional Termination*], entitle the Employer to copy, use and communicate the Contractor's Documents for which the Contractor has received payment

for the purpose of completing the Works and/or arranging for any other entities to do so.

The Contractor's Documents (and other design documents, if any, made by (or on behalf of) the Contractor) shall not, without the Contractor's prior consent, be used, copied or communicated to a third party by (or on behalf of) the Employer for purposes other than those permitted under this Sub-Clause.

1.11 Contractor's Use of Employer's Documents

As between the Parties, the Employer shall retain the copyright and other intellectual property rights in the Specification and Drawings and other documents made by (or on behalf of) the Employer. The Contractor may, at the Contractor's cost, copy, use and communicate these documents for the purposes of the Contract.

These documents (in whole or in part) shall not, without the Employer's prior consent, be copied, used or communicated to a third party by the Contractor, except as necessary for the purposes of the Contract.

但是，如果**工程师**未能发出是由于**承包商**的错误或拖延，包括**承包商文件**中的错误或提交拖延造成的，**承包商**无权要求此类**竣工时间**的延长和/或**成本加利润**。

1.10 雇主使用承包商文件

就**双方**而言，由**承包商**编制的**承包商文件**［以及由**承包商**（或其代表）编制的其他设计文件，如果有］，其版权和其他知识产权应归**承包商**所有。

承包商（通过签署**合同协议书**）应被认为已给予**雇主**无限期的、可转让的、不排他的、免版税的许可，复制、使用和传送**承包商文件**（以及此类其他设计文件，如果有），包括对其做出的修改和使用。这项许可应：

（a）适用于**工程**相关部分的实际或预期寿命期（取较长的）；

（b）允许具有**工程**相关部分正当占有权的任何人，为了完成、运行、维护、更改、调整、修复和拆除**工程**的目的，复制、使用和传送**承包商文件**（以及此类其他设计文件，如果有）；

（c）如果在**承包商文件**（以及此类其他设计文件，如果有）是电子或数字文件、计算机程序和其他软件形式，允许它们在**现场**和/或**合同**规定的和/或**雇主**、**工程师**的地点的其他场所的任何计算机上使用；（以及）

（d）在合同终止时：

（i）根据第15.2款［*因承包商违约的终止*］的规定，**雇主**有权复制、使用和传送**承包商文件**（以及由**承包商**或为**承包商**编写的其他设计文件，如果有）；（或）

（ii）根据第15.5款［*为雇主便利的终止*］、第16.2款［*由承包商终止*］和/或第18.5款［*自主选择终止*］的规定，**雇主**有权复制、使用和传送**承包商**已收到付款的**承包商文件**。

为完成**工程**和/或安排其他实体同样使用这些文件。

未经**承包商**事先同意，**雇主**（或其代表）不得在本款允许以外，为其他目的使用、复制**承包商文件**［以及由**承包商**（或其代表）编制的其他设计文件，如果有］，或将其传送给第三方。

1.11 承包商使用雇主文件

就**双方**而言，由**雇主**（或其代表）编制的**规范要求**、**图纸**和其他文件，其版权和其他知识产权应归**雇主**所有。**承包商**因合同的目的，可自费复制、使用和传送上述文件。

除**合同**需要外，未经**雇主**事先同意，**承包商**不得复制、使用这些文件（全部或部分），或将其传送给第三方。

1.12 Confidentiality

The Contractor shall disclose all such confidential and other information as the Engineer may reasonably require in order to verify the Contractor's compliance with the Contract.

The Contractor shall treat all documents forming the Contract as confidential, except to the extent necessary to carry out the Contractor's obligations under the Contract. The Contractor shall not publish, permit to be published, or disclose any particulars of the Contract in any trade or technical paper or elsewhere without the Employer's prior consent.

The Employer and the Engineer shall treat all information provided by the Contractor and marked "confidential", as confidential. The Employer shall not disclose or permit to be disclosed any such information to third parties, except as may be necessary when exercising the Employer's rights under Sub-Clause 15.2 [*Termination for Contractor's Default*].

A Party's obligation of confidentiality under this Sub-Clause shall not apply where the information:

(a) was already in that Party's possession without an obligation of confidentiality before receipt from the other Party;
(b) becomes generally available to the public through no breach of these Conditions; or
(c) is lawfully obtained by the Party from a third party which is not bound by an obligation of confidentiality.

1.13 Compliance with Laws

The Contractor and the Employer shall, in performing the Contract, comply with all applicable Laws. Unless otherwise stated in the Specification:

(a) the Employer shall have obtained (or shall obtain) the planning, zoning or building permit or similar permits, permissions, licences and/or approvals for the Permanent Works, and any other permits, permissions, licenses and/or approvals described in the Specification as having been (or being) obtained by the Employer. The Employer shall indemnify and hold the Contractor harmless against and from the consequences of any delay or failure to do so, unless the failure is caused by the Contractor's failure to comply with sub-paragraph (c) below;
(b) the Contractor shall give all notices, pay all taxes, duties and fees, and obtain all other permits, permissions, licences and/or approvals, as required by the Laws in relation to the execution of the Works. The Contractor shall indemnify and hold the Employer harmless against and from the consequences of any failure to do so unless the failure is caused by the Employer's failure to comply with Sub-Clause 2.2 [*Assistance*];
(c) within the time(s) stated in the Specification the Contractor shall provide such assistance and all documentation, as described in the Specification or otherwise reasonably required by the Employer, so as to allow the Employer to obtain any permit, permission, licence or approval under sub-paragraph (a) above; and
(d) the Contractor shall comply with all permits, permissions, licences and/or approvals obtained by the Employer under sub-paragraph (a) above.

If, having complied with sub-paragraph (c) above, the Contractor suffers delay and/or incurs Cost as a result of the Employer's delay or failure to obtain any permit, permission, licence or approval under sub-paragraph (a) above, the Contractor shall be entitled subject to Sub-Clause 20.2 [*Claims For Payment and/or EOT*] to EOT and/or payment of such Cost Plus Profit.

1.12 保密

对**工程师**为了证实**承包商**遵守**合同**的情况，合理需要的所有秘密和其他信息，**承包商**应当透露。

承包商应将构成合同的所有文件视为保密文件，但履行合同规定的**承包商**义务所必需的文件除外。未经雇主事先同意，**承包商**不应在任何行业或技术文件或其他地方发布、允许发布或透露合同的任何细节。

雇主和**工程师**应将**承包商**提供的所有标明"保密"的信息视为保密资料。**雇主**不应向第三方透露或允许透露任何此类信息，除根据第 15.2 款［*因承包商违约的终止*］的规定，**雇主**在履行其必要的权利时。

本**款**规定的一方保密义务不适用于以下情况：

（a） 该方从另一方收到该方事先已拥有的、没有保密义务的；

（b） 在不违反本**条件**的情况下，普遍向公众提供的；（或）

（c） 由该方从不受保密义务约束的第三方合法取得的。

1.13 遵守法律

承包商和**雇主**在履行合同期间，应遵守所有适用**法律**。除非**规范要求**中另有规定：

（a） **雇主**应已（或将）为**永久工程**取得规划、区域划定或施工许可证，或类似的许可证、准许、执照和／或批准以及**规范要求**中所述的**雇主**已（或将）取得的任何其他许可、许可证、执照和／或批准；**雇主**应保障和保持**承包商**免受未能完成上述工作带来的延误或损害；除非该延误或损害是由于**承包商**不能遵守下述（c）段的规定造成的；

（b） **承包商**应发出所有通知，缴纳各项税费、关税和费用，按照**法律**关于工程设计、实施和竣工等方面的要求，办理并领取所需要的所有其他许可、准许、执照和／或批准；**承包商**应保障和保持**雇主**免受因未能完成上述工作带来的损害；除非该损害是由于**雇主**未能遵守第 2.2 款［*协助*］的规定造成的；

（c） 在**规范要求**规定的时间内，**承包商**应按照**规范要求**中的规定或**雇主**的合理要求提供此类协助和所有文件，以便**雇主**获得上述（a）段所述的任何许可证、准许、执照或批准；（以及）

（d） **承包商**应遵守**雇主**根据上述（a）段获得的所有许可证、准许、执照和／或批准。

如果**承包商**遵守了上述（c）段的规定，由于**雇主**的延误或未能根据上述（a）段的规定获得任何许可证、准许、执照和／或批准，使**承包商**遭受延误和／或招致增加**费用**，**承包商**应根据第 20.2 款［*付款和／或竣工时间延长的索赔*］的规定，有权获得**竣工时间**的延长和／或此类**成本加利润**的支付。

If the Employer incurs additional costs as a result of the Contractor's failure to comply with:

(i) sub-paragraph (c) above; or
(ii) sub-paragraph (b) or (d) above, provided that the Employer shall have complied with Sub-Clause 2.2 [*Assistance*],

the Employer shall be entitled subject to Sub-Clause 20.2 [*Claims For Payment and/or EOT*] to payment of these costs by the Contractor.

1.14 Joint and Several Liability

If the Contractor is a Joint Venture:

(a) the members of the JV shall be jointly and severally liable to the Employer for the performance of the Contractor's obligations under the Contract;
(b) the JV leader shall have authority to bind the Contractor and each member of the JV; and
(c) neither the members nor (if known) the scope and parts of the Works to be carried out by each member nor the legal status of the JV shall be altered without the prior consent of the Employer (but such consent shall not relieve the altered JV from any liability under sub-paragraph (a) above).

1.15 Limitation of Liability

Neither Party shall be liable to the other Party for loss of use of any Works, loss of profit, loss of any contract or for any indirect or consequential loss or damage which may be suffered by the other Party in connection with the Contract, other than under:

(a) Sub-Clause 8.8 [*Delay Damages*];
(b) sub-paragraph (c) of Sub-Clause 13.3.1 [*Variation by Instruction*];
(c) Sub-Clause 15.7 [*Payment after Termination for Employer's Convenience*];
(d) Sub-Clause 16.4 [*Payment after Termination by Contractor*];
(e) Sub-Clause 17.3 [*Intellectual and Industrial Property Rights*];
(f) the first paragraph of Sub-Clause 17.4 [*Indemnities by Contractor*]; and
(g) Sub-Clause 17.5 [*Indemnities by Employer*].

The total liability of the Contractor to the Employer under or in connection with the Contract, other than:

(i) under Sub-Clause 2.6 [*Employer-Supplied Materials and Employer's Equipment*];
(ii) under Sub-Clause 4.19 [*Temporary Utilities*];
(iii) under Sub-Clause 17.3 [*Intellectual and Industrial Property Rights*]; and
(iv) under the first paragraph of Sub-Clause 17.4 [*Indemnities by Contractor*],

shall not exceed the sum stated in the Contract Data or (if a sum is not so stated) the Accepted Contract Amount.

This Sub-Clause shall not limit liability in any case of fraud, gross negligence, deliberate default or reckless misconduct by the defaulting Party.

如果**雇主**因**承包商**未能遵守下述规定而招致产生了额外费用：

（i） 按照上述（c）段；（或）
（ii） 按照上述（b）或（d）段，**雇主**已遵守第2.2款[*协助*]的规定。

雇主应有权根据第20.2款[*付款和/或竣工时间延长的索赔*]的规定，由**承包商**支付这些费用。

1.14 共同的和各自的责任

如果**承包商**是**联营体**：

（a） **联营体**成员应对**雇主**负有履行**合同**规定的**承包商**义务的共同的和各自的责任；

（b） **联营体**负责人有权约束**承包商**及其**联营体**的每个成员；（以及）

（c） 未经**雇主**事先同意，任何成员或（已知的）每个成员将要实施的**工程**的范围和部分，以及**联营体**的法律地位均不得改变[但此类同意不应免除变更后的**联营体**在上述（a）段规定的任何责任]。

1.15 责任限度

任一方均不对另一方根据本**合同**可能遭受的任何**工程**的使用损失、利润损失、任何合同的损失，或任何间接的或引起严重后果的损失或损害承担责任，除以下规定外：

（a） 第8.8款[*误期损害赔偿费*]；
（b） 第13.3.1项[*指示变更*]（c）段；
（c） 第15.7款[*为雇主便利终止后的付款*]；
（d） 第16.4款[*由承包商终止后的付款*]；
（e） 第17.3款[*知识产权和工业产权*]；
（f） 第17.4款[*由承包商保障*]第一段；（以及）
（g） 第17.5款[*由雇主保障*]。

以下规定以外的，**承包商**根据本**合同**或与本**合同**有关的规定对**雇主**的全部责任，包括：

（i） 第2.6款[*雇主提供的材料和雇主设备*]规定的；

（ii） 第4.19款[*临时公用设施*]规定的；

（iii） 第17.3款[*知识产权和工业产权*]规定的；（以及）

（iv） 第17.4款[*由承包商保障*]第1段规定的；

均不应超过**合同数据**中规定的金额，或（如果未规定金额）**中标合同金额**。

本款不应限制追究违约方在任何欺骗、重大过失、故意违约或轻率不当行为情况下的责任。

1.16
Contract Termination

Subject to any mandatory requirements under the governing law of the Contract, termination of the Contract under any Sub-Clause of these Conditions shall require no action of whatsoever kind by either Party other than as stated in the Sub-Clause.

2 The Employer

2.1
Right of Access to the Site

The Employer shall give the Contractor right of access to, and possession of, all parts of the Site within the time (or times) stated in the Contract Data. The right and possession may not be exclusive to the Contractor. If, under the Contract, the Employer is required to give (to the Contractor) possession of any foundation, structure, plant or means of access, the Employer shall do so in the time and manner stated in the Specification. However, the Employer may withhold any such right or possession until the Performance Security has been received.

If no such time is stated in the Contract Data, the Employer shall give the Contractor right of access to, and possession of, those parts of the Site within such times as may be required to enable the Contractor to proceed in accordance with the Programme or, if there is no Programme at that time, the initial programme submitted under Sub-Clause 8.3 [*Programme*].

If the Contractor suffers delay and/or incurs Cost as a result of a failure by the Employer to give any such right or possession within such time, the Contractor shall be entitled subject to Sub-Clause 20.2 [*Claims For Payment and/or EOT*] to EOT and/or payment of such Cost Plus Profit.

However, if and to the extent that the Employer's failure was caused by any error or delay by the Contractor, including an error in, or delay in the submission of, any of the applicable Contractor's Documents, the Contractor shall not be entitled to such EOT and/or Cost Plus Profit.

2.2
Assistance

If requested by the Contractor, the Employer shall promptly provide reasonable assistance to the Contractor so as to allow the Contractor to obtain:

(a) copies of the Laws of the Country which are relevant to the Contract but are not readily available; and

(b) any permits, permissions, licences or approvals required by the Laws of the Country (including information required to be submitted by the Contractor in order to obtain such permits, permissions, licences or approvals):

  (i) which the Contractor is required to obtain under Sub-Clause 1.13 [*Compliance with Laws*];
  (ii) for the delivery of Goods, including clearance through customs; and
  (iii) for the export of Contractor's Equipment when it is removed from the Site.

1.16 合同终止

根据本合同适用法律的任何强制性要求，按照本条件的任何条款终止**合同**，除按照本**条款**的规定外，不要求任何一方采取任何形式的行动。

2 雇主

2.1 现场进入权

雇主应在**合同数据**规定的时间（或几个时间）内，给予**承包商**进入和占用**现场**各部分的权利。此项进入和占用权可不为**承包商**独享。如果根据合同，要求**雇主**（向**承包商**）提供任何地基、结构、生产设备的占用权或进场方法，**雇主**应按**规范要求**规定的时间和方式提供。但是，**雇主**在收到**履约担保**前，可暂不给予上述任何进入或占用权。

如果**合同数据**中没有规定上述时间，**雇主**应在要求的时间内，给予**承包商**进入和占用部分**现场**的权利，使**承包商**能够按照**进度计划**进行施工，或如果当时没有**进度计划**，按第 8.3 款 *[进度计划]* 的规定提交的初步进度计划进行。

如果**雇主**未能及时给予**承包商**上述进入和占用的权利，使**承包商**遭受延误和/或招致增加**费用**，**承包商**应有权根据第 20.2 款 *[付款和/或竣工时间延长的索赔]* 的规定，获得**竣工时间**的延长和/或此类**成本加利润**的支付。

但是，如果**雇主**的违约是由于**承包商**的任何错误或延误，包括在任何可适用的**承包商文件**中的错误或提交延误造成的情况，**承包商**应无权获得此类**竣工时间**的延长和/或**成本加利润**。

2.2 协助

如果**承包商**提出请求，**雇主**应迅速对其提供合理的协助，以便**承包商**可获得：

(a) 与合同有关、但不易得到的**工程所在国**的**法律**文本；（以及）

(b) **工程所在国法律**要求的任何许可证、准许、执照或批准（包括为取得此类许可证、准许、执照或批准，**承包商**需要提交的信息）：

（i）根据第 1.13 款 *[遵守法律]* 的规定，**承包商**需要得到的；

（ii）为运送**货物**，包括清关需要的；（以及）

（iii）**承包商设备**运离**现场**时出口需要的。

2.3 Employer's Personnel and Other Contractors

The Employer shall be responsible for ensuring that the Employer's Personnel and the Employer's other contractors (if any) on or near the Site:

(a) co-operate with the Contractor's efforts under Sub-Clause 4.6 [*Co-operation*]; and
(b) comply with the same obligations which the Contractor is required to comply with under sub-paragraphs (a) to (e) of Sub-Clause 4.8 [*Health and Safety Obligations*] and under Sub-Clause 4.18 [*Protection of the Environment*].

The Contractor may require the Employer to remove (or cause to be removed) any person of the Employer's Personnel or of the Employer's other contractors (if any) who is found, based on reasonable evidence, to have engaged in corrupt, fraudulent, collusive or coercive practice.

2.4 Employer's Financial Arrangements

The Employer's arrangements for financing the Employer's obligations under the Contract shall be detailed in the Contract Data.

If the Employer intends to make any material change (affecting the Employer's ability to pay the part of the Contract Price remaining to be paid at that time as estimated by the Engineer) to these financial arrangements, or has to do so because of changes in the Employer's financial situation, the Employer shall immediately give a Notice to the Contractor with detailed supporting particulars.

If the Contractor:

(a) receives an instruction to execute a Variation with a price greater than ten percent (10%) of the Accepted Contract Amount, or the accumulated total of Variations exceeds thirty percent (30%) of the Accepted Contract Amount;
(b) does not receive payment in accordance with Sub-Clause 14.7 [*Payment*]; or
(c) becomes aware of a material change in the Employer's financial arrangements of which the Contractor has not received a Notice under this Sub-Clause,

the Contractor may request and the Employer shall, within 28 days after receiving this request, provide reasonable evidence that financial arrangements have been made and are being maintained which will enable the Employer to pay the part of the Contract Price remaining to be paid at that time (as estimated by the Engineer).

2.5 Site Data and Items of Reference

The Employer shall have made available to the Contractor for information, before the Base Date, all relevant data in the Employer's possession on the topography of the Site and on sub-surface, hydrological, climatic and environmental conditions at the Site. The Employer shall promptly make available to the Contractor all such data which comes into the Employer's possession after the Base Date.

The original survey control points, lines and levels of reference (the "items of reference" in these Conditions) shall be specified on the Drawings and/or in the Specification or issued to the Contractor by a Notice from the Engineer.

2.3 雇主人员和其他承包商

雇主应负责保证在**现场**或附近的**雇主人员**和**雇主**的其他**承包商**（如果有）做到：

（a） 根据**第 4.6 款**［*合作*］的规定，与**承包商**努力合作；（以及）

（b） 要求**承包商**遵守与**第 4.8 款**［*健康和安全义务*］（a）至（e）段和**第 4.18 款**［*环境保护*］规定的相同义务。

承包商可要求**雇主**撤换（或安排撤换）根据合理的证据，发现参与了腐败、欺诈、串通或胁迫行为的任何**雇主人员**或**雇主**的其他**承包商**人员（如果有）。

2.4 雇主的资金安排

合同数据中应详细说明**雇主**为履行**合同**义务所提供的资金安排。

如果**雇主**拟对这些资金安排做任何重要变更（影响**雇主**支付**工程师**当时估算还需支付的部分**合同价格**的能力），或由于**雇主**的资金状况发生变化而必须这样做，**雇主**应立刻向**承包商**发出**通知**，并提供详细的证明资料。

如果**承包商**：

（a） 接到指示，执行价格超过**中标合同金额** 10% 的变更，或累计变更总额超过**中标合同金额** 30% 的变更；

（b） 未收到按照**第 14.7 款**［*付款*］规定的付款；（或）

（c） 知道**雇主**的资金安排发生重大变化，而**承包商**还未收到本款规定的**通知**；

承包商可提出要求，**雇主**应在收到**承包商**的要求后 28 天内，提供其已做并将维持的资金安排的合理证据，说明**雇主**能够支付当时还需支付的（**工程师**估算的）部分**合同价格**。

2.5 现场数据和参考事项

雇主应在**基准日期**前，向**承包商**提供其拥有的**现场**地形和地下、水文、气候及环境条件方面的所有相关数据，供**承包商**参考。**雇主**应立即向**承包商**提供在**基准日期**后其拥有的所有此类数据。

原始测量控制点、基准线和基准标高（本**条件**中的"参考事项"）应在**图纸**和/或**规范要求**中规定，或由**工程师**向**承包商**发出**通知**。

2.6 Employer-Supplied Materials and Employer's Equipment

If Employer-Supplied Materials and/or Employer's Equipment are listed in the Specification for the Contractor's use in the execution of the Works, the Employer shall make such materials and/or equipment available to the Contractor in accordance with the details, times, arrangements, rates and prices stated in the Specification.

The Contractor shall be responsible for each item of Employer's Equipment whilst any of the Contractor's Personnel is operating it, driving it, directing it, using it, or in control of it.

3 The Engineer

3.1 The Engineer

The Employer shall appoint the Engineer, who shall carry out the duties assigned to the Engineer in the Contract.

The Engineer shall be vested with all the authority necessary to act as the Engineer under the Contract.

If the Engineer is a legal entity, a natural person employed by the Engineer shall be appointed and authorised to act on behalf of the Engineer under the Contract.

The Engineer (or, if a legal entity, the natural person appointed to act on its behalf) shall be:

(a) a professional engineer having suitable qualifications, experience and competence to act as the Engineer under the Contract; and
(b) shall be fluent in the ruling language defined in Sub-Clause 1.4 [*Law and Language*].

Where the Engineer is a legal entity, the Engineer shall give a Notice to the Parties of the natural person (or any replacement) appointed and authorised to act on its behalf. The authority shall not take effect until this Notice has been received by both Parties. The Engineer shall similarly give a Notice of any revocation of such authority.

3.2 Engineer's Duties and Authority

Except as otherwise stated in these Conditions, whenever carrying out duties or exercising authority, specified in or implied by the Contract, the Engineer shall act as a skilled professional and shall be deemed to act for the Employer.

The Engineer shall have no authority to amend the Contract or, except as otherwise stated in these Conditions, to relieve either Party of any duty, obligation or responsibility under or in connection with the Contract.

The Engineer may exercise the authority attributable to the Engineer as specified in or necessarily to be implied from the Contract. If the Engineer is required to obtain the consent of the Employer before exercising a specified authority, the requirements shall be as stated in the Particular Conditions. There shall be no requirement for the Engineer to obtain the Employer's consent before the Engineer exercises his/her authority under Sub-Clause 3.7 [*Agreement or Determination*]. The Employer shall not impose further constraints on the Engineer's authority.

2.6 雇主提供的材料和雇主设备

如果**规范要求**中列出了**雇主提供的材料**和／或**雇主设备**，供**承包商**在**工程**实施中使用，**雇主**应按照**规范要求**中规定的细节、时间、安排、费率和价格向**承包商**提供此类材料和／或可用设备。

当任何**承包商人员**操作、驾驶、指挥、使用或控制每项**雇主设备**时，**承包商**应对其负责。

3 工程师

3.1 工程师

雇主应任命**工程师**，**工程师**应履行合同中指派给他的任务。

工程师应享有合同规定中作为**工程师**所必需的一切权利。

如果**工程师**是法人实体，**工程师**雇用的自然人应根据合同规定任命和授权其代表**工程师**执行。

工程师（或如是法人实体，则为其任命、代表其行事的自然人），应该是：

(a) 具有适当资格、经验和能力，根据合同规定担任专业**工程师**；（以及）
(b) 应能流利地使用**第1.4款**［*法律和语言*］所规定的主导语言。

如果**工程师**是法人实体，**工程师**应向双方的自然人（或任何替代人员）发出**通知**，任命和授权代表其执行。该授权应在双方收到**通知**后生效。同样，撤销任何此类授权，**工程师**也应发出**通知**。

3.2 工程师的任务和权利

除本**条件**中另有说明外，每当**工程师**履行或行使**合同**规定或隐含的任务或权利时，应作为熟练的专业人员和视为代表**雇主**执行；

除本**条件**中另有说明外，**工程师**无权修改合同，或解除合同规定的或与合同有关的任一方的任何任务、义务或职责。

工程师可行使合同中规定或必然隐含的应属于**工程师**的权利。如果要求**工程师**在行使规定权利前须取得**雇主**同意，这些要求应在专用条件中写明。在**工程师**根据**第3.7款**［*商定或确定*］的规定行使权利前，不要求取得**雇主**的同意。**雇主**将不对**工程师**的权利做进一步的限制。

However, whenever the Engineer exercises a specified authority for which the Employer's consent is required, then (for the purposes of the Contract) such consent shall be deemed to have been given.

Any acceptance, agreement, approval, check, certificate, comment, consent, disapproval, examination, inspection, instruction, Notice, No-objection, record(s) of meeting, permission, proposal, record, reply, report, request, Review, test, valuation, or similar act (including the absence of any such act) by the Engineer, the Engineer's Representative or any assistant shall not relieve the Contractor from any duty, obligation or responsibility the Contractor has under or in connection with the Contract.

3.3 The Engineer's Representative

The Engineer may appoint an Engineer's Representative and delegate to him/her in accordance with Sub-Clause 3.4 [*Delegation by the Engineer*] the authority necessary to act on the Engineer's behalf at the Site, except to replace the Engineer's Representative.

The Engineer's Representative (if appointed) shall comply with sub-paragraphs (a) and (b) of Sub-Clause 3.1 [*The Engineer*] and shall be based at the Site for the whole time that the Works are being executed at the Site. If the Engineer's Representative is to be temporarily absent from the Site during the execution of the Works, an equivalently qualified, experienced and competent replacement shall be appointed by the Engineer, and the Contractor shall be given a Notice of such replacement.

3.4 Delegation by the Engineer

The Engineer may from time to time assign duties and delegate authority to assistants, and may also revoke such assignment or delegation, by giving a Notice to the Parties, describing the assigned duties and the delegated authority of each assistant. The assignment, delegation or revocation shall not take effect until this Notice has been received by both Parties. However, the Engineer shall not delegate the authority to:

(a) act under Sub-Clause 3.7 [*Agreement or Determination*]; and/or
(b) issue a Notice to Correct under Sub-Clause 15.1 [*Notice to Correct*].

Assistants shall be suitably qualified natural persons, who are experienced and competent to carry out these duties and exercise this authority, and who are fluent in the language for communications defined in Sub-Clause 1.4 [*Law and Language*].

Each assistant, to whom duties have been assigned or authority has been delegated, shall only be authorised to issue instructions to the Contractor to the extent defined by the Engineer's Notice of delegation under this Sub-Clause. Any act by an assistant, in accordance with the Engineer's Notice of delegation, shall have the same effect as though the act had been an act of the Engineer. However, if the Contractor questions any instruction or Notice given by an assistant, the Contractor may by giving a Notice refer the matter to the Engineer. The Engineer shall be deemed to have confirmed the assistant's instruction or Notice if the Engineer does not respond, within 7 days after receiving the Contractor's Notice, reversing or varying the assistant's instruction or Notice (as the case may be).

3.5 Engineer's Instructions

The Engineer may issue to the Contractor (at any time) instructions which may be necessary for the execution of the Works, all in accordance with the Contract. The Contractor shall only take instructions from the Engineer, or from the Engineer's Representative (if appointed) or an assistant to whom the appropriate authority to give instruction has been delegated under Sub-Clause 3.4 [*Delegation by the Engineer*].

但是，每当**工程师**行使需由**雇主**同意的规定权利时，则（就本**合同**而言）应视为**雇主**已予同意。

工程师、**工程师代表**或任何助手的任何接收、协议、批准、校核、证明、评论、同意、不批准、检查、检验、指示、**通知**、**不满意**、会议记录、许可、建议、记录、答复、报告、要求、**审核**、试验、评估或类似行动（包括无此类行动），不应解除**承包商**根据**合同**有关规定应承担的任何任务、义务或职责。

3.3 工程师代表

工程师可根据第 3.4 款 [*由工程师付托*] 的规定，任命**工程师代表**，并授权其在**现场**代表**工程师**行使所需的权利，但替换**工程师代表**的情况除外。

工程师代表（如有任命）应遵守第 3.1 款 [*工程师*]（a）和（b）段的规定，在整个**工程**施工期间常驻**现场**。如果**工程师代表**在**工程**施工期间临时离开**现场**，**工程师**应指定一名同等资格、经验丰富和能胜任的替换人员，并应向**承包商**发出此类替换**通知**。

3.4 由工程师付托

工程师有时可向其助手指派任务和付托权利，也可撤销这种指派或付托，可通过向双方发出**通知**，说明分配给每个助手的任务和授权。指派、付托或撤销应在双方收到**通知**后生效。但是，**工程师**不应将权利付托给：

(a) 根据第 3.7 款 [*商定或确定*] 的规定行事；(和 / 或)
(b) 根据第 15.1 款 [*通知改正*] 的规定发出**改正通知**。

助手应为具有适当资质的自然人，具备履行这些义务，行使此项权利，并能流利地使用第 1.4 款 [*法律和语言*] 规定的交流语言。

应只授权被指派任务或付托权利的每个助手，在**工程师**按照本款发出的付托**通知**规定的范围内向**承包商**发出指示。助手按照**工程师**的付托**通知**做出的任何行动，应与**工程师**做出的行动具有同等的效力。但是，如果**承包商**对助手的指示或**通知**提出质疑，**承包商**可通过发出**通知**将此事项提交**工程师**，如果**工程师**在收到**承包商**的**通知**后 7 天内，不能对该助手的指示或**通知**进行答复、取消或变更（视情况而定），**工程师**应被视为确认了该助手的指示或**通知**。

3.5 工程师的指示

工程师可（在任何时候）按照**合同**规定向**承包商**发出实施**工程**可能需要的指示。**承包商**仅应接受**工程师**或**工程师代表**（如有任命），或根据第 3.4 款 [*由工程师付托*] 的规定被付托适当权利的助手的指示。

Subject to the following provisions of this Sub-Clause, the Contractor shall comply with the instructions given by the Engineer or the Engineer's Representative (if appointed) or delegated assistant, on any matter related to the Contract.

If an instruction states that it constitutes a Variation, Sub-Clause 13.3.1 [*Variation by Instruction*] shall apply.

If not so stated, and the Contractor considers that the instruction:

(a) constitutes a Variation (or involves work that is already part of an existing Variation); or
(b) does not comply with applicable Laws or will reduce the safety of the Works or is technically impossible

the Contractor shall immediately, and before commencing any work related to the instruction, give a Notice to the Engineer with reasons. If the Engineer does not respond within 7 days after receiving this Notice, by giving a Notice confirming, reversing or varying the instruction, the Engineer shall be deemed to have revoked the instruction. Otherwise the Contractor shall comply with and be bound by the terms of the Engineer's response.

3.6 Replacement of the Engineer

If the Employer intends to replace the Engineer, the Employer shall, not less than 42 days before the intended date of replacement, give a Notice to the Contractor of the name, address and relevant experience of the intended replacement Engineer.

If the Contractor does not respond within 14 days after receiving this Notice, by giving a Notice stating an objection to such replacement with reasons, the Contractor shall be deemed to have accepted the replacement.

The Employer shall not replace the Engineer with a person (whether a legal entity or a natural person) against whom the Contractor has raised reasonable objection by a Notice under this Sub-Clause.

If the Engineer is unable to act as a result of death, illness, disability or resignation (or, in the case of an entity, the Engineer becomes unable or unwilling to carry out any of its duties, other than for a cause attributable to the Employer) the Employer shall be entitled to immediately appoint a replacement by giving a Notice to the Contractor with reasons and the name, address and relevant experience of the replacement. This appointment shall be treated as a temporary appointment until this replacement is accepted by the Contractor, or another replacement is appointed, under this Sub-Clause.

3.7 Agreement or Determination

When carrying out his/her duties under this Sub-Clause, the Engineer shall act neutrally between the Parties and shall not be deemed to act for the Employer.

Whenever these Conditions provide that the Engineer shall proceed under this Sub-Clause to agree or determine any matter or Claim, the following procedure shall apply:

3.7.1 Consultation to reach agreement

The Engineer shall consult with both Parties jointly and/or separately, and shall encourage discussion between the Parties in an endeavour to reach agreement. The Engineer shall commence such consultation promptly to allow adequate time to comply with the time limit for agreement under Sub-Clause 3.7.3 [*Time limits*]. Unless otherwise proposed by the Engineer and agreed by both Parties, the Engineer shall provide both Parties with a record of the consultation.

在符合本款下列规定的情况下，**承包商**应遵守**工程师**或**工程师代表**（如有任命）或付托助手对**合同**有关的任何事项发出的指示。

如指示说明构成一项**变更**，则**第 13.3.1 项**[*指示变更*]应适用。

如未说明，**承包商**则认为该指示：

（a） 构成一项**变更**（或已成为现有**变更**一部分的工作）；（或）

（b） 不符合适用的**法律**或将降低**工程**的安全性或技术上不可行。

承包商应在开始与本指示有关的任何工作之前，立即向**工程师**发出说明原因的**通知**。如**工程师**在收到**通知**后 7 天内未能做出答复，发出确认、取消或变更该指示的**通知**，**工程师**应被视为取消了该项指示。否则，**承包商**应遵守**工程师**的答复条件并受其约束。

3.6 工程师的替代

如果**雇主**拟替代**工程师**，**雇主**应在拟替代日期 42 天前**通知承包商**，告知拟替代**工程师**的姓名、地址和相关经验。

如果**承包商**在收到**通知**后 14 天内，不能对该**通知**进行答复，并在通知中说明对替换**工程师**的反对意见和理由，**承包商**应被视为已接受该替代。

对于**承包商**根据本款发出的通知，提出合理反对的人（无论是法人还是自然人），**雇主**不得用其替代**工程师**。

如果**工程师**因死亡、疾病、残疾或辞职而不能工作（或如为实体，**工程师**不能或不愿意履行其任何职责，而非**雇主**的原因），**雇主**应有权立即任命一名替代人员，向**承包商**发出**通知**说明替代人员的原因、姓名、地址和相关经验。此项任命应视为临时任命，直到**承包商**接受该替代人员，或根据**本款**的规定任命另一名替代人员。

3.7 商定或确定

在履行**本款**规定的任务时，**工程师**应在双方之间保持中立，不应被视为代表**雇主**行事。

本**条件**规定**工程师**应按照**本款**对任何事项或**索赔**进行商定或确定时，以下程序应适用：

3.7.1 协商达成协议

工程师应与双方共同和／或单独协商，并鼓励双方进行讨论，尽量达成协议。**工程师**应立即开始此类协商，以便有足够时间遵守根据**第 3.7.3 项**[*时限*]商定的时限。除非**工程师**另有提议并经双方商定，**工程师**应向双方提供协商会议记录。

If agreement is achieved, within the time limit for agreement under Sub-Clause 3.7.3 [*Time limits*], the Engineer shall give a Notice to both Parties of the agreement, which agreement shall be signed by both Parties. This Notice shall state that it is a "Notice of the Parties' Agreement" and shall include a copy of the agreement.

If:

(a) no agreement is achieved within the time limit for agreement under Sub-Clause 3.7.3 [*Time limits*]; or
(b) both Parties advise the Engineer that no agreement can be achieved within this time limit

whichever is the earlier, the Engineer shall give a Notice to the Parties accordingly and shall immediately proceed as specified under Sub-Clause 3.7.2 [*Engineer's Determination*].

3.7.2 Engineer's Determination

The Engineer shall make a fair determination of the matter or Claim, in accordance with the Contract, taking due regard of all relevant circumstances.

Within the time limit for determination under Sub-Clause 3.7.3 [*Time limits*], the Engineer shall give a Notice to both Parties of his/her determination. This Notice shall state that it is a "Notice of the Engineer's Determination", and shall describe the determination in detail with reasons and detailed supporting particulars.

3.7.3 Time limits

The Engineer shall give the Notice of agreement, if agreement is achieved, within 42 days or within such other time limit as may be proposed by the Engineer and agreed by both Parties (the "time limit for agreement" in these Conditions), after:

(a) in the case of a matter to be agreed or determined (not a Claim), the date of commencement of the time limit for agreement as stated in the applicable Sub-Clause of these Conditions;
(b) in the case of a Claim under sub-paragraph (c) of Sub-Clause 20.1 [*Claims*], the date the Engineer receives a Notice under Sub-Clause 20.1 from the claiming Party; or
(c) in the case of a Claim under sub-paragraph (a) or (b) of Sub-Clause 20.1 [*Claims*], the date the Engineer receives:

  (i) a fully detailed Claim under Sub-Clause 20.2.4 [*Fully Detailed Claim*]; or
  (ii) in the case of a Claim under Sub-Clause 20.2.6 [*Claims of continuing effect*], an interim or final fully detailed Claim (as the case may be).

The Engineer shall give the Notice of his/her determination within 42 days or within such other time limit as may be proposed by the Engineer and agreed by both Parties (the "time limit for determination" in these Conditions), after the date corresponding to his/her obligation to proceed under the last paragraph of Sub-Clause 3.7.1 [*Consultation to reach agreement*].

If the Engineer does not give the Notice of agreement or determination within the relevant time limit:

(i) in the case of a Claim, the Engineer shall be deemed to have given a determination rejecting the Claim; or

如果在**第3.7.3项**[*时限*]规定的商定期限内达成协议，**工程师**应将协议**通知**双方，由双方签署该协议。应说明该**通知**是一份"双方商定的**通知**"，并应包括一份副本。

如果：

（a） 在**第3.7.3项**[*时限*]规定的商定时限内未能达成协议；（或）

（b） 双方告知**工程师**未能在此时限内达成协议。

以较早的时限为准，**工程师**应相应地**通知**双方，并应立即按照**第3.7.2项**[*工程师的确定*]的规定执行。

3.7.2 工程师的确定

工程师应根据合同，在适当考虑所有相关情况下，对此事项或**索赔**做出公正的确定。

在**第3.7.3项**[*时限*]规定的确定时限内，**工程师**应将其确定**通知**双方。该**通知**应说明是"**工程师**确定的**通知**"，详细说明确定的理由并附详细证明资料。

3.7.3 时限

如果达成商定，**工程师**应在42天内或由**工程师**提议并经双方商定的其他时限（**本条件**中的"商定时限"）内，在下列时间之后发出商定**通知**：

（a） 如为待商定或确定的事项（不是**索赔**），**本条件**适用**条款**中规定的商定时限的开始日期；

（b） 如为**第20.1款**[*索赔*]（c）段规定的某项**索赔**，**工程师**收到**第20.1款**规定的**索赔**方通知的日期；（或）

（c） 如为**第20.1款**[*索赔*]（a）或（b）段规定的某项**索赔**，**工程师**收到（下述索赔）的日期：

　　（i） **第20.2.4项**[*充分详细的索赔*]规定的充分详细的**索赔**；（或）
　　（ii） **第20.2.6项**[*具有持续影响的索赔*]规定的**索赔**，临时或最终充分详细的**索赔**（视情况而定）。

工程师应在42天内或由**工程师**提议并经双方商定的其他时限（**本条件**中的"确定时限"）内发出其确定**通知**，在其根据**第3.7.1项**[*协商达成协议*]最后一段规定的履行其义务的相应日期之后。

如果**工程师**不能在相关时限内发出商定或确定**通知**，则：

（i） 在**索赔**的情况下，**工程师**应被视为已做出拒绝该项**索赔**的确定；（或）

(ii) in the case of a matter to be agreed or determined, the matter shall be deemed to be a Dispute which may be referred by either Party to the DAAB for its decision under Sub-Clause 21.4 [*Obtaining DAAB's Decision*] without the need for a NOD (and Sub-Clause 3.7.5 [*Dissatisfaction with Engineer's determination*] and sub-paragraph (a) of Sub-Clause 21.4.1 [*Reference of a Dispute to the DAAB*] shall not apply).

3.7.4 Effect of the agreement or determination

Each agreement or determination shall be binding on both Parties (and shall be complied with by the Engineer) unless and until corrected under this Sub-Clause or, in the case of a determination, it is revised under Clause 21 [*Disputes and Arbitration*].

If an agreement or determination concerns the payment of an amount from one Party to the other Party, the Contractor shall include such an amount in the next Statement and the Engineer shall include such amount in the Payment Certificate that follows that Statement.

If, within 14 days after giving or receiving the Engineer's Notice of agreement or determination, any error of a typographical or clerical or arithmetical nature is found:

(a) by the Engineer: then he/she shall immediately advise the Parties accordingly; or
(b) by a Party: then that Party shall give a Notice to the Engineer, stating that it is given under this Sub-Clause 3.7.4 and clearly identifying the error. If the Engineer does not agree there was an error, he/she shall immediately advise the Parties accordingly.

The Engineer shall within 7 days of finding the error, or receiving a Notice under sub-paragraph (b) above (as the case may be), give a Notice to both Parties of the corrected agreement or determination. Thereafter, the corrected agreement or determination shall be treated as the agreement or determination for the purpose of these Conditions.

3.7.5 Dissatisfaction with Engineer's determination

If either Party is dissatisfied with a determination of the Engineer:

(a) the dissatisfied Party may give a NOD to the other Party, with a copy to the Engineer;
(b) this NOD shall state that it is a "Notice of Dissatisfaction with the Engineer's Determination" and shall set out the reason(s) for dissatisfaction;
(c) this NOD shall be given within 28 days after receiving the Engineer's Notice of the determination under Sub-Clause 3.7.2 [*Engineer's Determination*] or, if applicable, his/her Notice of the corrected determination under Sub-Clause 3.7.4 [*Effect of the agreement or determination*] (or, in the case of a deemed determination rejecting the Claim, within 28 days after the time limit for determination under Sub-Clause 3.7.3 [*Time limits*] has expired); and
(d) thereafter, either Party may proceed under Sub-Clause 21.4 [*Obtaining DAAB's Decision*].

If no NOD is given by either Party within the period of 28 days stated in sub-paragraph (c) above, the determination of the Engineer shall be deemed to have been accepted by both Parties and shall be final and binding on them.

If the dissatisfied Party is dissatisfied with only part(s) of the Engineer's determination:

(i) this part(s) shall be clearly identified in the NOD;

(ii) 在待商定或确定的事项的情况下，该事项应视为是一项**争端**，任一方均可根据**第 21.4 款**［*取得争端避免/裁决委员会的决定*］的规定，将该**争端**提交**争端避免/裁决委员会**做出决定，而不必发出**不满意通知**（**第 3.7.5 项**［*对工程师的确定不满意*］和**第 21.4.1 项**［*争端提交给争端避免/裁决委员会*］（a）段将不适用）。

3.7.4
商定和确定的效力

每项商定或确定对双方均具有约束力（**工程师**应遵守），除非并直到根据本**款**进行了改正，或在某项确定的情况下，根据**第 21 条**［*争端和仲裁*］的规定做出了修改。

如果某项商定或确定涉及一方向另一方支付一笔金额，**承包商**应将此金额列入下一份**报表**，**工程师**应随该**报表**将此项金额列入**付款证书**。

在发出或收到**工程师**商定或确定的**通知**后 14 天内，如发现有任何印刷、文书或运算性的错误：

（a） 如由**工程师**发现：则由其立即**通知**双方；（或）

（b） 如由某一方发现：则该方应根据**第 3.7.4 项**的规定向**工程师**发出**通知**，并明确指出错误之处。如果**工程师**不同意有错误，其应视情立即**通知**双方。

工程师应在发现错误 7 天内，或收到上述（b）段（视情况而定）规定的**通知**后，向双方发出改正后的商定或确定**通知**。此后，此改正后的商定或确定应视为本**条件**的商定或确定。

3.7.5
对工程师的确定不满意

如果任一方对**工程师**的确定不满意：

（a） 不满意的一方可向另一方发出**不满意通知**，并抄送**工程师**；

（b） 此份**不满意通知**应说明是"对**工程师**的确定不满意的**通知**"，并应阐述不满意的理由；

（c） 此份**不满意通知**，应在收到**工程师**根据**第 3.7.2 项**［*工程师的确定*］的规定做出的确定**通知**，或如适用，在**工程师**根据**第 3.7.4 项**［*商定或确定的效力*］的规定发出的改正**通知**后 28 天内发出（或如被视为拒绝**索赔**的确定，在根据**第 3.7.3 项**［*时限*］的规定做出确定的时限期满后 28 天内）；（以及）

（d） 此后，任一方均可根据**第 21.4 款**［*取得争端避免/裁决委员会的决定*］的规定执行。

如果任一方在上述（c）段所规定的 28 天内未能发出**不满意通知**，**工程师**的确定应被视为已由双方接受的最终确定，并对双方均具有约束力。

如果不满意的一方仅对**工程师**的部分确定不满意：

（i） 该部分应在**不满意通知**中明确说明；

(ii) this part(s), and any other parts of the determination that are affected by such part(s) or rely on such part(s) for completeness, shall be deemed to be severable from the remainder of the determination; and

(iii) the remainder of the determination shall become final and binding on both Parties as if the NOD had not been given.

In the event that a Party fails to comply with an agreement of the Parties under this Sub-Clause 3.7 or a final and binding determination of the Engineer, the other Party may, without prejudice to any other rights it may have, refer the failure itself directly to arbitration under Sub-Clause 21.6 [*Arbitration*] in which case the first and the third paragraphs of Sub-Clause 21.7 [*Failure to Comply with DAAB's Decision*] shall apply to such reference in the same manner as these paragraphs apply to a final and binding decision of the DAAB.

3.8 Meetings

The Engineer or the Contractor's Representative may require the other to attend a management meeting to discuss arrangements for future work and/or other matters in connection with execution of the Works.

The Employer's other contractors, the personnel of legally constituted public authorities and/or private utility companies, and/or Subcontractors may attend any such meeting, if requested by the Engineer or the Contractor's Representative.

The Engineer shall keep a record of each management meeting and supply copies of the record to those attending and to the Employer. At any such meeting, and in the record, responsibilities for any actions to be taken shall be in accordance with the Contract.

The Contractor

4.1 Contractor's General Obligations

The Contractor shall execute the Works in accordance with the Contract. The Contractor undertakes that the execution of the Works and the completed Works will be in accordance with the documents forming the Contract, as altered or modified by Variations.

The Contractor shall provide the Plant (and spare parts, if any) and Contractor's Documents specified in the Contract, and all Contractor's Personnel, Goods, consumables and other things and services, whether of a temporary or permanent nature, required to fulfil the Contractor's obligations under the Contract.

The Contractor shall be responsible for the adequacy, stability and safety of all the Contractor's operations and activities, of all methods of construction and of all the Temporary Works. Except to the extent specified in the Contract, the Contractor:

(i) shall be responsible for all Contractor's Documents, Temporary Works, and such design of each item of Plant and Materials as is required for the item to be in accordance with the Contract; and

(ii) shall not otherwise be responsible for the design or specification of the Permanent Works.

（ii） 该部分以及受该部分影响或依赖其完整性的、确定的任何其他部分，应视为与确定的剩余部分可分割；（以及）

（iii） 确定的剩余部分应成为双方最终的并具有约束力的确定，如同未发出**不满意通知**。

如果一方未能遵守双方根据**第3.7款**达成的商定，或**工程师**做出的最终和具有约束力的确定，另一方可在不损害其可能拥有的任何其他权利的情况下，根据**第21.6款**[**仲裁**]的规定将不遵守的事项直接提交仲裁。在这种情况下，**第21.7款**[**未能遵守争端避免/裁决委员会的决定**]的第1段和第3段应适用于此类提交，提交方式与这些段落适用于**争端避免/裁决委员会**的最终和具有约束力的决定相同。

3.8 会议

工程师或**承包商代表**可要求其他人员参加管理会议，讨论进一步工作的安排和/或**工程**实施相关的其他事项。

应**工程师**或**承包商代表**的要求，**雇主**的其他承包商、合法成立的公共部门和/或私营公用事业公司的人员和/或**分包商**可以参加任何此类会议。

工程师应保存每次管理会议的记录，并将记录副本提供给与会人员和**雇主**。在任何此类会议和记录中，拟采取任何措施的责任应符合**合同**规定。

4 承包商

4.1 承包商的一般义务

承包商应按照合同实施**工程**。**承包商**承诺，**工程**的实施和竣工的**工程**将符合经过**变更**做出更改或修正而构成的**合同**文件的要求。

承包商应提供合同规定的**生产设备**（以及备件，如果有）和**承包商文件**，以及履行合同规定的**承包商**义务所需的所有临时性或永久性的**承包商人员**、**货物**、消耗品及其他物品和服务。

承包商应对所有**承包商**作业和活动、所有施工方法和所有**临时工程**的完备性、稳定性和安全性承担责任。除非**合同**另有规定，**承包商**：

（i） 应对所有**承包商文件**、**临时工程**，以及合同规定的每项**生产设备**和**材料**的设计承担责任；（以及）

（ii） 不应对其他**永久工程**的设计或规范要求负责。

The Contractor shall, whenever required by the Engineer, submit details of the arrangements and methods which the Contractor proposes to adopt for the execution of the Works. No significant alteration to these arrangements and methods shall be made without this alteration having been submitted to the Engineer.

If the Contract specifies that the Contractor shall design any part of the Permanent Works, then unless otherwise stated in the Particular Conditions:

(a) the Contractor shall prepare, and submit to the Engineer for Review, the Contractor's Documents for this part (and any other documents necessary to complete and implement the design during the execution of the Works and to instruct the Contractor's Personnel);

(b) these Contractor's Documents shall be in accordance with the Specification and Drawings and shall include additional information required by the Engineer to add to the Drawings for co-ordination of each Party's designs. If the Engineer instructs that further Contractor's Documents are reasonably required to demonstrate that the Contractor's design complies with the Contract, the Contractor shall prepare and submit them promptly to the Engineer at the Contractor's cost;

(c) construction of this part shall not commence until a Notice of No-objection is given (or is deemed to have been given) by the Engineer under sub-paragraph (i) of Sub-Clause 4.4.1 [*Preparation and Review*] for all the Contractor's Documents which are relevant to its design, and construction of such part shall be in accordance with these Contractor's Documents;

(d) the Contractor may modify any design or Contractor's Documents which have previously been submitted for Review, by giving a Notice to the Engineer with reasons. If the Contractor has commenced construction of the part of the Works to which such design or Contractor's Documents are relevant, work on this part shall be suspended, the provisions of Sub-Clause 4.4.1 [*Preparation and Review*] shall apply as if the Engineer had given a Notice in respect of the Contractor's Documents under sub-paragraph (ii) of Sub-Clause 4.4.1, and work shall not resume until a Notice of No-objection is given (or is deemed to have been given) by the Engineer for the revised documents;

(e) the Contractor shall be responsible for this part and it shall, when the Works are completed, be fit for such purpose(s) for which the part is intended as are specified in the Contract (or, where no purpose(s) are so defined and described, fit for their ordinary purpose(s));

(f) in addition to the Contractor's undertaking above, the Contractor undertakes that the design and the Contractor's Documents for this part will comply with the technical standards stated in the Specification and Laws (in force when the Works are taken over under Clause 10 [*Employer's Taking Over*]) and in accordance with the documents forming the Contract, as altered or modified by Variations;

(g) if Sub-Clause 4.4.2 [*As-built Records*] and/or Sub-Clause 4.4.3 [*Operation and Maintenance Manuals*] apply, the Contractor shall submit to the Engineer the Contractor's Documents for this part in accordance with such Sub-Clause(s) and in sufficient detail for the Employer to operate, maintain, dismantle, reassemble, adjust and repair this part; and

(h) if Sub-Clause 4.5 [*Training*] applies, the Contractor shall carry out training of the Employer's Personnel in the operation and maintenance of this part.

当工程师提出要求时，**承包商**应提交其拟采用的详细**工程**施工安排和方法。事先在未向**工程师**提交此类改变之前，不得对这些安排和方法进行重要改变。

如果**合同**规定**承包商**设计**永久工程**的任何部分，除非在**专用条件**中另有规定，则：

(a) **承包商**应编写并向**工程师**提交该部分的**承包商文件**（以及在**工程**施工期间完成和实施设计的和指导**承包商人员**所需的任何其他文件）进行**审核**；

(b) 这些**承包商文件**应符合**规范要求**和**图纸**的要求，并应包括**工程师**要求添加到**图纸**中的附加资料，以便协调各方的设计。如果**工程师**指示，合理要求进一步的**承包商文件**以证明**承包商**的设计符合**合同**要求，**承包商**应编写并及时提交**工程师**，费用由**承包商**承担；

(c) 在**工程师**根据第 4.4.1 项［*编制和审核*］(i) 段的规定，对**承包商**所有有关设计和该部分施工的**承包商文件**发出（或被视为已发出）**不反对通知**之前，不得开始该部分的施工；

(d) **承包商**可向**工程师**发出**通知**并说明理由，修改已提交送审的任何设计或**承包商文件**。如果**承包商**已开始与此类设计或**承包商文件**相关的部分**工程**的施工，则应暂停该部分**工程**的工作，第 4.4.1 项［*编制和审核*］的规定应适用，如同**工程师**已根据第 4.4.1 项 (ii) 段的规定就**承包商文件**发出了**通知**一样。在**工程师**就已修正的文件发出（或视为已发出）**不反对通知**之前，不得恢复工作；

(e) **承包商**应对**工程**竣工后该部分**工程**符合**合同**规定的预期目的负责（或如目的未如此定义和描述，则适用于其普通目的）；

(f) 除上述**承包商**的承诺外，**承包商**还承诺该部分**工程**的设计和**承包商文件**，将符合**规范要求**和**法律**中规定的技术标准（根据第 10 条［*雇主的接收*］的规定，接收**工程**时有效），并符合经过**变更**做出更改或修正后构成的**合同**文件的要求；

(g) 如第 4.4.2 项［*竣工记录*］和/或第 4.4.3 项［*操作和维护手册*］适用，**承包商**应按照该条款向**工程师**提交该部分足够详细的**承包商文件**，使**雇主**能操作、维护、拆卸、再组装、调整和修复该部分**工程**；（以及）

(h) 如果第 4.5 款［*培训*］适用，**承包商**应对**雇主**人员进行该部分的操作和维护方面的培训。

4.2 Performance Security

The Contractor shall obtain (at the Contractor's cost) a Performance Security to secure the Contractor's proper performance of the Contract, in the amount and currencies stated in the Contract Data. If no amount is stated in the Contract Data, this Sub-Clause shall not apply.

4.2.1 Contractor's obligations

The Contractor shall deliver the Performance Security to the Employer, with a copy to the Engineer, within 28 days after receiving the Letter of Acceptance. The Performance Security shall be issued by an entity and from within a country (or other jurisdiction) to which the Employer gives consent and shall be in the form annexed to the Particular Conditions, or in another form agreed by the Employer (but such consent and/or agreement shall not relieve the Contractor from any obligation under this Sub-Clause).

The Contractor shall ensure that the Performance Security remains valid and enforceable until the issue of the Performance Certificate and the Contractor has complied with Sub-Clause 11.11 [*Clearance of Site*]. If the terms of the Performance Security specify an expiry date, and the Contractor has not become entitled to receive the Performance Certificate by the date 28 days before the expiry date, the Contractor shall extend the validity of the Performance Security until the issue of the Performance Certificate and the Contractor has complied with Sub-Clause 11.11 [*Clearance of Site*].

Whenever Variations and/or adjustments under Clause 13 [*Variations and Adjustments*] result in an accumulative increase or decrease of the Contract Price by more than twenty percent (20%) of the Accepted Contract Amount:

(a) in the case of such an increase, at the Employer's request the Contractor shall promptly increase the amount of the Performance Security in that currency by a percentage equal to the accumulative increase. If the Contractor incurs Cost as a result of this Employer's request, Sub-Clause 13.3.1 [*Variation by Instruction*] shall apply as if the increase had been instructed by the Engineer; or

(b) in the case of such a decrease, subject to the Employer's prior consent the Contractor may decrease the amount of the Performance Security in that currency by a percentage equal to the accumulative decrease.

4.2.2 Claims under the Performance Security

The Employer shall not make a claim under the Performance Security, except for amounts to which the Employer is entitled under the Contract in the event of:

(a) failure by the Contractor to extend the validity of the Performance Security, as described in this Sub-Clause, in which event the Employer may claim the full amount (or, in the case of previous reduction(s), the full remaining amount) of the Performance Security;

(b) failure by the Contractor to pay the Employer an amount due, as agreed or determined under Sub-Clause 3.7 [*Agreement or Determination*] or agreed or decided under Clause 21 [*Disputes and Arbitration*], within 42 days after the date of the agreement or determination or decision or arbitral award (as the case may be);

(c) failure by the Contractor to remedy a default stated in a Notice given under Sub-Clause 15.1 [*Notice to Correct*] within 42 days or other time (if any) stated in the Notice;

(d) circumstances which entitle the Employer to terminate the Contract under Sub-Clause 15.2 [*Termination for Contractor's Default*], irrespective of whether a Notice of termination has been given; or

4.2 履约担保

承包商（承包商承担费用）应取得**履约担保**以确保**承包商**恰当履行合同，保证金额和币种应符合**合同数据**中的规定。如**合同数据**中未明确保证金额，本款应不适用。

4.2.1 承包商的义务

承包商应在收到**中标函**后 28 天内向**雇主**提交**履约担保**，并向**工程师**送一份副本。**履约担保**应由**雇主**同意的国家（或其他司法管辖区）内的实体提供，并采用**专用条件**所附格式或**雇主**同意的其他格式（但这种同意和／或商定不应解除**承包商**根据本款承担的任何义务）。

承包商应在其**履约证书**签发和**承包商**履行了第 11.11 款 [*现场清理*] 规定前，确保**履约担保**保持有效和可执行。如果在**履约担保**的条款中规定了其期满日期，而**承包商**在该期满日期前 28 天尚无权拿到**履约证书**，**承包商**应将**履约担保**的有效期延长至**履约证书**签发和**承包商**按照第 11.11 款 [*现场清理*] 的规定完成清理时为止。

根据第 13 条 [*变更和调整*] 的规定，当**变更**和／或**调整**导致**合同价格**累计增加或减少超过**中标合同金额**的 20% 时：

（a）在此类增加的情况下，应**雇主**的要求，**承包商**应立即按累计增加额的百分比，增加该货币的**履约担保**金额。如果**承包商**因**雇主**的要求而增加**费用**，则第 13.3.1 项 [*指示变更*] 应适用，如同增加**费用**是**工程师**指示的一样；（或）

（b）在此类减少的情况下，经**雇主**事先同意，**承包商**可按累计减少额的百分比，减少该货币的**履约担保**金额。

4.2.2 根据履约担保的索赔

除以下**雇主**根据合同规定有权获得的金额情况外，**雇主**不应根据**履约担保**提出索赔：

（a）**承包商**未能按本款所述的要求延长**履约担保**的有效期，这时**雇主**可以索赔**履约担保**的全部金额（或在先前减少的情况下，全部余额）；

（b）**承包商**未能在商定或确定，或决定或仲裁裁决后 42 天内，按照第 3.7 款 [*商定或确定*] 或第 21 条 [*争端和仲裁*] 规定的商定或确定，向**雇主**支付到期金额；

（c）**承包商**未能在收到根据第 15.1 款 [*通知改正*] 的规定发出的**通知**后 42 天内或**通知**中规定的其他时间内纠正违约；

（d）**雇主**有权根据第 15.2 款 [*因承包商违约的终止*] 的规定，无论是否已发出终止通知都应终止**合同**的情况；（或）

(e) if under Sub-Clause 11.5 [*Remedying of Defective Work off Site*] the Contractor removes any defective or damaged Plant from the Site, failure by the Contractor to repair such Plant, return it to the Site, reinstall it and retest it by the date of expiry of the relevant duration stated in the Contractor's Notice (or other date agreed by the Employer).

The Employer shall indemnify and hold the Contractor harmless against and from all damages, losses and expenses (including legal fees and expenses) resulting from a claim under the Performance Security to the extent that the Employer was not entitled to make the claim.

Any amount which is received by the Employer under the Performance Security shall be taken into account:

(i) in the Final Payment Certificate under Sub-Clause 14.13 [*Issue of FPC*]; or
(ii) if the Contract is terminated, in payment due to the Contractor under Sub-Clause 15.4 [*Payment after Termination for Contractor's Default*], Sub-Clause 15.7 [*Payment after Termination for Employer's Convenience*], Sub-Clause 16.4 [*Payment after Termination by Contractor*], Sub-Clause 18.5 [*Optional Termination*], or Sub-Clause 18.6 [*Release from Performance under the Law*] (as the case may be).

4.2.3 Return of the Performance Security

The Employer shall return the Performance Security to the Contractor:

(a) within 21 days after the issue of the Performance Certificate and the Contractor has complied with Sub-Clause 11.11 [*Clearance of Site*]; or
(b) promptly after the date of termination if the Contract is terminated in accordance with Sub-Clause 15.5 [*Termination for Employer's Convenience*], Sub-Clause 16.2 [*Termination by Contractor*], Sub-Clause 18.5 [*Optional Termination*] or Sub-Clause 18.6 [*Release from Performance under the Law*].

4.3 Contractor's Representative

The Contractor shall appoint the Contractor's Representative and shall give him/her all authority necessary to act on the Contractor's behalf under the Contract, except to replace the Contractor's Representative.

The Contractor's Representative shall be qualified, experienced and competent in the main engineering discipline applicable to the Works and fluent in the language for communications defined in Sub-Clause 1.4 [*Law and Language*].

Unless the Contractor's Representative is named in the Contract, the Contractor shall, before the Commencement Date, submit to the Engineer for consent the name and particulars of the person the Contractor proposes to appoint as Contractor's Representative. If consent is withheld or subsequently revoked, or if the appointed person fails to act as Contractor's Representative, the Contractor shall similarly submit the name and particulars of another suitable replacement for such appointment. If the Engineer does not respond within 28 days after receiving this submission, by giving a Notice to the Contractor objecting to the proposed person or replacement, the Engineer shall be deemed to have given his/her consent.

The Contractor shall not, without the Engineer's prior consent, revoke the appointment of the Contractor's Representative or appoint a replacement (unless the Contractor's Representative is unable to act as a result of death, illness, disability or resignation, in which case his/her appointment shall be deemed to have been revoked with immediate effect and the appointment of a replacement shall be treated as a temporary appointment until the Engineer gives his/her consent to this replacement, or another replacement is appointed, under this Sub-Clause).

(e) 如果根据第 11.5 款［*现场外缺陷工程的修补*］的规定，**承包商**将任何有缺陷或损害的**生产设备**移出**现场**，如**承包商**未能修复此项**设备**，则将其退回**现场**重新安装，并在**承包商通知**中规定的相关期满日期（或**雇主**同意的其他日期）前重新试验。

雇主应保障和使**承包商**免受因**雇主**无权提出索赔的**履约担保**规定的索赔，而产生的所有损害赔偿费、损失和开支（包括法律费用和开支）的损害。

根据**履约担保**的规定，**雇主**收到的任何款项，应考虑：

(i) 第 14.13 款［*最终付款证书的签发*］规定的**最终付款证书**；（或）

(ii) 如果合同终止，则根据第 15.4 款［*因承包商违约终止后的付款*］、第 15.7 款［*为雇主便利终止后的付款*］、第 16.4 款［*由承包商终止后的付款*］、第 18.5 款［*自主选择终止*］或第 18.6 款［*依法解除履约*］（视情况而定）规定的，应向**承包商**支付的应付款项。

4.2.3
履约担保的退还

雇主应在下述时间将**履约担保**退还**承包商**：

(a) **履约证书**签发后 21 天内，及**承包商**已遵守第 11.11 款［*现场清理*］的规定；（或）
(b) 如根据第 15.5 款［*为雇主便利的终止*］、第 16.2 款［*由承包商终止*］、第 18.5 款［*自主选择终止*］或第 18.6 款［*依法解除履约*］的规定终止合同，则在合同终止日期后。

4.3
承包商代表

承包商应任命**承包商代表**，并授予其代表**承包商**根据合同采取行动所需的全部权利，但替代**承包商代表**除外。

承包商代表应具备适用于本**工程**的主要工程专业的资格、经验和能力，并能流利使用第 1.4 款［*法律和语言*］规定的交流语言。

除非合同中已写明了**承包商代表**的姓名，**承包商**应在**开工日期**前，将其拟任命为**承包商代表**的人员姓名和详细资料提交**工程师**取得同意。如未获得同意，或随后撤销了同意，或任命的人员不能担任**承包商代表**，**承包商**应同样提交另外适合人选的姓名、详细资料，以取得该项任命。如果**工程师**在收到此项提交后 28 天内，未能向**承包商**发出对拟推荐人选和替代人员的反对**通知**，**工程师**应被视为已同意。

未经**工程师**事先同意，**承包商**不应撤销**承包商代表**的任命，或任命替代人员（除非**承包商代表**因死亡、疾病、残疾或辞职无法行事，在这种情况下，其任命应视为已被立即撤销，以及该替代人员的任命应视为临时任命，直到**工程师**同意或根据本款的规定任命另一名替代人员）。

The whole time of the Contractor's Representative shall be given to directing the Contractor's performance of the Contract. The Contractor's Representative shall act for and on behalf of the Contractor at all times during the performance of the Contract, including issuing and receiving all Notices and other communications under Sub-Clause 1.3 [*Notices and Other Communications*] and for receiving instructions under Sub-Clause 3.5 [*Engineer's Instructions*].

The Contractor's Representative shall be based at the Site for the whole time that the Works are being executed at the Site. If the Contractor's Representative is to be temporarily absent from the Site during the execution of the Works, a suitable replacement shall be temporarily appointed, subject to the Engineer's prior consent.

The Contractor's Representative may delegate any powers, functions and authority except:

(a) the authority to issue and receive Notices and other communications under Sub-Clause 1.3 [*Notices and Other Communications*]; and
(b) the authority to receive instructions under Sub-Clause 3.5 [*Engineer's Instructions*],

to any suitably competent and experienced person and may at any time revoke the delegation. Any delegation or revocation shall not take effect until the Engineer has received a Notice from the Contractor's Representative, naming the person, specifying the powers, functions and authority being delegated or revoked, and stating the timing of the delegation or revocation.

All these persons shall be fluent in the language for communications defined in Sub-Clause 1.4 [*Law and Language*].

4.4
Contractor's Documents

4.4.1
Preparation and Review

The Contractor's Documents shall comprise the documents:

(a) stated in the Specification;
(b) required to satisfy all permits, permissions, licences and other regulatory approvals which are the Contractor's responsibility under Sub-Clause 1.13 [*Compliance with Laws*];
(c) described in Sub-Clause 4.4.2 [*As-Built Records*] and Sub-Clause 4.4.3 [*Operation and Maintenance Manuals*], where applicable; and
(d) required under sub-paragraph (a) of Sub-Clause 4.1 [*Contractor's General Obligations*], where applicable.

Unless otherwise stated in the Specification, the Contractor's Documents shall be written in the language for communications defined in Sub-Clause 1.4 [*Law and Language*].

The Contractor shall prepare all Contractor's Documents and the Employer's Personnel shall have the right to inspect the preparation of all these documents, wherever they are being prepared.

If the Specification or these Conditions specify that a Contractor's Document is to be submitted to the Engineer for Review, it shall be submitted accordingly, together with a Notice from the Contractor stating that the Contractor's Document is ready for Review and that it complies with the Contract.

承包商代表应将其全部时间用于指导**承包商**履行**合同**。**承包商代表**应在履行**合同**期间始终代表**承包商**行事，包括根据第 1.3 款［*通知和其他通信交流*］的规定签发和接收所有**通知**和其他通信交流，并根据第 3.5 款［*工程师的指示*］的规定接收指示。

承包商代表应在整个**工程**施工期间常驻**现场**。如果**承包商代表**在工程施工期间临时离开**现场**，应事先征得**工程师**的同意，临时指定一名合适替代人员。

承包商代表可将任何职权、任务和权利付托，以下情况除外：

(a) 根据第 1.3 款［*通知和其他通信交流*］的规定签发和接收**通知**和其他通信交流的权利；(以及)
(b) 根据第 3.5 款［*工程师的指示*］接收指示的权利。

给有适当能力和经验的人员，并可随时撤销付托。任何付托或撤销应在**工程师**收到**承包商代表**发出的指明人员姓名，并说明付托或撤销的职权、职能和权利的**通知**后生效。

所有这些人员应能流利地使用第 1.4 款［*法律和语言*］规定的交流语言。

4.4
承包商文件

4.4.1
编制和审核

承包商文件应包括的文件：

(a) 在**规范要求**中规定的；
(b) 满足第 1.13 款［*遵守法律*］规定的**承包商**负责的所有许可证、准许、执照或其他监管批准要求的；
(c) 适用情况下，第 4.4.2 项［*竣工记录*］和第 4.4.3 项［*操作和维护手册*］中所述的；(以及)
(d) 适用情况下，第 4.1 款［*承包商的一般义务*］(a) 段的规定要求的。

除非**规范要求**另有规定，**承包商文件**应按照第 1.4 款［*法律和语言*］规定的交流语言编写。

承包商应编写所有**承包商文件**，**雇主人员**应有权检查所有这些文件的编写，无论这些文件在何处编写。

如果**规范要求**或本条件规定**承包商文件**应提交**工程师**审核，则应将其连同**承包商**的**通知**一并提交，说明**承包商文件**已编写好供**审核**并符合**合同**要求。

The Engineer shall, within 21 days after receiving the Contractor's Document and this Notice from the Contractor, give a Notice to the Contractor:

(i) of No-objection (which may include comments concerning minor matters which will not substantially affect the Works); or
(ii) that the Contractor's Document fails (to the extent stated) to comply with the Contract, with reasons.

If the Engineer gives no Notice within this period of 21 days, the Engineer shall be deemed to have given a Notice of No-objection to the Contractor's Document.

After receiving a Notice under sub-paragraph (ii), above, the Contractor shall revise the Contractor's Document and resubmit it to the Engineer for Review in accordance with this Sub-Clause and the period of 21 days for Review shall be calculated from the date that the Engineer receives it.

4.4.2 As-Built Records

If no as-built records to be prepared by the Contractor are stated in the Specification, this Sub-Clause shall not apply.

The Contractor shall prepare, and keep up-to-date, a complete set of "as-built" records of the execution of the Works, showing the exact as-built locations, sizes and details of the work as executed by the Contractor. The format, referencing system, system of electronic storage and other relevant details of the as-built records shall be as stated in the Specification (if not stated, as acceptable to the Engineer). These records shall be kept on the Site and shall be used exclusively for the purposes of this Sub-Clause.

The as-built records shall be submitted to the Engineer for Review, and the Works shall not be considered to be completed for the purposes of taking-over under Sub-Clause 10.1 [*Taking Over the Works and Sections*] until the Engineer has given (or is deemed to have given) a Notice of No-objection under sub-paragraph (i) of Sub-Clause 4.4.1 [*Preparation and Review*].

The number of copies of as-built records to be submitted by the Contractor under this Sub-Clause shall be as required under Sub-Clause 1.8 [*Care and Supply of Documents*].

4.4.3 Operation and Maintenance Manuals

If no operation and maintenance manuals to be prepared by the Contractor are stated in the Specification, this Sub-Clause shall not apply.

The Contractor shall prepare, and keep up-to-date, the operation and maintenance manuals in the format and other relevant details as stated in the Specification.

The operation and maintenance manuals shall be submitted to the Engineer for Review, and the Works shall not be considered to be completed for the purposes of taking-over under Sub-Clause 10.1 [*Taking Over the Works and Sections*] until the Engineer has given (or is deemed to have given) a Notice of No-objection under sub-paragraph (i) of Sub-Clause 4.4.1 [*Preparation and Review*].

4.5 Training

If no training of employees of the Employer (and/or other identified personnel) by the Contractor is stated in the Specification, this Sub-Clause shall not apply.

工程师应在收到承包商文件和承包商通知后 21 天内，向承包商发出通知：

(i) 不反对（可能包括不会对工程造成实际性影响的非重要事项的意见）；（或）

(ii) 承包商文件（在规定的范围内）不符合合同规定，并说明原因。

如果工程师在 21 天内未能发出通知，工程师应被视为对承包商文件发出不反对通知。

在收到上述第（ii）段规定的通知后，承包商应根据本款修改承包商文件，并重新提交给工程师审核，审核期限为 21 天，从工程师收到通知的日期算起。

4.4.2 竣工记录

如果在规范要求中未规定承包商准备的竣工记录，本款应不适用。

承包商应编写并随时更新一套完整的工程施工"竣工"记录，如实记载竣工的准确位置、尺寸和承包商已实施工作的详细说明。竣工记录的格式、参考系统、电子存储系统和其他相关细节应按规范要求中的规定（如没有规定，按工程师所接受的）。这些记录应保存在现场，并仅限用于本款的需要。

竣工记录应提交给工程师审核，在工程师根据第 4.4.1 项 [*编制和审核*]（i）段的规定发出（或视为已发出）不反对通知之前，该工程不应被认为已按第 10.1 款 [*工程和分项工程的接收*] 规定的接收要求竣工。

承包商根据本款规定提交竣工记录副本的数量应符合第 1.8 款 [*文件的照管和提供*] 的要求。

4.4.3 操作和维护手册

如果在规范要求中没有规定由承包商编写操作和维护手册，本款应不适用。

承包商应按照规范要求规定的格式和其他相关细节，编写和更新操作和维护手册。

操作和维护手册应提交工程师审核，在工程师根据第 4.4.1 项 [*编制和审查*]（i）段的规定发出（或视为已发出）不反对通知之前，该工程不应被认为已按第 10.1 款 [*工程和分项工程的接收*] 规定的接收要求竣工。

4.5 培训

如果规范要求中未规定承包商对雇主的雇员（和/或其他确定的人员）的培训，本款应不适用。

The Contractor shall carry out training of the Employer's employees (and/or other personnel identified in the Specification) in the operation and maintenance of the Works, and any other aspect of the Works, to the extent stated in the Specification. The timing of the training shall be as stated in the Specification (if not stated, as acceptable to the Employer). The Contractor shall provide qualified and experienced training staff, training facilities and all training materials as necessary and/or as stated in the Specification.

If the Specification specifies training which is to be carried out before taking over, the Works shall not be considered to be completed for the purposes of taking over under Sub-Clause 10.1 [*Taking Over the Works and Sections*] until this training has been completed in accordance with the Specification.

4.6 Co-operation

The Contractor shall, as stated in the Specification or as instructed by the Engineer, co-operate with and allow appropriate opportunities for carrying out work by:

(a) the Employer's Personnel;
(b) any other contractors employed by the Employer; and
(c) the personnel of any legally constituted public authorities and private utility companies,

who may be employed in the carrying out, on or near the Site, of any work not included in the Contract. Such appropriate opportunities may include the use of Contractor's Equipment, Temporary Works, access arrangements which are the responsibility of the Contractor, and/or other Contractor's facilities or services on the Site.

The Contractor shall be responsible for the Contractor's construction activities on the Site, and shall use all reasonable endeavours to co-ordinate these activities with those of other contractors to the extent (if any) stated in the Specification or as instructed by the Engineer.

If the Contractor suffers delay and/or incurs Cost as a result of an instruction under this Sub-Clause, to the extent (if any) that co-operation, allowance of opportunities and coordination was Unforeseeable having regard to that stated in the Specification, the Contractor shall be entitled subject to Sub-Clause 20.2 [*Claims For Payment and/or EOT*] to EOT and/or payment of such Cost Plus Profit.

4.7 Setting Out

The Contractor shall set out the Works in relation to the items of reference under Sub-Clause 2.5 [*Site Data and Items of Reference*].

4.7.1 Accuracy

The Contractor shall:

(a) verify the accuracy of all these items of reference before they are used for the Works;
(b) promptly deliver the results of each verification to the Engineer;
(c) rectify any error in the positions, levels, dimensions or alignment of the Works; and
(d) be responsible for the correct positioning of all parts of the Works.

4.7.2 Errors

If the Contractor finds an error in any items of reference, the Contractor shall give a Notice to the Engineer describing it:

承包商应按照**规范要求**中规定的范围，对**雇主**的雇员（和/或**规范要求**中确定的其他人员）进行**工程**操作和维护以及**工程**任何其他方面的培训。培训时间应按**规范要求**中的规定（如未规定，按**雇主**可接受的）。**承包商**应提供合格和经验丰富的培训人员、培训设施和/或**规范要求**中规定的所有必要的培训教材。

如果**规范要求**规定了**工程**接收前要进行培训，在按照**规范要求**的规定完成培训之前，不应认为**工程**已按照第 10.1 款［*工程和分项工程的接收*］规定的接收要求竣工。

4.6 合作

承包商应按照**规范要求**的规定或**工程师**的指示，与下列人员合作并为其工作提供适当的机会：

（a） **雇主**人员；
（b） **雇主**雇用的任何其他承包商；（以及）
（c） 任何合法成立的公共部门和私营公用事业公司的人员。

他们可能被雇用在**现场**或**现场**附近从事本合同中未计划进行的任何工作。此类适当的机会可包括使用**承包商设备**、**临时工程**、由**承包商**负责的进出安排和/或**现场**的其他**承包商**设施或服务。

承包商应负责其在**现场**的施工活动，并应在本**规范要求**或**工程师**指示中的规定范围内（如果有），尽一切合理的努力与其他**承包商**协调这些活动。

如果**承包商**因本款规定的指示而遭受延误和/或增加**费用**，在考虑到本**规范要求**中规定的合作、提供的机会和协调的范围（如果有）是不可预见的情况下，**承包商**应有权根据第 20.2 款［*付款和/或竣工时间延长的索赔*］的规定，获得**竣工时间**的延长和/或此类**成本加利润**的支付。

4.7 放线

承包商应按照第 2.5 款［*现场数据和参考事项*］的规定对与参考事项有关的工程放线。

4.7.1 准确性

承包商应当：

（a） 在**工程**使用前，核实所有这些参考事项的准确性；
（b） 及时将核实的每项结果送交**工程师**；
（c） 纠正在**工程**的位置、标高、尺寸或定线中的任何错误；（以及）

（d） 负责对**工程**的所有部分正确定位。

4.7.2 错误

如果**承包商**在任何参考事项中发现错误，**承包商**应向**工程师**发出**通知**，按以下要求对错误进行说明：

(a) within the period stated in the Contract Data (if not stated, 28 days) calculated from the Commencement Date, if the items of reference are specified on the Drawings and/or in the Specification; or

(b) as soon as practicable after receiving the items of reference, if they are issued by the Engineer under Sub-Clause 2.5 [*Site Data and Items of Reference*].

4.7.3 Agreement or Determination of rectification measures, delay and/or Cost

After receiving a Notice from the Contractor under Sub-Clause 4.7.2 [*Errors*], the Engineer shall proceed under Sub-Clause 3.7 [*Agreement or Determination*] to agree or determine:

(a) whether or not there is an error in the items of reference;

(b) whether or not (taking account of cost and time) an experienced contractor exercising due care would have discovered such an error

- when examining the Site, the Drawings and the Specification before submitting the Tender; or
- if the items of reference are specified on the Drawings and/or in the Specification and the Contractor's Notice is given after the period stated in sub-paragraph (a) of Sub-Clause 4.7.2; and

(c) what measures (if any) the Contractor is required to take to rectify the error

(and, for the purpose of Sub-Clause 3.7.3 [*Time limits*], the date the Engineer receives the Contractor's Notice under Sub-Clause 4.7.2 [*Errors*] shall be the date of commencement of the time limit for agreement under Sub-Clause 3.7.3).

If, under sub-paragraph (b) above, an experienced contractor would not have discovered the error:

(i) Sub-Clause 13.3.1 [*Variation by Instruction*] shall apply to the measures that the Contractor is required to take (if any); and

(ii) if the Contractor suffers delay and/or incurs Cost as a result of the error, the Contractor shall be entitled subject to Sub-Clause 20.2 [*Claims For Payment and/or EOT*] to EOT and/or payment of such Cost Plus Profit.

4.8 Health and Safety Obligations

The Contractor shall:

(a) comply with all applicable health and safety regulations and Laws;

(b) comply with all applicable health and safety obligations specified in the Contract;

(c) comply with all directives issued by the Contractor's health and safety officer (appointed under Sub-Clause 6.7 [*Health and Safety of Personnel*]);

(d) take care of the health and safety of all persons entitled to be on the Site and other places (if any) where the Works are being executed;

(e) keep the Site, Works (and the other places (if any) where the Works are being executed) clear of unnecessary obstruction so as to avoid danger to these persons;

(f) provide fencing, lighting, safe access, guarding and watching of:

(i) the Works, until the Works are taken over under Clause 10 [*Employer's Taking Over*]; and

(ii) any part of the Works where the Contractor is executing outstanding works or remedying any defects during the DNP; and

(a) 如果在**图纸和/或规范要求**中规定了参考事项，从**开工日期**算起的、**合同数据**中规定的期限内（如未规定，则为 28 天）；（或）

(b) 如果参考事项是由**工程师**根据**第 2.5 款**[*现场数据和参考事项*]的规定签发的，在可行的范围内收到参考事项后。

4.7.3
对整改措施、延误和/或成本的商定或确定

在收到**承包商**根据**第 4.7.2 项**[*错误*]规定发出的**通知**后，**工程师**应根据**第 3.7 款**[*商定或确定*]的规定就以下事项做出商定或确定：

(a) 参考事项中是否有错误；
(b) 谨慎行事、经验丰富的**承包商**是否会（考虑到成本和时间）发现这样的错误：

- 在提交**投标书**前，对**现场**、**图纸**和**规范要求**进行审核时；或

- 在第 4.7.2 项（a）段规定的期限内检查参考事项时，如果**图纸和/或规范要求**中规定了参考事项；（以及）⊖

(c) **承包商**需要采取什么措施（如果有）改正错误。

（以及，就**第 3.7.3 项**[*时限*]而言，**工程师**根据**第 4.7.2 项**[*错误*]的规定，收到**承包商通知**的日期应是**第 3.7.3 项**规定的商定时限的开始日期）。

根据以上（b）段所述，如果经验丰富的**承包商**不能发现错误：

(i) **第 13.3.1 项**[*指示变更*]的规定应适用于要求**承包商**采取的措施（如果有）；（以及）

(ii) 如果**承包商**因错误而遭受延误和/或招致增加**费用**，**承包商**应根据**第 20.2 款**[*付款和/或竣工时间延长的索赔*]的规定，有权获得**竣工时间的延长**和/或此类**成本加利润**的支付。

4.8
健康和安全义务

承包商应：

(a) 遵守所有适用的健康和安全的规则和**法律**；
(b) 遵守**合同**中规定的所有适用的健康和安全义务；
(c) 遵守**承包商**健康和安全官员（根据**第 6.7 款**[*人员的健康和安全*]的规定任命）发布的所有指令；
(d) 保护所有有权在施工**现场**和其他地方（如果有）工作人员的健康和安全；
(e) 保持**现场**、**工程**[以及正在实施**工程**的其他地方（如果有）]清洁，清除不需要的障碍物，以避免对这些人员造成危险；
(f) 提供围栏、照明、安全通道、保卫和看守：

(i) 在**工程**按照**第 10 条**[*雇主的接收*]的规定移交前；（以及）

(ii) **承包商**在**缺陷通知期限**正在执行扫尾工作或修补任何缺陷的任何工程部分；（以及）

⊖ 此处中文按照勘误表改正后的英文翻译。——译者注

(g) provide any Temporary Works (including roadways, footways, guards and fences) which may be necessary, because of the execution of the Works, for the use and protection of the public and of owners and occupiers of adjacent land and property.

Within 21 days of the Commencement Date and before commencing any construction on the Site, the Contractor shall submit to the Engineer for information a health and safety manual which has been specifically prepared for the Works, the Site and other places (if any) where the Contractor intends to execute the Works. This manual shall be in addition to any other similar document required under applicable health and safety regulations and Laws.

The health and safety manual shall set out all the health and safety requirements:

(i) stated in the Specification;
(ii) that comply with all the Contractor's health and safety obligations under the Contract; and
(iii) that are necessary to effect and maintain a healthy and safe working environment for all persons entitled to be on the Site and other places (if any) where the Works are being executed.

This manual shall be revised as necessary by the Contractor or the Contractor's health and safety officer, or at the reasonable request of the Engineer. Each revision of the manual shall be submitted promptly to the Engineer.

In addition to the reporting requirement of sub-paragraph (g) of Sub-Clause 4.20 [*Progress Reports*], the Contractor shall submit to the Engineer details of any accident as soon as practicable after its occurrence and, in the case of an accident causing serious injury or death, shall inform the Engineer immediately.

The Contractor shall, as stated in the Specification and as the Engineer may reasonably require, maintain records and make reports (in compliance with the applicable health and safety regulations and Laws) concerning the health and safety of persons and any damage to property.

4.9 Quality Management and Compliance Verification Systems

4.9.1 Quality Management System

The Contractor shall prepare and implement a QM System to demonstrate compliance with the requirements of the Contract. The QM System shall be specifically prepared for the Works and submitted to the Engineer within 28 days of the Commencement Date. Thereafter, whenever the QM System is updated or revised, a copy shall promptly be submitted to the Engineer.

The QM System shall be in accordance with the details stated in the Specification (if any) and shall include the Contractor's procedures:

(a) to ensure that all Notices and other communications under Sub-Clause 1.3 [*Notices and Other Communications*], Contractor's Documents, as-built records (if applicable), operation and maintenance manuals (if applicable), and contemporary records can be traced, with full certainty, to the Works, Goods, work, workmanship or test to which they relate;

(g) 因实施**工程**，为了公众和邻近土地和财产的所有人、占用人的使用并对其保护，提供可能需要的任何**临时工程**（包括道路、人行道、防护物和围栏等）。

在**开工日期**后 21 天内，**现场**开始任何施工前，**承包商**应向**工程师**提交一份专门为**工程**和**承包商**拟实施工程的**现场**和其他地方（如果有）编写的健康和安全手册，以供参考。该手册应是对适用的健康和安全规则和**法律**所要求的任何其他类似文件的补充。

健康和安全手册应规定所有健康和安全要求：

（i） **规范要求**中规定的；
（ii） 符合**合同**规定的所有**承包商**健康和安全义务的；（以及）
（iii） 对所有有权在**现场**和正在实施**工程**的其他地方（如果有）的所有人员创造和保持健康和安全的工作环境所需的。

该手册应在必要时由**承包商**或**承包商**的健康和安全官员进行修订，或根据**工程师**的合理要求进行修订。手册的每次修订都应及时提交给**工程师**。

除根据第 4.20 款 [*进度报告*]（g）段规定的报告要求外，**承包商**应在事故发生后尽快向**工程师**提交事故的详细情况，如果事故造成人员重伤或死亡，应立即通知**工程师**。

承包商应按照**规范要求**中的规定和**工程师**的合理要求，保存有关人员的健康和安全以及财产损失的记录和报告（符合适用的健康和安全规则和**法律**）。

4.9 质量管理和合规验证体系

4.9.1 质量管理体系

承包商应制定和实施**质量管理体系**，以证明其符合**合同**要求。**质量管理体系**应为**工程**量身定制，并应在**工程**开工之日起 28 天内提交给**工程师**。此后，无论何时更新或修订**质量管理体系**，应立即向**工程师**提交一份副本。

质量管理体系应符合**规范要求**（如果有）的详细规定，并应包括**承包商**的程序：

(a) 确保根据第 1.3 款 [*通知和其他通信交流*] 规定发出的所有**通知**和通信交流、**承包商文件**、竣工记录（如适用）、操作和维护手册（如适用）和同期记录均能完全确定地追溯到与其相关的**工程**、**货物**、工作、工艺或试验；

(b) to ensure proper coordination and management of interfaces between the stages of execution of the Works, and between Subcontractors; and

(c) for the submission of Contractor's Documents to the Engineer for Review.

The Engineer may Review the QM System and may give a Notice to the Contractor stating the extent to which it does not comply with the Contract. Within 14 days after receiving this Notice, the Contractor shall revise the QM System to rectify such non-compliance. If the Engineer does not give such a Notice within 21 days of the date of submission of the QM System, the Engineer shall be deemed to have given a Notice of No-objection.

The Engineer may, at any time, give a Notice to the Contractor stating the extent to which the Contractor is failing to correctly implement the QM System to the Contractor's activities under the Contract. After receiving this Notice, the Contractor shall immediately remedy such failure.

The Contractor shall carry out internal audits of the QM System regularly, and at least once every 6 months. The Contractor shall submit to the Engineer a report listing the results of each internal audit within 7 days of completion. Each report shall include, where appropriate, the proposed measures to improve and/or rectify the QM System and/or its implementation.

If the Contractor is required by the Contractor's quality assurance certification to be subject to external audit, the Contractor shall immediately give a Notice to the Engineer describing any failing(s) identified in any external audit. If the Contractor is a JV, this obligation shall apply to each member of the JV.

4.9.2 Compliance Verification System

The Contractor shall prepare and implement a Compliance Verification System to demonstrate that the design (if any), Materials, Employer-Supplied Materials (if any), Plant, work and workmanship comply in all respects with the Contract.

The Compliance Verification System shall be in accordance with the details stated in the Specification (if any) and shall include a method for reporting the results of all inspections and tests carried out by the Contractor. In the event that any inspection or test identifies a non-compliance with the Contract, Sub-Clause 7.5 [*Defects and Rejection*] shall apply.

The Contractor shall prepare and submit to the Engineer a complete set of compliance verification documentation for the Works or Section (as the case may be), fully compiled and collated in the manner described in the Specification or, if not so described, in a manner acceptable to the Engineer.

4.9.3 General provision

Compliance with the QM System and/or Compliance Verification System shall not relieve the Contractor from any duty, obligation or responsibility under or in connection with the Contract.

4.10 Use of Site Data

The Contractor shall be responsible for interpreting all data referred to under Sub-Clause 2.5 [*Site Data and Items of Reference*].

To the extent which was practicable (taking account of cost and time), the Contractor shall be deemed to have obtained all necessary information as to risks, contingencies and other circumstances which may influence or affect the Tender or Works. To the same extent, the Contractor shall be deemed to have inspected and examined the Site, access to the Site, its surroundings, the above data and other available information, and to have been satisfied before submitting the Tender as to all matters relevant to the execution of the Works, including:

(b) 确保**工程**实施阶段之间以及**分包商**之间衔接的适当协调和管理；（以及）

(c) 将**承包商**文件提交给**工程师**审核。

工程师可审核**质量管理体系**并可向**承包商**发出**通知**，说明其不符合合同规定的程度。承包商应在收到**通知**后 14 天内，修改**质量管理体系**，以纠正此类不合规情况。如果**工程师**未在**质量管理体系**提交之日起 21 天内发出此类**通知**，**工程师**应被视为已发出**不反对通知**。

工程师可随时向**承包商**发出**通知**，说明**承包商**未能按照合同的规定，在**承包商**的活动中正确实施**质量管理体系**的程度。收到该**通知**后，**承包商**应立即改正此类问题。

承包商应定期进行**质量管理体系**的内部审计，至少每 6 个月进行一次。**承包商**应在审计结束后 7 天内向**工程师**提交一份审计报告，列出每次内部审计的结果。适当时，每份报告均应包括在改进和 / 或纠正**质量管理体系**和 / 或其实施方面拟采取的措施。

如果**承包商**质量保证证书要求**承包商**接受外部审计，**承包商**应立即向**工程师**发出**通知**，说明在任何外部审计中发现的任何问题。如果**承包商**是**联营体**，则该义务应适用于**联营体**的每个成员。

4.9.2
合规验证体系

承包商应制定并实施**合规验证体系**，以证明设计（如果有）、**材料**、雇主提供的材料（如果有）、**生产设备**、工作和工艺在所有方面均符合合同要求。

合规验证体系应符合**规范要求**（如果有）的详细规定，并应包括**承包商**进行的所有检测和试验结果的报告方法。如果任何检测或试验发现合同要求的不符合项，则第 7.5 款[*缺陷和拒收*]应适用。

承包商应编写并向**工程师**提交一套完整的**工程**或分项工程（视情况而定）的合规验证文件，并按**规范要求**规定的方式进行充分汇总和整理，如未规定，则按**工程师**可接受的方式进行。

4.9.3
一般规定

遵守**质量管理体系**和 / 或**合规验证体系**，不应免除合同规定的或与合同有关的**承包商**的任何任务、义务或职责。

4.10
现场数据的使用

承包商应负责解释根据第 2.5 款[*现场数据和参考事项*]提供的所有数据。

在实际可行（考虑费用和时间）的范围内，**承包商**应被认为已取得可能对**投标书**或**工程**产生影响或作用的有关风险、偶发事件和其他情况的所有必要资料。同样，**承包商**应被认为在提交**投标书**前，已检验和检查了**现场**、进入**现场**的通道、周围环境、上述数据和其他得到的资料，并认为与实施**工程**相关的所有事项满足要求，包括：

(a) the form and nature of the Site, including sub-surface conditions;
(b) the hydrological and climatic conditions, and the effects of climatic conditions at the Site;
(c) the extent and nature of the work and Goods necessary for the execution of the Works;
(d) the Laws, procedures and labour practices of the Country; and
(e) the Contractor's requirements for access, accommodation, facilities, personnel, power, transport, water and any other utilities or services.

4.11 Sufficiency of the Accepted Contract Amount

The Contractor shall be deemed to:

(a) have satisfied himself/herself as to the correctness and sufficiency of the Accepted Contract Amount; and
(b) have based the Accepted Contract Amount on the data, interpretations, necessary information, inspections, examinations and satisfaction as to all relevant matters described in Sub-Clause 4.10 [*Use of Site Data*].

Unless otherwise stated in the Contract, the Accepted Contract Amount shall be deemed to cover all the Contractor's obligations under the Contract and all things necessary for the proper execution of the Works in accordance with the Contract.

4.12 Unforeseeable Physical Conditions

In this Sub-Clause, "physical conditions" means natural physical conditions and physical obstructions (natural or man-made) and pollutants, which the Contractor encounters at the Site during execution of the Works, including sub-surface and hydrological conditions but excluding climatic conditions at the Site and the effects of those climatic conditions.

If the Contractor encounters physical conditions which the Contractor considers to have been Unforeseeable and that will have an adverse effect on the progress and/or increase the Cost of the execution of the Works, the following procedure shall apply:

4.12.1 Contractor's Notice

After discovery of such physical conditions, the Contractor shall give a Notice to the Engineer, which shall:

(a) be given as soon as practicable and in good time to give the Engineer opportunity to inspect and investigate the physical conditions promptly and before they are disturbed;
(b) describe the physical conditions, so that they can be inspected and/or investigated promptly by the Engineer;
(c) set out the reasons why the Contractor considers the physical conditions to be Unforeseeable; and
(d) describe the manner in which the physical conditions will have an adverse effect on the progress and/or increase the Cost of the execution of the Works.

4.12.2 Engineer's inspection and investigation

The Engineer shall inspect and investigate the physical conditions within 7 days, or a longer period agreed with the Contractor, after receiving the Contractor's Notice.

The Contractor shall continue execution of the Works, using such proper and reasonable measures as are appropriate for the physical conditions and to enable the Engineer to inspect and investigate them.

(a) **现场**的状况和性质，包括地下条件；
(b) 水文和气候条件，以及**现场**气候条件的影响；
(c) 实施**工程**所需的工作和**货物**的范围和性质；
(d) **工程**所在国的**法律**、程序和劳务惯例；（以及）
(e) **承包商**对进场、食宿、设施、人员、电力、运输、水和其他公共设施或服务的要求。

4.11 中标合同金额的充分性

承包商应被认为：

(a) 已确信**中标合同金额**的正确性和充分性；（以及）

(b) 已将**中标合同金额**建立在根据第 4.10 款 [*现场数据的使用*] 中提到的所有相关事项的数据、解释、必要的资料、检验、检查和满意的基础上。

除非**合同**另有规定，**中标合同金额**应被视为包括**承包商**根据合同规定应承担的全部义务，以及按照**合同**规定为正确地实施**工程**所需的全部有关事项的费用。

4.12 不可预见的物质条件

本款中的"物质条件"系指**承包商**在**现场**施工期间遇到的自然物质条件，及物质障碍（自然的或人为的）和污染物，包括地下和水文条件，但不包括**现场**的气候条件以及这些气候条件的影响。

如果**承包商**遇到其认为**不可预见**的，并将对**工程**进度有不利影响和/或增加**工程**实施**费用**的物质条件，以下程序应适用。

4.12.1 承包商通知

发现此类物质条件后，**承包商**应**通知工程师**，**工程师**应：

(a) 在实际可行的情况下，适时给**工程师**机会在物质条件被干扰前，进行适当检查和调查；

(b) 描述物质条件，以便**工程师**及时进行检查和/或调查；

(c) 提出**承包商**认为物质条件是**不可预见**的理由；（以及）

(d) 说明物质条件对**工程**施工进度的不利影响和/或增加**成本**的方式。

4.12.2 工程师检验和调查

工程师应在收到**承包商通知**后 7 天内，或与**承包商**商定的更长时间内，对物质条件进行检验和调查。

承包商应采取与物质条件相适应的适当、合理的措施继续施工，并使**工程师**能够检验和调查。

4.12.3
Engineer's instructions

The Contractor shall comply with any instructions which the Engineer may give for dealing with the physical conditions and, if such an instruction constitutes a Variation, Sub-Clause 13.3.1 [*Variation by Instruction*] shall apply.

4.12.4
Delay and/or Cost

If and to the extent that the Contractor suffers delay and/or incurs Cost due to these physical conditions, having complied with Sub-Clauses 4.12.1 to 4.12.3 above, the Contractor shall be entitled subject to Sub-Clause 20.2 [*Claims For Payment and/or EOT*] to EOT and/or payment of such Cost.

4.12.5
Agreement or Determination of Delay and/or Cost

The agreement or determination, under Sub-Clause 20.2.5 [*Agreement or determination of the Claim*], of any Claim under Sub-Clause 4.12.4 [*Delay and/or Cost*] shall include consideration of whether and (if so) to what extent the physical conditions were Unforeseeable.

The Engineer may also review whether other physical conditions in similar parts of the Works (if any) were more favourable than could reasonably have been foreseen by the Base Date. If and to the extent that these more favourable conditions were encountered, the Engineer may take account of the reductions in Cost which were due to these conditions in calculating the additional Cost to be agreed or determined under this Sub-Clause 4.12.5. However, the net effect of all additions and reductions under this Sub-Clause 4.12.5 shall not result in a net reduction in the Contract Price.

The Engineer may take account of any evidence of the physical conditions foreseen by the Contractor by the Base Date, which the Contractor may include in the supporting particulars for the Claim under Sub-Clause 20.2.4 [*Fully detailed Claim*], but shall not be bound by any such evidence.

4.13
Rights of Way and Facilities

The Contractor shall bear all costs and charges for special and/or temporary rights-of-way which may be required for the purposes of the Works, including those for access to the Site.

The Contractor shall also obtain, at the Contractor's risk and cost, any additional facilities outside the Site which may be required for the purposes of the Works.

4.14
Avoidance of Interference

The Contractor shall not interfere unnecessarily or improperly with:

(a) the convenience of the public; or
(b) the access to and use and occupation of all roads and footpaths, irrespective of whether they are public or in the possession of the Employer or of others.

The Contractor shall indemnify and hold the Employer harmless against and from all damages, losses and expenses (including legal fees and expenses) resulting from any such unnecessary or improper interference.

4.15
Access Route

The Contractor shall be deemed to have been satisfied, at the Base Date, as to the suitability and availability of the access routes to the Site. The Contractor shall take all necessary measures to prevent any road or bridge from being damaged by the Contractor's traffic or by the Contractor's Personnel. These measures shall include the proper use of appropriate vehicles (conforming to legal load and width limits (if any) and any other restrictions) and routes.

4.12.3
工程师的指示

承包商应遵守工程师为应对物质条件而发出的任何指示,如果这种指示构成变更,第 13.3.1 项 [*指示变更*] 应适用。

4.12.4
延误和/或费用

如果**承包商**在遵守上述第 **4.12.1** 项至第 **4.12.3** 项的规定后,因这些**物质条件**而遭受延误和/或招致增加**费用**,**承包商**应有权根据第 **20.2** 款 [*付款和/或竣工时间延长的索赔*] 的规定,有权获得**竣工时间**的延长和/或此类**费用**的支付。

4.12.5
延误和/或费用的商定或确定

根据第 **20.2.5** 项 [*索赔的商定或确定*] 和第 **4.12.4** 项 [*延误和/或费用*] 的规定,对任何**索赔**的商定或确定,应包括考虑**物质条件**是否**不可预见**,以及(如果是)此类**物质条件不可预见**的程度。

工程师还可审核**工程**的类似部分中(如果有)其他**物质条件**是否比**基准日期**前能合理预见的更为有利。如果并在一定程度上遇到这些更为有利的条件,**工程师**在计算根据本第 **4.12.5** 项商定或确定额外**费用**时,可考虑由于这些条件而引起的**费用**减少额。但是,根据本第 **4.12.5** 项规定的所有增加和减少额的净作用,不应造成**合同价格**净减少的结果。

工程师可以考虑**承包商**在**基准日期**前所预见的任何**物质条件**的证据,**承包商**根据第 **20.2.4** 项 [*充分详细的索赔*] 的规定,可以在**索赔**的证明资料中包括这些证据,但不应受任何此类证据的约束。

4.13
道路通行权和设施

承包商应为**工程**目的所需的专用和/或临时道路,包括进场道路的通行权,承担全部费用和开支。

承包商还应自担风险和费用,取得为**工程**目的可能需要的**现场**以外的任何附加设施。

4.14
避免干扰

承包商应避免对以下事项产生不必要或不当的干扰:

(a) 公众的便利;(或)
(b) 所有道路和人行道的进入、使用和占用,不论它们是公共的,或是**雇主**或其他人所有的。

承包商应保障和保持使**雇主**免受因任何此类不必要或不当的干扰造成的任何损害赔偿费、损失和开支(包括法律费用和开支)。

4.15
进场通路

承包商应被认为已对在**基准日期现场**的进入通路的适宜性和可用性感到满意。**承包商**应采取一切必要的措施,防止任何道路或桥梁因**承包商**的通行或**承包商人员**受到损坏。这些措施应包括正确使用适宜的车辆 [符合法定的负载和宽度限制(如果有)以及任何其他限制] 和通路。

Except as otherwise stated in these Conditions:

(a) the Contractor shall (as between the Parties) be responsible for repair of any damage caused to, and any maintenance which may be required for the Contractor's use of, access routes;

(b) the Contractor shall provide all necessary signs or directions along access routes, and shall obtain any permissions or permits which may be required from the relevant authorities, for the Contractor's use of routes, signs and directions;

(c) the Employer shall not be responsible for any third party claims which may arise from the Contractor's use or otherwise of any access route;

(d) the Employer does not guarantee the suitability or availability of particular access routes; and

(e) all Costs due to non-suitability or non-availability, for the use required by the Contractor, of access routes shall be borne by the Contractor.

To the extent that non-suitability or non-availability of an access route arises as a result of changes to that access route by the Employer or a third party after the Base Date and as a result the Contractor suffers delay and/or incurs Cost, the Contractor shall be entitled subject to Sub-Clause 20.2 [*Claims For Payment and/or EOT*] to EOT and/or payment of such Cost.

4.16 Transport of Goods

The Contractor shall:

(a) give a Notice to the Engineer not less than 21 days before the date on which any Plant, or a major item of other Goods (as stated in the Specification), will be delivered to the Site;

(b) be responsible for packing, loading, transporting, receiving, unloading, storing and protecting all Goods and other things required for the Works;

(c) be responsible for customs clearance, permits, fees and charges related to the import, transport and handling of all Goods, including all obligations necessary for their delivery to the Site; and

(d) indemnify and hold the Employer harmless against and from all damages, losses and expenses (including legal fees and expenses) resulting from the import, transport and handling of all Goods, and shall negotiate and pay all third party claims arising from their import, transport and handling.

4.17 Contractor's Equipment

The Contractor shall be responsible for all Contractor's Equipment. When brought on to the Site, Contractor's Equipment shall be deemed to be exclusively intended for the execution of the Works. The Contractor shall not remove from the Site any major items of Contractor's Equipment without the Engineer's consent. However, consent shall not be required for vehicles transporting Goods or Contractor's Personnel off Site.

In addition to any Notice given under Sub-Clause 4.16 [*Transport of Goods*], the Contractor shall give a Notice to the Engineer of the date on which any major item of Contractor's Equipment has been delivered to the Site. This Notice shall be given within 7 days of the delivery date, shall identify whether the item of Contractor's Equipment is owned by the Contractor or a Subcontractor or another person and, if rented or leased, shall identify the rental or leasing entity.

4.18 Protection of the Environment

The Contractor shall take all necessary measures to:

除本**条件**另有规定外：

（a） **承包商**应（就**双方**而言）负责维修对进场通路造成的损害，并维护其使用的进场通路；

（b） **承包商**应提供进场通路所有必需的标识或方向指示，还应为其使用这些通路、标识和方向指示，取得必要的有关部门的任何许可或许可证；

（c） **雇主**不应对因**承包商**使用或以其他方式使用进场通路而引起的任何第三方索赔负责；

（d） **雇主**不保证特定进场通路的适宜性和可用性；（以及）

（e） 因**承包商**需要使用进场通路而产生的不适宜性和不可用性的所有**费用**均应由**承包商**承担。

如果因**雇主**或第三方在**基准日期**后更改通路而导致进场通路不适宜或不可用，而**承包商**因此遭受延误和／或招致增加**费用**，**承包商**应有权根据第 20.2 款［*付款和／或竣工时间延长的索赔*］的规定，获得**竣工**时间的延长和／或此类**费用**的支付。

4.16 货物运输

承包商应：

（a） 在不少于 21 天前，将任何**生产设备**或每项其他主要**货物**（如本**规范要求**所述）运到**现场**的日期，**通知工程师**；

（b） 负责**工程**需要的所有**货物**和其他物品的包装、装货、运输、接收、卸货、存储和保护；

（c） 负责所有**货物**的进口、运输和办理有关的通关、许可证、费用和杂费，并承担将货物交付到**现场**所需的所有义务；（以及）

（d） 保障并保持**雇主**免受因进口、运输和装卸所有**货物**引起的所有损害赔偿费、损失和开支（包括法律费用和开支），并应协商和支付由于**货物**进口、运输和装卸而引起的所有第三方索赔。

4.17 承包商设备

承包商应负责所有**承包商设备**。**承包商设备**运到**现场**后，应被视作准备为**工程**施工专用。未经**工程师**同意，**承包商**不得从**现场**运走任何主要**承包商设备**。但运送**货物**或**承包商**人员离开**现场**的车辆，无须经过同意。

除了根据第 4.16 款［*货物运输*］的规定发出的任何**通知**外，**承包商**还应向**工程师**发出**通知**，说明**承包商设备**的任何主要部件已交付到**现场**的日期。本通知应在交货日期后 7 天内发出，应确定**承包商设备**部件是否归**承包商**或**分包商**或其他人所有，如果是租用或租赁的，应明确租用或租赁实体。

4.18 环境保护

承包商应采取一切必要措施：

(a) protect the environment (both on and off the Site);
(b) comply with the environmental impact statement for the Works (if any); and
(c) limit damage and nuisance to people and property resulting from pollution, noise and other results of the Contractor's operations and/ or activities.

The Contractor shall ensure that emissions, surface discharges, effluent and any other pollutants from the Contractor's activities shall exceed neither the values indicated in the Specification, nor those prescribed by applicable Laws.

4.19 Temporary Utilities

The Contractor shall, except as stated below, be responsible for the provision of all temporary utilities, including electricity, gas, telecommunications, water and any other services the Contractor may require for the execution of the Works.

The following provisions of this Sub-Clause shall only apply if, as stated in the Specification, the Employer is to provide utilities for the Contractor's use. The Contractor shall be entitled to use, for the purposes of the Works, the utilities on the Site for which details and prices are given in the Specification. The Contractor shall, at the Contractor's risk and cost, provide any apparatus necessary for the Contractor's use of these services and for measuring the quantities consumed. The apparatus provided for measuring quantities consumed shall be subject to the Engineer's consent. The quantities consumed (if any) during each period of payment stated in the Contract Data (if not stated, each month) shall be measured by the Contractor, and the amount to be paid by the Contractor for such quantities (at the prices stated in the Specification) shall be included in the relevant Statement.

4.20 Progress Reports

Monthly progress reports, in the format stated in the Specification (if not stated, in a format acceptable to the Engineer) shall be prepared by the Contractor and submitted to the Engineer. Each progress report shall be submitted in one paper-original, one electronic copy and additional paper copies (if any) as stated in the Contract Data. The first report shall cover the period up to the end of the first month following the Commencement Date. Reports shall be submitted monthly thereafter, each within 7 days after the last day of the month to which it relates.

Reporting shall continue until the Date of Completion of the Works or, if outstanding work is listed in the Taking-Over Certificate, the date on which such outstanding work is completed. Unless otherwise stated in the Specification, each progress report shall include:

(a) charts, diagrams and detailed descriptions of progress, including each stage of (design by the Contractor, if any) Contractor's Documents, procurement, manufacture, delivery to Site, construction, erection and testing;
(b) photographs and/or video recordings showing the status of manufacture and of progress on and off the Site;
(c) for the manufacture of each main item of Plant and Materials, the name of the manufacturer, manufacture location, percentage progress, and the actual or expected dates of:

(i) commencement of manufacture,
(ii) Contractor's inspections,
(iii) tests, and

(a) 保护（**现场**内外）环境；

(b) 遵守**工程**的环境影响说明（如果有）；（以及）

(c) 限制由**承包商**施工作业和／或活动引起的污染、噪声和其他后果对公众和财产造成的损害和妨害。

承包商应确保因其活动产生的气体排放、地面排水、排污及任何其他污染物等，不得超过**规范要求**规定的数值，也不得超过适用**法律**规定的数值。

4.19 临时公用设施

除下述情况外，**承包商**应负责提供所有临时公用设施，包括所有电、天然气、电信、水和**承包商**施工可能需要的其他服务。

本**款**的下列规定仅适用于**规范要求**中规定的**雇主**提供**承包商**使用的公用设施的情况。**承包商**应有权因**工程**的需要使用**现场**公用设施，其详细规定和价格见**规范要求**。**承包商**应自担风险和费用，提供其使用这些服务及计量所需要的任何仪器。用于测量消耗数量的设备应征得**工程师**的同意。在**合同数据**中规定的每个付款期间（如未规定，按每个月）的消耗数量（如果有）应由**承包商**测量，**承包商**为此类数量支付的金额（按**规范要求**中规定的价格）应包含在相关**报表**中。

4.20 进度报告

承包商应按照**规范要求**中规定的格式（如未规定，按照**工程师**可接受的格式）编制月度进度报告，并提交给**工程师**。每份进度报告应按照**合同数据**的规定，提交一份纸质原件、一份电子副本和附加纸质副本（如果有）。第一次报告所包含的时期，应自开**工日期**起至当月月底止。此后，应每月提交一次报告，在每次报告月最后一天后 7 日内报出。

报告应持续到**工程竣工**的日期，或如果**工程接收证书**中列出了未完成扫尾工作，报告应持续到此类扫尾工作完成为止。除非**规范要求**中另有规定，每份进度报告应包括：

(a) 图表、示意图和详细进度说明，包括（**承包商**设计，如果有）**承包商文件**、采购、制造、货物送达**现场**、施工、安装、试验的每个阶段；

(b) 反映制造和**现场**内外进展情况的照片和／或录像；

(c) 关于每项主要**生产设备**和**材料**的制造、制造商的名称、制造地点、进度百分比，以及下列事项的实际或预计日期：

（i）开始制造；
（ii）**承包商**检验；
（iii）试验；（以及）

(iv) shipment and arrival at the Site;

(d) the details described in Sub-Clause 6.10 [*Contractor's Records*];
(e) copies of quality management documents, inspection reports, test results, and compliance verification documentation (including certificates of Materials);
(f) a list of Variations, and any Notices given (by either Party) under Sub-Clause 20.2.1 [*Notice of Claim*];
(g) health and safety statistics, including details of any hazardous incidents and activities relating to environmental aspects and public relations; and
(h) comparisons of actual and planned progress, with details of any events or circumstances which may adversely affect the completion of the Works in accordance with the Programme and the Time for Completion, and the measures being (or to be) adopted to overcome delays.

However, nothing stated in any progress report shall constitute a Notice under a Sub-Clause of these Conditions.

4.21 Security of the Site

The Contractor shall be responsible for the security of the Site, and:

(a) for keeping unauthorised persons off the Site; and
(b) authorised persons shall be limited to the Contractor's Personnel, the Employer's Personnel, and to any other personnel identified as authorised personnel (including the Employer's other contractors on the Site), by a Notice from the Employer or the Engineer to the Contractor.

4.22 Contractor's Operations on Site

The Contractor shall confine the Contractor's operations to the Site, and to any additional areas which may be obtained by the Contractor and acknowledged by the Engineer as working areas. The Contractor shall take all necessary precautions to keep Contractor's Equipment and Contractor's Personnel within the Site and these additional areas, and to keep them off adjacent land.

At all times, the Contractor shall keep the Site free from all unnecessary obstruction, and shall properly store or remove from the Site any Contractor's Equipment (subject to 4.17 [*Contractor's Equipment*]) and/or surplus materials. The Contractor shall promptly clear away and remove from the Site any wreckage, rubbish, hazardous waste and Temporary Works which are no longer required.

Promptly after the issue of a Taking-Over Certificate, the Contractor shall clear away and remove, from that part of the Site and Works to which the Taking-Over Certificate refers, all Contractor's Equipment, surplus material, wreckage, rubbish, hazardous waste and Temporary Works. The Contractor shall leave that part of the Site and the Works in a clean and safe condition. However, the Contractor may retain at locations on the Site agreed with the Engineer, during the DNP, such Goods as are required for the Contractor to fulfil obligations under the Contract.

4.23 Archaeological and Geological Findings

All fossils, coins, articles of value or antiquity, and structures and other remains or items of geological or archaeological interest found on the Site shall be placed under the care and authority of the Employer. The Contractor shall take all reasonable precautions to prevent Contractor's Personnel or other persons from removing or damaging any of these findings.

　　　　　　（ⅳ）发货和运抵现场。

(d) 第 6.10 款 [*承包商的记录*] 中所述的细节；
(e) 质量管理文件、检测报告、试验结果及合规验证文件（包括**材料**证书）的副本；
(f) **变更**清单；（以及）（由任一方）根据第 20.2.1 项 [*索赔通知*] 的规定发出的任何**通知**；
(g) 健康和安全统计，包括对环境方面和公共关系有危害的任何事件和活动的详细情况；（以及）
(h) 实际进度与计划进度的对比，包括可能影响按照**计划**和**竣工时间**完成**工程**的任何不利事件或情况的详情；（以及）为消除延误正在（或准备）采取的措施。

但是，任何进度报告中的任何内容都不应构成本**条件条款**的**通知**。

4.21 现场安保

承包商应负责**现场**的安全，和

(a) 阻止未经授权的人员进入**现场**；（以及）
(b) 由**雇主**或**工程师**向**承包商**发出**通知**，授权人员应仅限于**承包商人员**、**雇主人员**和被确定为授权人员的任何其他人员（包括**雇主**在**现场**的其他承包商）。

4.22 承包商的现场作业

承包商应将其作业范围限制在**现场**，以及**承包商**可得到并经**工程师**认可作为工作场地的任何其他区域内。**承包商**应采取一切必要的预防措施，以保持**承包商设备**和**承包商人员**处在**现场**和此类其他区域内，避免其进入邻近区域。

在任何时间，**承包商**应保持**现场**没有一切不必要的障碍物，并应妥善存放和移除**承包商设备**（按照第 4.17 款 [*承包商设备*]）和 / 或剩余的材料。**承包商**应立即从**现场**清除并运走任何残物、垃圾、危险废物和不再需要的**临时工程**。⊖

在签发一项**接收证书**后，**承包商**应立即从**接收证书**涉及的**现场**和**工程**部分，清除并运走所有**承包商设备**、剩余材料、残物、垃圾、危险废物和**临时工程**。**承包商**应使该部分**现场**和**工程**处于清洁和安全的状态。但是，在**缺陷通知期限**内，**承包商**可在与**工程师**商定的**现场**位置保留其按照**合同**完成规定义务所需要的此类**货物**。

4.23 考古和地质发现

在**现场**发现的所有化石、硬币、有价值的物品或古物；以及具有地质或考古价值的结构物和其他遗迹或物品，应置于**雇主**的照管和权限下。**承包商**应采取一切合理的预防措施，防止**承包商人员**或其他人员移动或损坏任何此类发现物。

⊖ 此处中文按照勘误表改正后的英文翻译。——译者注

The Contractor shall, as soon as practicable after discovery of any such finding, give a Notice to the Engineer in good time to give the Engineer opportunity to promptly inspect and/or investigate the finding before it is disturbed. This Notice shall describe the finding and the Engineer shall issue instructions for dealing with it.

If the Contractor suffers delay and/or incurs Cost from complying with the Engineer's instructions, the Contractor shall be entitled subject to Sub-Clause 20.2 [*Claims For Payment and/or EOT*] to EOT and/or payment of such Cost.

5 Subcontracting

5.1 Subcontractors

The Contractor shall not subcontract:

(a) works with a total accumulated value greater than the percentage of the Accepted Contract Amount stated in the Contract Data (if not stated, the whole of the Works); or

(b) any part of the Works for which subcontracting is not permitted as stated in the Contract Data.

The Contractor shall be responsible for the work of all Subcontractors, for managing and coordinating all the Subcontractors' works, and for the acts or defaults of any Subcontractor, any Subcontractor's agents or employees, as if they were the acts or defaults of the Contractor.

The Contractor shall obtain the Engineer's prior consent to all proposed Subcontractors, except:

(i) suppliers of Materials; or

(ii) a subcontract for which the Subcontractor is named in the Contract.

Where the Contractor is required to obtain the Engineer's consent to a proposed Subcontractor, the Contractor shall submit the name, address, detailed particulars and relevant experience of such a Subcontractor and the work intended to be subcontracted to the Engineer and further information which the Engineer may reasonably require. If the Engineer does not respond within 14 days after receiving this submission (or further information if requested), by giving a Notice objecting to the proposed Subcontractor, the Engineer shall be deemed to have given his/her consent.

The Contractor shall give a Notice to the Engineer not less than 28 days before the intended date of the commencement of each Subcontractor's work, and of the commencement of such work on the Site.

5.2 Nominated Subcontractors

5.2.1 Definition of "nominated Subcontractor"

In this Sub-Clause, "nominated Subcontractor" means a Subcontractor named as such in the Specification or whom the Engineer, under Sub-Clause 13.4 [*Provisional Sums*], instructs the Contractor to employ as a Subcontractor.

发现任何此类物品后，**承包商**应在可行的范围内尽快**通知工程师**，以便**工程师**有机会在发现任何此类物品受到干扰前，及时进行检查和/或调查。该**通知**应说明这一发现，**工程师**应就处理发现的物品发出指示。

如**承包商**因执行**工程师**指示遭受延误和/或招致增加**费用**，**承包商**应有权根据第 20.2 款 [*付款和/或竣工时间延长的索赔*] 的规定，要求获得**竣工时间的延长**和/或此类**费用**的支付。

---

5 分包

5.1 分包商

承包商不得分包：

(a) 累计总价值大于**合同数据**中规定的**中标合同金额**百分比的工程（如未规定，则为全部工程）；（或）

(b) **合同数据**中规定不允许分包的任何**工程**部分。

承包商应负责所有**分包商**的工作，管理和协调所有**分包商**的工程，并对任何**分包商**、任何**分包商**代理或雇员的行为或违约负责，如同其是**承包商**自己的行为或违约一样。

承包商应事先获得**工程师**对所有拟定的**分包商**的同意，但下列情况除外：

(i) **材料**供应商；（或）
(ii) 合同中指定**分包商**的分包合同。

如果要求**承包商**获得**工程师**对拟定的**分包商**的同意，**承包商**应向**工程师**提交该**分包商**的姓名、地址、详细资料和相关经验，以及拟分包的工作，和**工程师**可能合理要求的更多信息。如果**工程师**在收到这份提交（或如果要求提供更多信息）后 14 天内未做出答复，未对拟定的**分包商**发出**不同意通知**，应视为**工程师**已同意。

承包商应在每个**分包商**工作拟定开工日期前不少于 28 天，向**工程师**发出**通知**并在**现场**开始此类工作。

5.2 指定分包商

5.2.1 "指定分包商"的定义

在本款中，"指定分包商"系指在**规范要求**中指定的**分包商**，或工程师根据第 13.4 款 [*暂列金额*] 的规定指示**承包商**雇用的**分包商**。

5.2.2
Objection to Nomination

The Contractor shall not be under any obligation to employ a Subcontractor whom the Engineer instructs and against whom the Contractor raises reasonable objection by giving a Notice to the Engineer, with detailed supporting particulars, no later than 14 days after receiving the Engineer's instruction. An objection shall be deemed reasonable if it arises from (among other things) any of the following matters, unless the Employer agrees to indemnify the Contractor against and from the consequences of the matter:

(a) there are reasons to believe that the Subcontractor nominated does not have sufficient competence, resources or financial strength;

(b) the subcontract does not specify that the nominated Subcontractor shall indemnify the Contractor against and from any negligence or misuse of Goods by the nominated Subcontractor, the nominated Subcontractor's agents and employees; or

(c) the subcontract does not specify that, for the subcontracted work (including design, if any), the nominated Subcontractor shall:

   (i) undertake to the Contractor such obligations and liabilities as will enable the Contractor to discharge the Contractor's corresponding obligations and liabilities under the Contract, and

   (ii) indemnify the Contractor against and from all obligations and liabilities arising under or in connection with the Contract and from the consequences of any failure by the Subcontractor to perform these obligations or to fulfil these liabilities.

5.2.3
Payments to nominated Subcontractors

The Contractor shall pay to the nominated Subcontractor the amounts due in accordance with the subcontract. These amounts plus other charges shall be included in the Contract Price in accordance with sub-paragraph (b) of Sub-Clause 13.4 [*Provisional Sums*], except as stated in Sub-Clause 5.2.4 [*Evidence of Payments*].

5.2.4
Evidence of Payments

Before issuing a Payment Certificate which includes an amount payable to a nominated Subcontractor, the Engineer may request the Contractor to supply reasonable evidence that the nominated Subcontractor has received all amounts due in accordance with the previous Payment Certificates, less applicable deductions for retention or otherwise. Unless the Contractor:

(a) submits this reasonable evidence to the Engineer, or

(b) (i) satisfies the Engineer in writing that the Contractor is reasonably entitled to withhold or refuse to pay these amounts, and

   (ii) submits to the Engineer reasonable evidence that the nominated Subcontractor has been notified of the Contractor's entitlement,

then the Employer may (at the Employer's sole discretion) pay, directly to the nominated Subcontractor, part or all of such amounts previously certified (less applicable deductions) as are due to the nominated Subcontractor and for which the Contractor has failed to submit the evidence described in sub-paragraphs (a) or (b) above.

Thereafter, the Engineer shall give a Notice to the Contractor stating the amount paid directly to the nominated Subcontractor by the Employer and, in the next IPC after this Notice, shall include this amount as a deduction under sub-paragraph (b) of Sub-Clause 14.6.1 [*The IPC*].

5.2.2
反对指定

承包商不应有任何义务雇用工程师指示的指定分包商，承包商应在收到工程师指示后 14 天内，向工程师发出附有详细证明资料的通知，提出合理的反对意见。（其中）任何以下事项引起的反对，应认为是合理的，除非雇主同意保障承包商免受这些事项的影响：⊖

(a) 有理由相信，指定分包商没有足够的能力、资源或财力；

(b) 分包合同未明确规定，指定分包商应保障承包商不承担指定分包商及其代理和雇员的任何疏忽而误用货物的责任；（或）

(c) 对分包的工作（包括设计，如果有），分包合同未明确规定指定分包商应：

（i） 为承包商承担此项义务和责任，使承包商能履行其合同规定的相应义务和责任；（以及）

（ii） 保障承包商免除其合同规定的或与合同有关的所有义务和责任，以及因分包商未能履行这些义务或责任的影响而产生的义务和责任。

5.2.3
对指定分包商付款

承包商应按照分包合同规定的应付金额给指定分包商付款。除了第 5.2.4 项 [付款证据] 中所述的情况外，这些金额连同其他费用，应按照第 13.4 款 [暂列金额] 的规定，计入合同价格。

5.2.4
付款证据

工程师在发出包含应付给指定分包商金额的付款证书前，可要求承包商提供合理的证据，证明指定分包商已收到按照此前付款证书应付的、减去合理的保留金或其他扣除后的所有金额。除非承包商：

(a) 向工程师提交此项合理证据；（或）

(b) （i） 提出使工程师满意的、承包商合理有权暂扣或拒付该金额的书面说明；（以及）
（ii） 向工程师提交合理证据，证明指定分包商已被告知承包商的授权。

然后，雇主可（雇主自行决定）直接向指定分包商支付部分或全部以前已证明应付的金额（减去合理的扣减额），以及承包商未能提交上述 (a) 或 (b) 段所述证据的部分或全部金额。

此后，工程师应向承包商发出通知，说明雇主直接支付指定分包商的金额，并在该通知后的下一次期中付款证书中包含第 14.6.1 项 [期中付款证书] (b) 段规定的扣减额。

⊖ 此处中文按照勘误表改正后的英文翻译。——译者注

6 Staff and Labour

6.1 Engagement of Staff and Labour

Except as otherwise stated in the Specification, the Contractor shall make arrangements for the engagement of all Contractor's Personnel, and for their payment, accommodation, feeding, transport and welfare.

6.2 Rates of Wages and Conditions of Labour

The Contractor shall pay rates of wages, and observe conditions of labour, which comply with all applicable Laws and are not lower than those established for the trade or industry where the work is carried out.

If no established rates or conditions are applicable, the Contractor shall pay rates of wages and observe conditions which are not lower than the general level of wages and conditions observed locally by employers whose trade or industry is similar to that of the Contractor.

6.3 Recruitment of Persons

The Contractor shall not recruit, or attempt to recruit, staff and labour from amongst the Employer's Personnel.

Neither the Employer nor the Engineer shall recruit, or attempt to recruit, staff and labour from amongst the Contractor's Personnel.

6.4 Labour Laws

The Contractor shall comply with all the relevant labour Laws applicable to the Contractor's Personnel, including Laws relating to their employment (including wages and working hours), health, safety, welfare, immigration and emigration, and shall allow them all their legal rights.

The Contractor shall require the Contractor's Personnel to obey all applicable Laws, including those concerning health and safety at work.

6.5 Working Hours

No work shall be carried out on the Site on locally recognised days of rest, or outside the normal working hours stated in the Contract Data, unless:

(a) otherwise stated in the Contract;
(b) the Engineer gives consent; or
(c) the work is unavoidable or necessary for the protection of life or property or for the safety of the Works, in which case the Contractor shall immediately give a Notice to the Engineer with reasons and describing the work required.

6.6 Facilities for Staff and Labour

Except as otherwise stated in the Specification, the Contractor shall provide and maintain all necessary accommodation and welfare facilities for the Contractor's Personnel.

If such accommodation and facilities are to be located on the Site, except where the Employer has given the Contractor prior permission, they shall be located within the areas identified in the Contract. If any such accommodation or facilities are found elsewhere within the Site, the Contractor shall immediately remove them at the Contractor's risk and cost. The Contractor shall also provide facilities for the Employer's Personnel as stated in the Specification.

6 员工

6.1 员工的雇用

除**规范要求**中另有说明外，**承包商**应安排从当地或其他地方雇用所有的**承包商人员**，并负责他们的报酬、住宿、膳食、交通和福利。

6.2 工资标准和劳动条件

承包商应支付薪酬并遵守所有适用**法律**的劳动条件，应不低于工作所在地该工种或行业制定的标准和条件。

如果没有现成的适用标准和条件，**承包商**所付的工资标准和遵守的劳动条件，应不低于当地与**承包商**类似的工种或行业雇主所付的一般工资标准和遵守的劳动条件。

6.3 招聘人员

承包商不应从**雇主人员**中招聘或试图招聘员工。

雇主或**工程师**均不得从**承包商人员**中招聘或试图招聘员工。

6.4 劳动法

承包商应遵守所有适用于**承包商人员**的相关劳动**法**，包括有关他们的雇用（包括薪酬和工作时间）、健康、安全、福利、入境和出境等**法律**，并应允许他们享有所有合法权利。

承包商应要求**承包商人员**遵守所有适用的**法律**，包括有关工作健康和安全的法律。

6.5 工作时间

除非出现下列情况，否则在当地公认的休息日，或**合同数据**中规定的正常工作时间以外，不应在**现场**进行工作：

(a) 合同中另有规定；
(b) **工程师**同意；（或）
(c) 因保护生命或财产，或因**工程**安全而不可避免或必需的工作，在此情况下，**承包商**应立即向**工程师**发出**通知**，说明原因和所需的工作。

6.6 为员工提供设施

除**规范要求**中另有说明外，**承包商**应为**承包商人员**提供和保持一切必要的食宿和福利设施。

如果此类住宿和设施将位于**现场**，除非**雇主**事先许可，否则应位于合同规定的区域内。如果在**现场**其他地方发现任何此类住宿和设施，**承包商**应立即将其移除，风险和费用由**承包商**承担。**承包商**还应按照**规范要求**中的规定为**雇主人员**提供设施。

6.7 Health and Safety of Personnel

In addition to the requirements of Sub-Clause 4.8 [*Health and Safety Obligations*], the Contractor shall at all times take all necessary precautions to maintain the health and safety of the Contractor's Personnel. In collaboration with local health authorities, the Contractor shall ensure that:

(a) medical staff, first aid facilities, sick bay, ambulance services and any other medical services stated in the Specification are available at all times at the Site and at any accommodation for Contractor's and Employer's Personnel; and

(b) suitable arrangements are made for all necessary welfare and hygiene requirements and for the prevention of epidemics.

The Contractor shall appoint a health and safety officer at the Site, responsible for maintaining health, safety and protection against accidents. This officer shall:

(i) be qualified, experienced and competent for this responsibility; and

(ii) have the authority to issue directives for the purpose of maintaining the health and safety of all personnel authorised to enter and/or work on the Site and to take protective measures to prevent accidents.

Throughout the execution of the Works, the Contractor shall provide whatever is required by this person to exercise this responsibility and authority.

6.8 Contractor's Superintendence

From the Commencement Date until the issue of the Performance Certificate, the Contractor shall provide all necessary superintendence to plan, arrange, direct, manage, inspect, test and monitor the execution of the Works.

Superintendence shall be given by a sufficient number of persons:

(a) who are fluent in or have adequate knowledge of the language for communications (defined in Sub-Clause 1.4 [*Law and Language*]); and

(b) who have adequate knowledge of the operations to be carried out (including the methods and techniques required, the hazards likely to be encountered and methods of preventing accidents),

for the satisfactory and safe execution of the Works.

6.9 Contractor's Personnel

The Contractor's Personnel (including Key Personnel, if any) shall be appropriately qualified, skilled, experienced and competent in their respective trades or occupations.

The Engineer may require the Contractor to remove (or cause to be removed) any person employed on the Site or Works, including the Contractor's Representative and Key Personnel (if any), who:

(a) persists in any misconduct or lack of care;
(b) carries out duties incompetently or negligently;
(c) fails to comply with any provision of the Contract;
(d) persists in any conduct which is prejudicial to safety, health, or the protection of the environment;
(e) is found, based on reasonable evidence, to have engaged in corrupt, fraudulent, collusive or coercive practice; or
(f) has been recruited from the Employer's Personnel in breach of Sub-Clause 6.3 [*Recruitment of Persons*].

| 6.7 人员的健康和安全 | 除第4.8款[*健康和安全义务*]的要求外，**承包商**应始终采取一切必要的预防措施，维护**承包商人员**的健康和安全。**承包商**应与当地卫生部门合作，确保：

（a） 随时在**现场**，以及**承包商人员**和**雇主人员**的任何住地，配备医务人员、急救设施、病房、救护车服务及**规范要求**中规定的任何其他医疗服务；

（b） 对所有必要的福利和卫生要求，以及预防传染病做出适当安排。

承包商应指派一名健康和安全官员，负责**现场**的人身健康和安全及安全事故预防工作。该人员应：

（i） 具备资格、经验丰富并胜任此项工作；（以及）
（ii） 有权发布指示，维护被授权进入和/或在**现场**工作的所有人员的健康和安全，及采取防止事故的保护措施。

在整个**工程**实施过程中，**承包商**应提供该人员为履行其职责和权利所需的任何事项。

| 6.8 承包商的监督 | 从开工日期到**履约证书**的签发，**承包商**应对**工程**实施的规划、安排、指导、管理、检测、试验和监测，提供一切必要的监督。

此类监督应由足够数量的人员执行：

（a） 其应能流利使用或熟练掌握（第1.4款[*法律和语言*]所规定的）交流语言；（以及）

（b） 具有对要进行的各项作业所需的足够知识（包括所需的方法和技术、可能遇到的危险和预防事故的方法）。

以便**工程**能够令人满意和安全地实施。

| 6.9 承包商人员 | **承包商人员**（包括**关键人员**，如果有）都应是在他们各自工种或职业内，具有相应资质、技能、经验和能力的人员。

工程师可要求**承包商**撤换（或敦促撤换）受雇于**现场**或**工程**的、有下列行为的任何人员，也包括**承包商代表**和**关键人员**（如果有）：

（a） 经常行为不当，或工作漫不经心；
（b） 无能力履行义务或玩忽职守；
（c） 不遵守**合同**的任何规定；
（d） 坚持任何有损安全、健康或有损环境保护的行为；
（e） 根据合理证据被认定从事腐败、欺诈、串通或胁迫行为；（或）
（f） 违反第6.3款[*招聘人员*]的规定从**雇主人员**中招聘员工。

If appropriate, the Contractor shall then promptly appoint (or cause to be appointed) a suitable replacement. In the case of replacement of the Contractor's Representative, Sub-Clause 4.3 [*Contractor's Representative*] shall apply. In the case of replacement of Key Personnel (if any), Sub-Clause 6.12 [*Key Personnel*] shall apply.

6.10 Contractor's Records

Unless otherwise proposed by the Contractor and agreed by the Engineer, in each progress report under Sub-Clause 4.20 [*Progress Reports*], the Contractor shall include records of:

(a) occupations and actual working hours of each class of Contractor's Personnel;
(b) the type and actual working hours of each of the Contractor's Equipment;
(c) the types of Temporary Works used;
(d) the types of Plant installed in the Permanent Works; and
(e) the quantities and types of Materials used

for each work activity shown in the Programme, at each work location and for each day of work.

6.11 Disorderly Conduct

The Contractor shall at all times take all necessary precautions to prevent any unlawful, riotous or disorderly conduct by or amongst the Contractor's Personnel, and to preserve peace and protection of persons and property on and near the Site.

6.12 Key Personnel

If no Key Personnel are stated in the Specification this Sub-Clause shall not apply.

The Contractor shall appoint the natural persons named in the Tender to the positions of Key Personnel. If not so named, or if an appointed person fails to act in the relevant position of Key Personnel, the Contractor shall submit to the Engineer for consent the name and particulars of another person the Contractor proposes to appoint to such position. If consent is withheld or subsequently revoked, the Contractor shall similarly submit the name and particulars of a suitable replacement for such position.

If the Engineer does not respond within 14 days after receiving any such submission, by giving a Notice stating his/her objection to the appointment of such person (or replacement) with reasons, the Engineer shall be deemed to have given his/her consent.

The Contractor shall not, without the Engineer's prior consent, revoke the appointment of any of the Key Personnel or appoint a replacement (unless the person is unable to act as a result of death, illness, disability or resignation, in which case the appointment shall be deemed to have been revoked with immediate effect and the appointment of a replacement shall be treated as a temporary appointment until the Engineer gives his/her consent to this replacement, or another replacement is appointed, under this Sub-Clause).

All Key Personnel shall be based at the Site (or, where Works are being executed off the Site, at the location of the Works) for the whole time that the Works are being executed. If any of the Key Personnel is to be temporarily absent during execution of the Works, a suitable replacement shall be temporarily appointed, subject to the Engineer's prior consent.

如果适宜，**承包商**随后应立即指派（或敦促指派）合适的替代人员。如果替代**承包商代表**，第 4.3 款［*承包商代表*］应适用。如果替代**关键人员**（如果有），第 6.12 款［*关键人员*］应适用。

6.10 承包商的记录

除非**承包商**另有提议并经**工程师**同意，否则在根据**第 4.20 款**［*进度报告*］规定的每份进度报告中，**承包商**应包括以下记录：

（a）各类**承包商人员**的职业和实际工作时间；

（b）各项**承包商设备**的类型和实际工作时间；

（c）所使用的**临时工程**的类型；
（d）**永久工程**中安装**生产设备**的类型；（以及）
（e）所用**材料**的数量和类型。

对**进度计划**中所示的每项工作活动、在每个工作地点和每天的工作。

6.11 无序行为

承包商应始终采取各种必要的预防措施，防止**承包商人员**或其内部发生任何非法的、骚动的，或无序的行为，以保持安定、保护**现场**及邻近人员和财产的安全。

6.12 关键人员

如果**规范要求**中未规定**关键人员**，本**款**应不适用。

承包商应任命**投标书**中提名的自然人担任**关键人员**。如未提名，或如果任命的人员未能担任**关键人员**的相关职位，**承包商**应向**工程师**提交其拟任命担任该职位的另一人员的姓名和详细资料，以征得其同意。如果拒绝同意或随后撤销同意，**承包商**应同样地提交该职位的合适替代人员的姓名和详细资料。

如果**工程师**在收到任何此类提交后 14 天内未做出答复，发出**通知**说明其对该人员（或替代人员）的反对意见，并附说明理由，则应视为**工程师**已同意。

未经**工程师**事先同意，**承包商**不得撤销任何**关键人员**的任命，或任命替代人员（除非该人员因死亡、疾病、残疾或辞职无法工作，在这种情况下，该任命应被视为立即撤销并无效，在**工程师**同意或根据本**款**的规定任命另一名替代人员之前，该替代人员的任命应视为临时任命）。

所有**关键人员**应在整个**工程**施工期间常驻**现场**（或，在施工**现场**以外的地方，驻扎在**工程**所在地）。如果任何**关键人员**在**工程**施工期间临时离开**现场**，应事先征得**工程师**的同意，临时指定一名合适的替代人员。

All Key Personnel shall be fluent in the language for communications defined in Sub-Clause 1.4 [*Law and Language*].

7 Plant, Materials and Workmanship

7.1 Manner of Execution

The Contractor shall carry out the manufacture, supply, installation, testing and commissioning and/or repair of Plant, the production, manufacture, supply and testing of Materials, and all other operations and activities during the execution of the Works:

(a) in the manner (if any) specified in the Contract;
(b) in a proper workmanlike and careful manner, in accordance with recognised good practice; and
(c) with properly equipped facilities and non-hazardous Materials, except as otherwise specified in the Contract.

7.2 Samples

The Contractor shall submit the following samples of Materials, and relevant information, to the Engineer for consent prior to using the Materials in or for the Works:

(a) manufacturer's standard samples of Materials and samples specified in the Contract, all at the Contractor's cost, and
(b) additional samples instructed by the Engineer as a Variation.

Each sample shall be labelled as to origin and intended use in the Works.

7.3 Inspection

The Employer's Personnel shall, during all the normal working hours stated in the Contract Data and at all other reasonable times:

(a) have full access to all parts of the Site and to all places from which natural Materials are being obtained;
(b) during production, manufacture and construction (at the Site and elsewhere), be entitled to:

   (i) examine, inspect, measure and test (to the extent stated in the Specification) the Materials, Plant and workmanship,
   (ii) check the progress of manufacture of Plant and production and manufacture of Materials, and
   (iii) make records (including photographs and/or video recordings); and

(c) carry out other duties and inspections, as specified in these Conditions and the Specification.

The Contractor shall give the Employer's Personnel full opportunity to carry out these activities, including providing safe access, facilities, permissions and safety equipment.

The Contractor shall give a Notice to the Engineer whenever any Materials, Plant or work is ready for inspection, and before it is to be covered up, put out of sight, or packaged for storage or transport. The Employer's Personnel shall then either carry out the examination, inspection, measurement or testing without unreasonable delay, or the Engineer shall promptly give a Notice to the Contractor that the Employer's Personnel

所有关键人员应能流利地使用第 1.4 款 [*法律和语言*] 规定的交流语言。

7 生产设备、材料和工艺

7.1 实施方法

承包商应按以下方法进行制造、供应、安装、试验和调试，和 / 或维修**生产设备**、生产、制造、供应和检测**材料**，以及**工程**的所有其他实施作业和活动：

(a) 按照**合同**规定的方法（如果有）；
(b) 按照公认的良好惯例，使用恰当、精巧和仔细的方法；（以及）

(c) 除**合同**另有规定外，使用适当配备的设施和无危险的材料。

7.2 样品

承包商应在**工程**中或为**工程**使用**材料**前，向工程师提交以下**材料**样品和有关资料，以取得其同意：

(a) 制造商的**材料**标准样品和**合同**规定的样品，均由**承包商**自费提供；（以及）
(b) 由**工程师**指示的、作为**变更**的附加样品。

每种样品均应标明其原产地和在**工程**中的拟定用途。

7.3 检验

雇主人员应在**合同数据**中规定的所有正常工作时间和所有其他合理的时间内：

(a) 有充分机会进入**现场**的所有部分，以及获得天然**材料**的所有地点；
(b) 有权在生产、加工和施工期间（在**现场**和其他地方）：

　(i) 检查、检验、测量和试验（在**规范要求**中规定的范围内）所用**材料**、**生产设备**和工艺；
　(ii) 检查**生产设备**的制造和**材料**的生产加工进度；（以及）

　(iii) 做好记录（包括照片和 / 或录像）；（以及）

(c) 执行本**条件**和**规范要求**中规定的其他任务和检查。

承包商应为**雇主人员**进行上述活动提供一切机会，包括提供安全进入条件、设施、许可和安全装备。

每当任何**材料**、**生产设备**或工作已经准备好接受检验，在覆盖、掩蔽、包装以便储存或运输前，**承包商**应通知**工程师**。这时，**雇主人员**应及时进行检查、检验、测量或试验，不得无故拖延，否则**工程师**应立即

do not require to do so. If the Engineer gives no such Notice and/or the Employer's Personnel do not attend at the time stated in the Contractor's Notice (or such time as may be agreed with the Contractor), the Contractor may proceed with covering up, putting out of sight or packaging for storage or transport.

If the Contractor fails to give a Notice in accordance with this Sub-Clause, the Contractor shall, if and when required by the Engineer, uncover the work and thereafter reinstate and make good, all at the Contractor's risk and cost.

7.4 Testing by the Contractor

This Sub-Clause shall apply to all tests specified in the Contract, other than the Tests after Completion (if any).

The Contractor shall provide all apparatus, assistance, documents and other information, temporary supplies of electricity and water, equipment, fuel, consumables, instruments, labour, materials, and suitably qualified, experienced and competent staff, as are necessary to carry out the specified tests efficiently and properly. All apparatus, equipment and instruments shall be calibrated in accordance with the standards stated in the Specification or defined by applicable Laws and, if requested by the Engineer, the Contractor shall submit calibration certificates before carrying out testing.

The Contractor shall give a Notice to the Engineer, stating the time and place for the specified testing of any Plant, Materials and other parts of the Works. This Notice shall be given in reasonable time, having regard to the location of the testing, for the Employer's Personnel to attend.

The Engineer may, under Clause 13 [*Variations and Adjustments*], vary the location or timing or details of specified tests, or instruct the Contractor to carry out additional tests. If these varied or additional tests show that the tested Plant, Materials or workmanship is not in accordance with the Contract, the Cost and any delay incurred in carrying out this Variation shall be borne by the Contractor.

The Engineer shall give a Notice to the Contractor of not less than 72 hours of his/her intention to attend the tests. If the Engineer does not attend at the time and place stated in the Contractor's Notice under this Sub-Clause, the Contractor may proceed with the tests, unless otherwise instructed by the Engineer. These tests shall then be deemed to have been made in the Engineer's presence. If the Contractor suffers delay and/or incurs Cost from complying with any such instruction or as a result of a delay for which the Employer is responsible, the Contractor shall be entitled subject to Sub-Clause 20.2 [*Claims For Payment and/or EOT*] to EOT and/or payment of Cost Plus Profit.

If the Contractor causes any delay to specified tests (including varied or additional tests) and such delay causes the Employer to incur costs, the Employer shall be entitled subject to Sub-Clause 20.2 [*Claims For Payment and/or EOT*] to payment of these costs by the Contractor.

The Contractor shall promptly forward to the Engineer duly certified reports of the tests. When the specified tests have been passed, the Engineer shall endorse the Contractor's test certificate, or issue a test certificate to the Contractor, to that effect. If the Engineer has not attended the tests, he/she shall be deemed to have accepted the readings as accurate.

通知**承包商**，**雇主人员**无须进行这些工作。如果**工程师**没有发出此类**通知**和/或雇主人员没有在**承包商通知**中规定的时间（或与**承包商**商定的时间）内参加此类活动，**承包商**可继续覆盖、掩蔽、包装以便储存或运输。

如果**承包商**没有按照本款的规定发出此类**通知**，而当**工程师**提出要求时，**承包商**应除去物件上的覆盖，并在随后恢复完好，所有风险和费用由**承包商**承担。

7.4
由承包商试验

本款适用于**竣工后试验**（如果有）以外的合同规定的所有试验。

为有效和适当地进行规定的试验，**承包商**应提供所需的所有仪器、协助、文件和其他资料，临时供应电力和水、装备、燃料、消耗品、工具、劳动力、材料，以及具有适当资质、经验和胜任的工作人员。所有仪器、设备和仪表应按照**规范要求**中规定或适用**法律**规定的标准进行校准。如果**工程师**要求，**承包商**应在进行试验前提交校准证书。

承包商应向**工程师**发出**通知**，说明对任何**生产设备**、**材料**和**工程**其他部分进行规定试验的时间和地点。该**通知**应在考虑试验地点的合理时间内发出，以便**雇主人员**参加。

根据第 13 条［*变更和调整*］的规定，**工程师**可以改变进行规定试验的位置、时间或细节，或指示**承包商**进行附加试验。如果这些改变的或附加的试验证明经过试验的**生产设备**、**材料**或工艺不符合**合同**要求，**承包商**应承担进行本项**变更**和任何延误产生的**费用**。

工程师应至少提前 72 小时将参加试验的意图**通知承包商**。如果**工程师**没有在**承包商**根据本款发出的**通知**中规定的时间和地点参加试验，除非**工程师**另有指示，**承包商**可自行进行试验，这些试验应被视为是在**工程师**在场情况下进行的。如果由于遵守这些指示或因**雇主**应负责的原因延误，使**承包商**遭受延误和/或招致增加**费用**，**承包商**应有权根据第 20.2 款［*付款和/或竣工时间延长的索赔*］的规定，要求获得竣工时间的延长和/或成本加利润的支付。

如果**承包商**对规定的试验（包括变更或附加试验）造成任何延误，且此类延误导致**雇主**产生费用，**雇主**有权根据第 20.2 款［*付款和/或竣工时间延长的索赔*］的规定，要求**承包商**支付这些费用。

承包商应迅速向**工程师**提交充分证实的试验报告。当规定的试验通过时，**工程师**应在**承包商**的试验证书上签字，或向**承包商**签发等效的证书。如果**工程师**未参加试验，其应被视为已经认可试验读数是准确的。

Sub-Clause 7.5 [*Defects and Rejection*] shall apply in the event that any Plant, Materials and other parts of the Works fails to pass a specified test.

7.5 Defects and Rejection

If, as a result of an examination, inspection, measurement or testing, any Plant, Materials, Contractor's design (if any) or workmanship is found to be defective or otherwise not in accordance with the Contract, the Engineer shall give a Notice to the Contractor describing the item of Plant, Materials, design or workmanship that has been found to be defective. The Contractor shall then promptly prepare and submit a proposal for necessary remedial work.

The Engineer may Review this proposal, and may give a Notice to the Contractor stating the extent to which the proposed work, if carried out, would not result in the Plant, Materials, Contractor's design (if any) or workmanship complying with the Contract. After receiving such a Notice the Contractor shall promptly submit a revised proposal to the Engineer. If the Engineer gives no such Notice within 14 days after receiving the Contractor's proposal (or revised proposal), the Engineer shall be deemed to have given a Notice of No-objection.

If the Contractor fails to promptly submit a proposal (or revised proposal) for remedial work, or fails to carry out the proposed remedial work to which the Engineer has given (or is deemed to have given) a Notice of No-objection, the Engineer may:

(a) instruct the Contractor under sub-paragraph (a) and/or (b) of Sub-Clause 7.6 [*Remedial Work*]; or
(b) reject the Plant, Materials, Contractor's design (if any) or workmanship by giving a Notice to the Contractor, with reasons, in which case sub-paragraph (a) of Sub-Clause 11.4 [*Failure to Remedy Defects*] shall apply.

After remedying defects in any Plant, Materials, design (if any) or workmanship, if the Engineer requires any such items to be retested, the tests shall be repeated in accordance with Sub-Clause 7.4 [*Testing by the Contractor*] at the Contractor's risk and cost. If the rejection and retesting cause the Employer to incur additional costs, the Employer shall be entitled subject to Sub-Clause 20.2 [*Claims For Payment and/or EOT*] to payment of these costs by the Contractor.

7.6 Remedial Work

In addition to any previous examination, inspection, measurement or testing, or test certificate or Notice of No-objection by the Engineer, at any time before the issue of the Taking-Over Certificate for the Works the Engineer may instruct the Contractor to:

(a) repair or remedy (if necessary, off the Site), or remove from the Site and replace any Plant or Materials which are not in accordance with the Contract;
(b) repair or remedy, or remove and re-execute, any other work which is not in accordance with the Contract; and
(c) carry out any remedial work which is urgently required for the safety of the Works, whether because of an accident, unforeseeable event or otherwise.

The Contractor shall comply with the instruction as soon as practicable and not later than the time (if any) specified in the instruction, or immediately if urgency is specified under sub-paragraph (c) above.

如果任何**生产设备**、**材料**和**工程**其他部分未能通过规定的试验，**第7.5款**[*缺陷和拒收*]应适用。

7.5 **缺陷和拒收**	如果检查、检验、测量或试验结果，发现任何**生产设备**、**材料**、**承包商**的设计（如果有）或工艺有缺陷，或不符合**合同**要求，**工程师**可向**承包商**发出**通知**，并说明被发现有缺陷的**生产设备**、**材料**、设计或工艺项目。**承包商**应迅速准备并提交必要补救工作的建议书。

工程师可审核该建议书，并可向**承包商**发出**通知**，说明拟定的工作如果实施，会在多大程度上导致**生产设备**、**材料**、**承包商**的设计（如果有）或工艺不符合**合同**规定。收到该**通知**后，**承包商**应立即向**工程师**提交修订的建议书。如果**工程师**在收到**承包商**的建议书（或修订的建议书）后14天内未发出此类**通知**，**工程师**应被视为已发出**不反对通知**。

如果**承包商**未能及时提交补救工作的建议书（或修订的建议书），或未能执行**工程师**已发出（或被视为已发出）的**不反对通知**中建议的补救工作，**工程师**可：

(a) 根据第7.6款[*修补工作*](a)段和/或(b)段指示**承包商**；(或)

(b) 向**承包商**发出**通知**并说明理由，拒收生产设备、**材料**、**承包商**的设计（如果有）或工艺，在这种情况下，**第11.4款**[*未能修补缺陷*](a)段应适用。

在修补任何**生产设备**、**材料**、**承包商**的设计（如果有）或工艺的缺陷后，如果**工程师**要求对任何此类项目再次进行试验，这些试验应按照**第7.4款**[*由承包商试验*]的规定重新进行，风险和费用由**承包商**承担。如果此项拒收和重新试验使**雇主**增加了额外费用，**雇主**应有权按照**第20.2款**[*付款和/或竣工时间延长的索赔*]的规定，要求**承包商**支付这笔费用。

7.6 **修补工作**	除**工程师**先前的任何检查、检验、测量或试验，或试验证书或**不反对通知**外，在签发**工程接收证书**之前的任何时候，**工程师**可指示**承包商**进行以下工作：

(a) 修理或修补（如有必要，在**现场**外）或从**现场**移除并更换不符合**合同**要求的任何**生产设备**或**材料**；

(b) 修理或修补，或移除并重新实施不符合**合同**规定的任何其他工作；(以及)

(c) 实施因意外、不可预见的事件或其他原因引起的、**工程**安全迫切需要的任何修补工作。

承包商应尽快在切实可行且不迟于指示中规定的时间（如果有）内执行该指示，或在上述(c)段规定的紧急情况下立即实施。

The Contractor shall bear the cost of all remedial work required under this Sub-Clause, except to the extent that any work under sub-paragraph (c) above is attributable to:

(i) any act by the Employer or the Employer's Personnel. If the Contractor suffers delay and/or incurs Cost in carrying out such work, the Contractor shall be entitled subject to Sub-Clause 20.2 [*Claims For Payment and/or EOT*] to EOT and/or payment of such Cost Plus Profit; or
(ii) an Exceptional Event, in which case Sub-Clause 18.4 [*Consequences of an Exceptional Event*] shall apply.

If the Contractor fails to comply with the Engineer's instruction, the Employer may (at the Employer's sole discretion) employ and pay other persons to carry out the work. Except to the extent that the Contractor would have been entitled to payment for work under this Sub-Clause, the Employer shall be entitled subject to Sub-Clause 20.2 [*Claims For Payment and/or EOT*] to payment by the Contractor of all costs arising from this failure. This entitlement shall be without prejudice to any other rights the Employer may have, under the Contract or otherwise.

7.7 Ownership of Plant and Materials

Each item of Plant and Materials shall, to the extent consistent with the mandatory requirements of the Laws of the Country, become the property of the Employer at whichever is the earlier of the following times, free from liens and other encumbrances:

(a) when it is delivered to the Site;
(b) when the Contractor is paid the value of the Plant and Materials under Sub-Clause 8.11 [*Payment for Plant and Materials after Employer's Suspension*]; or
(c) when the Contractor is paid the amount determined for the Plant and Materials under Sub-Clause 14.5 [*Plant and Materials intended for the Works*].

7.8 Royalties

Unless otherwise stated in the Specification, the Contractor shall pay all royalties, rents and other payments for:

(a) natural Materials obtained from outside the Site, and
(b) the disposal of material from demolitions and excavations and of other surplus material (whether natural or man-made), except to the extent that disposal areas within the Site are stated in the Specification.

8 Commencement, Delays and Suspension

8.1 Commencement of Works

The Engineer shall give a Notice to the Contractor stating the Commencement Date, not less than 14 days before the Commencement Date. Unless otherwise stated in the Particular Conditions, the Commencement Date shall be within 42 days after the Contractor receives the Letter of Acceptance.

The Contractor shall commence the execution of the Works on, or as soon as is reasonably practicable after, the Commencement Date and shall then proceed with the Works with due expedition and without delay.

承包商应承担本款规定的所有修补工作的费用，上述（c）段规定的任何工作由以下原因造成的除外：

(i) **雇主**或**雇主人员**的任何行为。如果**承包商**在进行此类工作中遭受延误和/或招致增加**费用**，**承包商**应有权按照**第 20.2 款**［*付款和/或竣工时间延长的索赔*］的规定，要求获得**竣工时间**的延长和/或**成本**加利润的支付；（或）

(ii) **例外事件**，在这种情况下，**第 18.4 款**［*例外事件的后果*］应适用。

如果**承包商**未能遵从指示，**雇主**可（**雇主**自行决定）雇用并付款给他人从事该工作。除**承包商**原有权根据本款规定从该工作所得付款外，**雇主**应有权按照**第 20.2 款**［*付款和/或竣工时间延长的索赔*］的规定，要求由**承包商**支付因其未履行指示招致的所有费用。该权利不应损害**雇主**根据**合同**或其他规定可能享有的任何其他权利。

7.7 生产设备和材料的所有权

从下列时间的较早者起，在符合工程所在国法律的强制性要求规定的范围内，每项**生产设备和材料**都应无抵押权和其他阻碍地成为**雇主**的财产：

(a) 当上述**生产设备和材料**运至**现场**时；
(b) 当根据**第 8.11 款**［*雇主暂停后对生产设备和材料的付款*］的规定，**承包商**得到按**生产设备和材料**价值的付款时；（或）
(c) 当根据**第 14.5 款**［*拟用于工程的生产设备和材料*］的规定，**承包商**得到**生产设备和材料**的确定金额的付款时。

7.8 土地（矿区）使用费

除非**规范要求**中另有说明，**承包商**应为以下事项支付所有的土地（矿区）使用费、租金和其他款项：

(a) 从**现场**以外地区得到的天然**材料**；（以及）
(b) 在**规范要求**规定的**现场**范围内的弃置地区以外，弃置拆除、开挖的材料和其他剩余材料（不论是天然的或人工的）。

8 开工、延误和暂停

8.1 工程的开工

工程师应在不少于 14 天前向**承包商**发出**开工日期**的通知。除非**专用条件**中另有说明，否则**开工日期**应在**承包商**收到**中标函**后 42 天内。

承包商应在**开工日期**或**开工日期**后，在合理可能的情况下尽早开始**工程**的实施，随后应以适当速度，不拖延地进行**工程**。

8.2
Time for Completion

The Contractor shall complete the whole of the Works, and each Section (if any), within the Time for Completion for the Works or Section (as the case may be), including completion of all work which is stated in the Contract as being required for the Works or Section to be considered to be completed for the purposes of taking over under Sub-Clause 10.1 [*Taking Over the Works and Sections*].

8.3
Programme

The Contractor shall submit an initial programme for the execution of the Works to the Engineer within 28 days after receiving the Notice under Sub-Clause 8.1 [*Commencement of Works*]. This programme shall be prepared using programming software stated in the Specification (if not stated, the programming software acceptable to the Engineer). The Contractor shall also submit a revised programme which accurately reflects the actual progress of the Works, whenever any programme ceases to reflect actual progress or is otherwise inconsistent with the Contractor's obligations.

The initial programme and each revised programme shall be submitted to the Engineer in one paper copy, one electronic copy and additional paper copies (if any) as stated in the Contract Data, and shall include:

(a) the Commencement Date and the Time for Completion, of the Works and of each Section (if any);

(b) the date right of access to and possession of (each part of) the Site is to be given to the Contractor in accordance with the time (or times) stated in the Contract Data. If not so stated, the dates the Contractor requires the Employer to give right of access to and possession of (each part of) the Site;

(c) the order in which the Contractor intends to carry out the Works, including the anticipated timing of each stage of design (if any), preparation and submission of Contractor's Documents, procurement, manufacture, inspection, delivery to Site, construction, erection, installation, work to be undertaken by any nominated Subcontractor (as defined in Sub-Clause 5.2 [*Nominated Subcontractors*]) and testing;

(d) the Review periods for any submissions stated in the Specification or required under these Conditions;

(e) the sequence and timing of inspections and tests specified in, or required by, the Contract;

(f) for a revised programme: the sequence and timing of the remedial work (if any) to which the Engineer has given a Notice of No-objection under Sub-Clause 7.5 [*Defects and Rejection*] and/or the remedial work (if any) instructed under Sub-Clause 7.6 [*Remedial Work*];

(g) all activities (to the level of detail stated in the Specification), logically linked and showing the earliest and latest start and finish dates for each activity, the float (if any), and the critical path(s);

(h) the dates of all locally recognised days of rest and holiday periods (if any);

(i) all key delivery dates of Plant and Materials;

(j) for a revised programme and for each activity: the actual progress to date, any delay to such progress and the effects of such delay on other activities (if any); and

(k) a supporting report which includes:

    (i) a description of all the major stages of the execution of the Works;

    (ii) a general description of the methods which the Contractor intends to adopt in the execution of the Works;

8.2 竣工时间

承包商应在工程或分项工程（如果有）的**竣工时间**内，完成整个工程和每个**分项工程**（视情况而定），包括完成合同规定的、工程或分项工程按照第 10.1 款 [*工程和分项工程的接收*] 的规定接收要求的竣工所需的全部工作。

8.3 进度计划

承包商应在收到根据第 8.1 款 [*工程的开工*] 规定发出的**通知**后 28 天内，向**工程师**提交一份实施**工程**的初步进度计划。该进度计划应使用**规范要求**中规定的编程软件编制（如未规定，使用**工程师**可接受的编程软件）。如果进度计划不能反映实际进度，或与**承包商**义务不相符，**承包商**还应提交一份准确反映**工程**实际进度的修订计划。

初步进度计划和每份修订的进度计划，应按**合同数据**中规定的一份纸质副本、一份电子副本和附加纸质副本（如果有），提交**工程师**，并应包括：

(a) **工程**和各**分项工程**（如果有）的**开工日期**和**竣工日期**；

(b) 根据**合同数据**规定的时间（或时间段），给予**承包商现场**（各部分）进入和占有权的日期。如果没有这样的规定，则为**承包商**要求**雇主**给予进入和占有**现场**（各部分）权的日期；

(c) **承包商**计划实施**工程**的工作顺序，包括设计（如果有）、**承包商文件**的编制和提交、采购、制造、检查、运到**现场**、施工、安装、由（第 5.2 款 [*指定分包商*] 规定的）任何指定**分包商**承担的工作和试验各个阶段的预期时间安排；

(d) 本**规范要求**规定的或本**条件**要求的任何提交文件的**审核期**；

(e) 合同中规定或要求的各项检验和试验的顺序和时间安排；

(f) 对修订的进度计划：**工程师**根据第 7.5 款 [*缺陷和拒收*] 的规定，发出**不反对通知**的修补工作（如果有），和 / 或第 7.6 款 [*修补工作*] 指示的修补工作（如果有）的顺序和时间安排；

(g) 所有活动（达到**规范要求**中规定的详细程度），在逻辑上衔接并显示每项活动的最早和最晚开始和结束日期、浮动时间（如果有）和关键路径；

(h) 所有当地公认的休息日和假期的日期（如果有）；

(i) **生产设备**和**材料**的所有关键交货日期；

(j) 对修订的进度计划和每项活动：迄今为止的实际进度情况、此类进度的任何延误以及延误对其他活动（如果有）的影响；（以及）

(k) 一份支持报告，内容包括：

(i) 对**工程**实施所有主要阶段的描述；

(ii) **承包商**在工程施工中拟采用的方法的一般描述；

(iii) details showing the Contractor's reasonable estimate of the number of each class of Contractor's Personnel, and of each type of Contractor's Equipment, required on the Site, for each major stage of the execution of the Works;
(iv) if a revised programme, identification of any significant change(s) to the previous programme submitted by the Contractor; and
(v) the Contractor's proposals to overcome the effects of any delay(s) on progress of the Works.

The Engineer shall Review the initial programme and each revised programme submitted by the Contractor and may give a Notice to the Contractor stating the extent to which it does not comply with the Contract or ceases to reflect actual progress or is otherwise inconsistent with the Contractor's obligations. If the Engineer gives no such Notice:

- within 21 days after receiving the initial programme; or
- within 14 days after receiving a revised programme

the Engineer shall be deemed to have given a Notice of No-objection and the initial programme or revised programme (as the case may be) shall be the Programme.

The Contractor shall proceed in accordance with the Programme, subject to the Contractor's other obligations under the Contract. The Employer's Personnel shall be entitled to rely on the Programme when planning their activities.

Nothing in any programme, the Programme or any supporting report shall be taken as, or relieve the Contractor of any obligation to give, a Notice under the Contract.

If, at any time, the Engineer gives a Notice to the Contractor that the Programme fails (to the extent stated) to comply with the Contract or ceases to reflect actual progress or is otherwise inconsistent with the Contractor's obligations, the Contractor shall within 14 days after receiving this Notice submit a revised programme to the Engineer in accordance with this Sub-Clause.

8.4 Advance Warning

Each Party shall advise the other and the Engineer, and the Engineer shall advise the Parties, in advance of any known or probable future events or circumstances which may:

(a) adversely affect the work of the Contractor's Personnel;
(b) adversely affect the performance of the Works when completed;
(c) increase the Contract Price; and/or
(d) delay the execution of the Works or a Section (if any).

The Engineer may request the Contractor to submit a proposal under Sub-Clause 13.3.2 [*Variation by Request for Proposal*] to avoid or minimise the effects of such event(s) or circumstance(s).

8.5 Extension of Time for Completion

The Contractor shall be entitled subject to Sub-Clause 20.2 [*Claims For Payment and/or EOT*] to Extension of Time if and to the extent that completion for the purposes of Sub-Clause 10.1 [*Taking Over the Works and Sections*] is or will be delayed by any of the following causes:

(a) a Variation (except that there shall be no requirement to comply with Sub-Clause 20.2 [*Claims For Payment and/or EOT*]);

（iii） 承包商对工程实施各主要阶段现场所需各级承包商人员和各类承包商设备合理估计数量的详细情况；

（iv） 如果是修订的进度计划，说明与承包商以前提交的进度计划相比所做的任何重大变化；（以及）

（v） 承包商关于克服任何延误对工程进度影响的建议。

工程师应审核承包商提交的初步进度计划和每个修订的进度计划，并可向承包商发出通知，指出其中不符合合同要求或不能反映实际进度或与承包商义务不一致的部分。如果工程师没有发出此类通知：

- 在收到初步进度计划后 21 天内；（或）
- 在收到修订的进度计划后 14 天内。

工程师应被视为已发出不反对通知，初步进度计划或修订的进度计划（视情况而定）应为进度计划。

承包商应按照进度计划进行工作，并应遵守合同规定的承包商其他义务。雇主人员应有权依照进度计划安排他们的活动。

任何进度计划的内容、进度计划或任何支持报告均不得视为合同规定的通知，或免除承包商履行根据合同的规定发出通知的任何义务。

如果任何时候工程师向承包商发出通知，指出进度计划（在规定的范围内）不符合合同要求，或不能反映实际进度或与承包商的义务不一致时，承包商应在收到该通知后 14 天内，按照本款的规定向工程师提交一份修订的进度计划。

8.4 预先警示	在任何已知的或可能将要发生的未来事件或情况之前，每一方应告知另一方和工程师，工程师应告知双方以下事项： （a） 对承包商人员的工作产生不利影响的； （b） 竣工后对工程的性能产生不利影响的； （c） 提高合同价格的；（和 / 或） （d） 延误工程或分项工程（如果有）施工的。 工程师可要求承包商根据第 13.3.2 项［建议书要求的变更］的规定提交建议书，以避免或尽量减少此类事件或情况的影响。
8.5 竣工时间的延长	如果由于下列任何原因，致使第 10.1 款［工程和分项工程的接收］要求的竣工受到或将受到延误，承包商应有权按照第 20.2 款［付款和 /或竣工时间延长的索赔］的规定获得竣工时间的延长： （a） 变更（但不要求遵守第 20.2 款［付款和/或竣工时间延长的索赔］的规定）；

(b) a cause of delay giving an entitlement to EOT under a Sub-Clause of these Conditions;

(c) exceptionally adverse climatic conditions, which for the purpose of these Conditions shall mean adverse climatic conditions at the Site which are Unforeseeable having regard to climatic data made available by the Employer under Sub-Clause 2.5 [*Site Data and Items of Reference*] and/or climatic data published in the Country for the geographical location of the Site;

(d) Unforeseeable shortages in the availability of personnel or Goods (or Employer-Supplied Materials, if any) caused by epidemic or governmental actions; or

(e) any delay, impediment or prevention caused by or attributable to the Employer, the Employer's Personnel, or the Employer's other contractors on the Site.

The Contractor shall be entitled subject to Sub-Clause 20.2 [*Claims For Payment and/or EOT*] to EOT if the measured quantity of any item of work in accordance with Clause 12 [*Measurement and Valuation*] is greater than the estimated quantity of this item in the Bill of Quantities or other Schedule by more than ten percent (10%) and such increase in quantities causes a delay to completion for the purposes of Sub-Clause 10.1 [*Taking Over the Works and Sections*]. The agreement or determination of any such Claim, under Sub-Clause 20.2.5 [*Agreement or determination of the Claim*], may include a review by the Engineer of measured quantities of other items of work which are significantly less (by more than 10%) than the corresponding estimated quantities in the Bill of Quantities or other Schedule. To the extent that there are such lesser measured quantities, the Engineer may take account of any favourable effect on the critical path of the Programme. However, the net effect of all such consideration shall not result in a net reduction in the Time for Completion

When determining each EOT under Sub-Clause 20.2 [*Claims For Payment and/or EOT*], the Engineer shall review previous determinations under Sub-Clause 3.7 [*Agreement or Determination*] and may increase, but shall not decrease, the total EOT.

If a delay caused by a matter which is the Employer's responsibility is concurrent with a delay caused by a matter which is the Contractor's responsibility, the Contractor's entitlement to EOT shall be assessed in accordance with the rules and procedures stated in the Special Provisions (if not stated, as appropriate taking due regard of all relevant circumstances).

8.6 Delays Caused by Authorities

If:

(a) the Contractor has diligently followed the procedures laid down by the relevant legally constituted public authorities or private utility entities in the Country;

(b) these authorities or entities delay or disrupt the Contractor's work; and

(c) the delay or disruption was Unforeseeable,

then this delay or disruption will be considered as a cause of delay under sub-paragraph (b) of Sub-Clause 8.5 [*Extension of Time for Completion*].

8.7 Rate of Progress

If, at any time:

(a) actual progress is too slow to complete the Works or a Section (if any) within the relevant Time for Completion; and/or

(b) 根据本**条件**某款，有权获得**竣工时间**的延长的原因；

(c) 异常不利的气候条件，就本**条件**而言，应指**雇主**根据第 2.5 款 [*现场数据和参考事项*] 提供的气候数据和/或**现场**地理位置所在国公布的气候数据、**不可预见**的**现场**不利气候条件；

(d) 由于流行病或政府行为造成可用的人员或**货物**（或**雇主**提供的**材料**，如果有）的**不可预见**的短缺；（或）

(e) 由**雇主**、**雇主人员**或在**现场**的**雇主**的其他**承包商**造成或引起的任何延误、妨碍或阻碍。

如果按照第 12 条 [*测量和估价*] 的规定测量的任何工作事项的数量超过**工程量清单**，或其他**资料表**中该项工作量的估算数量 10% 以上，而此类工作量的增加导致第 10.1 款 [*工程和分项工程的接收*] 中规定的竣工延误，**承包商**应有权按照第 20.2 款 [*付款和/或竣工时间延长的索赔*] 的规定提出**延长竣工时间**。根据第 20.2.5 项 [*索赔的商定或确定*] 的规定，对任何此类索赔的商定或确定，可包括工程师对其他工作事项测量数量的审核，这些测量数量明显少于（超过 10%）工程量清单或其他资料表中相应的估算数量。如果此类测量数量减少较小，**工程师**可考虑对**进度计划**的关键路径的任何有利作用。但是，所有这些考虑的净作用不应造成**竣工时间**净减少的结果。

工程师每次按照第 20.2 款 [*付款和/或竣工时间延长的索赔*] 的规定确定**竣工时间的延长**时，应对以前根据第 3.7 款 [*商定或确定*] 所做的确定进行审核，可以增加，但不得减少总的**竣工时间的延长**。

如果由**雇主**责任引起的延误和由**承包商**责任引起的延误同时发生，应按照**特别规定**中规定的规则和程序，评估**承包商**获得**竣工时间的延长**的权利（如未规定，应适当考虑所有相关情况）。

8.6
部门造成的延误

如果：

(a) **承包商**已努力遵守了**工程所在国**依法成立的有关公共部门或私有公用事业实体制定的程序；

(b) 这些部门或实体延误或扰乱了**承包商**的工作；（以及）

(c) 延误或中断是**不可预见**的。

则上述延误或中断可视为根据第 8.5 款 [*竣工时间的延长*]（b）段规定的延误或中断的原因。

8.7
工程进度

如果在任何时候：

(a) 实际工程进度过慢，未能在**竣工时间**内完成**工程**或**分项工程**（如果有）；（和/或）

(b) progress has fallen (or will fall) behind the Programme (or the initial programme if it has not yet become the Programme) under Sub-Clause 8.3 [*Programme*],

other than as a result of a cause listed in Sub-Clause 8.5 [*Extension of Time for Completion*], then the Engineer may instruct the Contractor to submit, under Sub-Clause 8.3 [*Programme*], a revised programme describing the revised methods which the Contractor proposes to adopt in order to expedite progress and complete the Works or a Section (if any) within the relevant Time for Completion.

Unless the Engineer gives a Notice to the Contractor stating otherwise, the Contractor shall adopt these revised methods, which may require increases in the working hours and/or in the numbers of Contractor's Personnel and/or the Goods, at the Contractor's risk and cost. If these revised methods cause the Employer to incur additional costs, the Employer shall be entitled subject to Sub-Clause 20.2 [*Claims For Payment and/or EOT*] to payment of these costs by the Contractor, in addition to Delay Damages (if any).

Sub-Clause 13.3.1 [*Variation by Instruction*] shall apply to revised methods, including acceleration measures, instructed by the Engineer to reduce delays resulting from causes listed under Sub-Clause 8.5 [*Extension of Time for Completion*].

8.8 Delay Damages

If the Contractor fails to comply with Sub-Clause 8.2 [*Time for Completion*], the Employer shall be entitled subject to Sub-Clause 20.2 [*Claims For Payment and/or EOT*] to payment of Delay Damages by the Contractor for this default. Delay Damages shall be the amount stated in the Contract Data, which shall be paid for every day which shall elapse between the relevant Time for Completion and the relevant Date of Completion of the Works or Section. The total amount due under this Sub-Clause shall not exceed the maximum amount of Delay Damages (if any) stated in the Contract Data.

These Delay Damages shall be the only damages due from the Contractor for the Contractor's failure to comply with Sub-Clause 8.2 [*Time for Completion*], other than in the event of termination under Sub-Clause 15.2 [*Termination for Contractor's Default*] before completion of the Works. These Delay Damages shall not relieve the Contractor from the obligation to complete the Works, or from any other duties, obligations or responsibilities which the Contractor may have under or in connection with the Contract.

This Sub-Clause shall not limit the Contractor's liability for Delay Damages in any case of fraud, gross negligence, deliberate default or reckless misconduct by the Contractor.

8.9 Employer's Suspension

The Engineer may at any time instruct the Contractor to suspend progress of part or all of the Works, which instruction shall state the date and cause of the suspension.

During such suspension, the Contractor shall protect, store and secure such part or all of the Works (as the case may be) against any deterioration, loss or damage.

To the extent that the cause of such suspension is the responsibility of the Contractor, Sub-Clauses 8.10 [*Consequences of Employer's Suspension*], 8.11 [*Payment for Plant and Materials after Employer's Suspension*] and 8.12 [*Prolonged Suspension*] shall not apply.

（b） 进度已（或将）落后于根据第 8.3 款 [*进度计划*] 的规定制定的**进度计划**（或还未成为**进度计划**的初步进度计划）。

除由于第 8.5 款 [*竣工时间的延长*] 中列举的某项原因造成的结果外，**工程师**可指示**承包商**根据第 8.3 款 [*进度计划*] 的规定提交一份修订的进度计划，说明**承包商**为加快进度在相关**竣工时间**内完成**工程**或某**分项工程**（如果有），而建议采用的修订方法。

除非**工程师**给**承包商**的**通知**中另有说明，否则**承包商**应采用这些修订方法，对可能需要增加工时和 / 或**承包商人员**和 / 或**货物**的数量，**承包商**应自行承担风险和费用。如果这些修订方法使**雇主**招致额外费用，**雇主**应有权根据第 20.2 款 [*付款和 / 或竣工时间延长的索赔*] 的规定，要求**承包商**支付这些费用，以及**误期损害赔偿费**（如果有）。

第 13.3.1 项 [*指示变更*] 应适用于**工程师**为减少因第 8.5 款 [*竣工时间的延长*] 中所列原因造成的延误，而指示的修订方法，包括加速措施。

8.8 误期损害赔偿费	如果**承包商**未能遵守第 8.2 款 [*竣工时间*] 的要求，**承包商**应当为其违约行为根据第 20.2 款 [*付款和 / 或竣工时间延长的索赔*] 的规定，向**雇主**支付**误期损害赔偿费**。**误期损害赔偿费**应按**合同数据**中规定的每天应付的金额，以**工程**或**分项工程**竣工日期超过相应**竣工时间**的天数计算。但按本**款**计算的赔偿总额不得超过**合同数据**中规定的**误期损害赔偿费**的最高限额（如果有）。

除在工程竣工前根据第 15.2 款 [*因承包商违约的终止*] 的规定终止的情况外，这些**误期损害赔偿费**应是**承包商**未能遵守第 8.2 款 [*竣工时间*] 的规定，为此类违约应付的唯一损害赔偿费。这些**误期损害赔偿费**不应免除**承包商**完成**工程**的义务，或**合同**规定的或与**合同**有关的其可能承担的任何其他任务、义务或职责。

本款不应限制**承包商**在任何欺诈、重大过失、故意违约或轻率不当行为情况下，对**误期损害赔偿费**的责任。 |
| 8.9 雇主暂停 | **工程师**可以随时指示**承包商**暂停**工程**某一部分或全部的施工进度，该项指示应说明暂停的日期和原因。

在暂停期间，**承包商**应保护、保管并保证该部分或全部**工程**（视情况而定）不致产生任何变质、损失或损害。

如果暂停的原因是由于**承包商**的责任造成的，则第 8.10 款 [*雇主暂停的后果*]、第 8.11 款 [*雇主暂停后对生产设备和材料的付款*] 和第 8.12 款 [*拖长的暂停*] 应不适用。 |

8.10 Consequences of Employer's Suspension

If the Contractor suffers delay and/or incurs Cost from complying with an Engineer's instruction under Sub-Clause 8.9 [*Employer's Suspension*] and/or from resuming the work under Sub-Clause 8.13 [*Resumption of Work*], the Contractor shall be entitled subject to Sub-Clause 20.2 [*Claims For Payment and/or EOT*] to EOT and/or payment of such Cost Plus Profit.

The Contractor shall not be entitled to EOT, or to payment of the Cost incurred, in making good:

(a) the consequences of the Contractor's faulty or defective (design, if any) workmanship, Plant or Materials; and/or

(b) any deterioration, loss or damage caused by the Contractor's failure to protect, store or secure in accordance with Sub-Clause 8.9 [*Employer's Suspension*].

8.11 Payment for Plant and Materials after Employer's Suspension

The Contractor shall be entitled to payment of the value (as at the date of suspension instructed under Sub-Clause 8.9 [*Employer's Suspension*]) of Plant and/or Materials which have not been delivered to Site, if:

(a) the work on Plant, or delivery of Plant and/or Materials, has been suspended for more than 28 days and

    (i) the Plant and/or Materials were scheduled, in accordance with the Programme, to have been completed and ready for delivery to the Site during the suspension period; and

    (ii) the Contractor provides the Engineer with reasonable evidence that the Plant and/or Materials comply with the Contract; and

(b) the Contractor has marked the Plant and/or Materials as the Employer's property in accordance with the Engineer's instructions.

8.12 Prolonged Suspension

If the suspension under Sub-Clause 8.9 [*Employer's Suspension*] has continued for more than 84 days, the Contractor may give a Notice to the Engineer requesting permission to proceed.

If the Engineer does not give a Notice under Sub-Clause 8.13 [*Resumption of Work*] within 28 days after receiving the Contractor's Notice under this Sub-Clause, the Contractor may either:

(a) agree to a further suspension, in which case the Parties may agree the EOT and/or Cost Plus Profit (if the Contractor incurs Cost), and/or payment for suspended Plant and/or Materials, arising from the total period of suspension;

or (and if the Parties fail to reach agreement under this sub-paragraph (a))

(b) after giving a (second) Notice to the Engineer, treat the suspension as an omission of the affected part of the Works (as if it had been instructed under Sub-Clause 13.3.1 [*Variation by Instruction*]) with immediate effect including release from any further obligation to protect, store and secure under Sub-Clause 8.9 [*Employer's Suspension*]. If the suspension affects the whole of the Works, the Contractor may give a Notice of termination under Sub-Clause 16.2 [*Termination by Contractor*].

8.10

雇主暂停的后果

如果**承包商**因执行**工程师**根据第 8.9 款［*雇主暂停*］的规定发出的指示，和 / 或因根据第 8.13 款［*复工*］的规定复工而遭受延误和 / 或招致增加**费用**，**承包商**应有权根据第 20.2 款［*付款和 / 或竣工时间延长的索赔*］的规定，要求获得**竣工时间**的延长和 / 或**成本加利润**的支付。

以下情况，**承包商**无权得到**竣工时间**的延长或所招致增加**费用**的支付：

(a) 因**承包商**有不合格或有缺陷（设计，如果有）的**工艺**、**生产设备**或**材料**而带来的后果；（和 / 或）

(b) 因**承包商**未能按照第 8.9 款［*雇主暂停*］的规定进行保护、保管或保证安全而带来的任何变质、损失或损害的后果。

8.11

雇主暂停后对生产设备和材料的付款

承包商应有权获得尚未交付**现场**的**生产设备**和 / 或**材料**（按照第 8.9 款［*雇主暂停*］指示的暂停开始的日期时）的价值的付款，如果：

(a) **生产设备**的生产或**生产设备**和 / 或**材料**的交付被暂停达 28 天以上；（以及）

(i) 按照**进度计划**，**生产设备**和 / 或**材料**已按计划完成，并准备在暂停期间交付到**现场**；（以及）

(ii) **承包商**向**工程师**提供合理的证据，证明**生产设备**和 / 或**材料**符合合同要求；（以及）

(b) **承包商**已按**工程师**的指示，标明上述**生产设备**和 / 或**材料**为**雇主**的财产。

8.12

拖长的暂停

如果第 8.9 款［*雇主暂停*］所述的暂停已持续 84 天以上，**承包商**可以向**工程师**发出**通知**，要求**工程师**允许继续施工。

如果**工程师**在收到**承包商**根据本款发出的**通知**后 28 天内，未能按照第 8.13 款［*复工*］规定发出通知，**承包商**可以：

(a) 同意继续暂停，在这种情况下，双方可就整个暂停期限内产生的**竣工时间**的延长和 / 或**成本加利润**（如果**承包商**产生了**费用**），和 / 或**生产设备**和 / 或**材料**暂停的支付达成商定一致；

或［双方未能根据本款（a）段规定达成商定一致］

(b) 在向**工程师**发出（第二次）**通知**后，将暂停视为受影响工程部分的删减（如同根据第 13.3.1 项［*指示变更*］的规定指示的），并立即生效，包括免除第 8.9 款［*雇主暂停*］规定的保护、保管或保证安全的任何进一步义务。如果暂停影响到整个工程，**承包商**可根据第 16.2 款［*由承包商终止*］的规定发出终止**通知**。

8.13
Resumption of Work

The Contractor shall resume work as soon as practicable after receiving a Notice from the Engineer to proceed with the suspended work.

At the time stated in this Notice (if not stated, immediately after the Contractor receives this Notice), the Contractor and the Engineer shall jointly examine the Works and the Plant and Materials affected by the suspension. The Engineer shall record any deterioration, loss, damage or defect in the Works or Plant or Materials which has occurred during the suspension and shall provide this record to the Contractor. The Contractor shall promptly make good all such deterioration, loss, damage or defect so that the Works, when completed, shall comply with the Contract.

9 Tests on Completion

9.1
Contractor's Obligations

The Contractor shall carry out the Tests on Completion in accordance with this Clause and Sub-Clause 7.4 [*Testing by the Contractor*], after submitting the documents under Sub-Clause 4.4.2 [*As-Built Records*] (if applicable) and Sub-Clause 4.4.3 [*Operation and Maintenance Manuals*] (if applicable).

The Contractor shall submit to the Engineer, not less than 42 days before the date the Contractor intends to commence the Tests on Completion, a detailed test programme showing the intended timing and resources required for these tests.

The Engineer may Review the proposed test programme and may give a Notice to the Contractor stating the extent to which it does not comply with the Contract. Within 14 days after receiving this Notice, the Contractor shall revise the test programme to rectify such non-compliance. If the Engineer gives no such Notice within 14 days after receiving the test programme (or revised test programme), the Engineer shall be deemed to have given a Notice of No-objection. The Contractor shall not commence the Tests on Completion until a Notice of No-objection is given (or is deemed to have been given) by the Engineer.

In addition to any date(s) shown in the test programme, the Contractor shall give a Notice to the Engineer, of not less than 21 days, of the date after which the Contractor will be ready to carry out each of the Tests on Completion. The Contractor shall commence the Tests on Completion within 14 days after this date, or on such day or days as the Engineer shall instruct, and shall proceed in accordance with the Contractor's test programme to which the Engineer has given (or is deemed to have given) a Notice of No-objection.

As soon as the Works or Section have, in the Contractor's opinion, passed the Tests on Completion, the Contractor shall submit a certified report of the results of these tests to the Engineer. The Engineer shall Review such a report and may give a Notice to the Contractor stating the extent to which the results of the tests do not comply with the Contract. If the Engineer does not give such a Notice within 14 days after receiving the results of the tests, the Engineer shall be deemed to have given a Notice of No-objection.

In considering the results of the Tests on Completion, the Engineer shall make allowances for the effect of any use of (any part of) the Works by the Employer on the performance or other characteristics of the Works.

8.13
复工

收到工程师发出的继续实施暂停工作的通知后，**承包商**应在切实可行的范围内尽快恢复工作。

在本**通知**规定的时间内（如未规定，则在**承包商**收到该**通知**后），**承包商**和**工程师**应共同对受暂停影响的**工程**、**生产设备**和**材料**进行检查。**工程师**应记录在暂停期间发生的**工程**、**生产设备**或**材料**中的任何变质、损失、损害或缺陷，并应将此记录提供给**承包商**。**承包商**应负责立即修复所有此类变质、损失、损害或缺陷，以便工程竣工时符合**合同**要求。

9 竣工试验

9.1 承包商的义务

承包商应在按照第 4.4.2 项 [*竣工记录*]（如适用）和第 4.4.3 项 [*操作和维护手册*]（如适用）的规定提供各种文件后，按照本条和第 7.4 款 [*由承包商试验*] 的要求进行**竣工试验**。

承包商应在其计划开始进行每项**竣工试验**的日期前至少 42 天，向**工程师**提交一份详细的试验进度计划，说明这些试验的预期时间和资源。

工程师可审核拟定的试验进度计划，并可向**承包商**发出**通知**，说明其与合同不符的程度。在收到此**通知**后 14 天内，**承包商**应修改试验进度计划，以纠正此类不合格的情况。如果**工程师**在收到试验进度计划（或修订的试验进度计划）后 14 天内没有发出此类**通知**，**工程师**应被视为已发出**不反对通知**。在**工程师**发出（或视为已发出）**不反对通知**之前，**承包商**不得开始进行**竣工试验**。

除试验进度计划中所示的任何日期外，**承包商**应提前 21 天向**工程师**发出**通知**，说明**承包商**将可以进行每项**竣工试验**的日期。**承包商**应在该日期后 14 天内，或在工程师指示的某日或某几日内进行，并应按照**工程师**已（或被视为已）发出**不反对通知**的**承包商**试验进度计划进行。

一旦**承包商**认为**工程**或某**分项工程**通过了**竣工试验**，**承包商**应向**工程师**提供一份此类试验结果的认证报告。**工程师**应审核此类报告，并可向**承包商**发出**通知**，说明试验结果与合同不符的程度。如果**工程师**在收到试验结果后 14 天内没有发出此类**通知**，**工程师**应视为已发出**不反对通知**。

工程师在考虑**竣工试验**结果时，应考虑到**雇主**对**工程**（任何部分）的任何使用对**工程**的性能或其他特性的影响。

9.2 Delayed Tests

If the Contractor has given a Notice under Sub-Clause 9.1 [*Contractor's Obligations*] that the Works or Section (as the case may be) are ready for Tests on Completion, and these tests are unduly delayed by the Employer's Personnel or by a cause for which the Employer is responsible, Sub-Clause 10.3 [*Interference with Tests on Completion*] shall apply.

If the Tests on Completion are unduly delayed by the Contractor, the Engineer may by giving a Notice to the Contractor require the Contractor to carry out the tests within 21 days after receiving the Notice. The Contractor shall carry out the tests on such day or days within this period of 21 days as the Contractor may fix, for which the Contractor shall give a prior Notice to the Engineer of not less than 7 days.

If the Contractor fails to carry out the Tests on Completion within this period of 21 days:

(a) after a second Notice is given by the Engineer to the Contractor, the Employer's Personnel may proceed with the tests;
(b) the Contractor may attend and witness these tests;
(c) within 28 days of these tests being completed, the Engineer shall send a copy of the test results to the Contractor; and
(d) if the Employer incurs additional costs as a result of such testing, the Employer shall be entitled subject to Sub-Clause 20.2 [*Claims For Payment and/or EOT*] to payment by the Contractor of the costs reasonably incurred.

Whether or not the Contractor attends, these Tests on Completion shall be deemed to have been carried out in the presence of the Contractor and the results of these tests shall be accepted as accurate.

9.3 Retesting

If the Works, or a Section, fail to pass the Tests on Completion, Sub-Clause 7.5 [*Defects and Rejection*] shall apply. The Engineer or the Contractor may require these failed tests, and the Tests on Completion on any related work, to be repeated under the same terms and conditions. Such repeated tests shall be treated as Tests on Completion for the purposes of this Clause.

9.4 Failure to Pass Tests on Completion

If the Works, or a Section, fail to pass the Tests on Completion repeated under Sub-Clause 9.3 [*Retesting*], the Engineer shall be entitled to:

(a) order further repetition of Tests on Completion under Sub-Clause 9.3 [*Retesting*];
(b) reject the Works if the effect of the failure is to deprive the Employer of substantially the whole benefit of the Works in which event the Employer shall have the same remedies as are provided in sub-paragraph (d) of Sub-Clause 11.4 [*Failure to Remedy Defects*];
(c) reject the Section if the effect of the failure is that the Section cannot be used for its intended purpose(s) under the Contract, in which event the Employer shall have the same remedy as is provided in sub-paragraph (c) of Sub-Clause 11.4 [*Failure to Remedy Defects*]; or
(d) issue a Taking-Over Certificate, if the Employer so requests.

In the event of sub-paragraph (d) above, the Contractor shall then proceed in accordance with all other obligations under the Contract, and the Employer shall be entitled subject to Sub-Clause 20.2 [*Claims For Payment and/or EOT*] to payment by the Contractor or a reduction in the Contract Price as described under sub-paragraph (b) of Sub-Clause 11.4 [*Failure to Remedy Defects*], respectively. This entitlement shall be without prejudice to any other rights the Employer may have, under the Contract or otherwise.

9.2 延误的试验

如果**承包商**已按照第 9.1 款 [*承包商的义务*] 的规定发出**通知**，说明**工程**或**分项工程**（视情况而定）已可以进行**竣工试验**，并且这些试验因**雇主人员**或**雇主**负责的原因而不当地延误，第 10.3 款 [*对竣工试验的干扰*] 应适用。

如果**承包商**不当地延误了**竣工试验**，**工程师**可向**承包商**发出**通知**，要求其在接到**通知**后 21 天内进行**竣工试验**。**承包商**应在规定的 21 天内其能确定的某日或某几日内进行**竣工试验**，并应提前不少于 7 天将该日期**通知工程师**。

如果**承包商**未在规定的 21 天内进行**竣工试验**：

(a) **工程师**向**承包商**发出第二次**通知**后，**雇主人员**可以继续进行试验；
(b) **承包商**可参加并见证这些试验；
(c) 在试验完成后 28 天内，**工程师**应将试验结果的副本发送给**承包商**；（以及）
(d) 如果**雇主**因此类试验而招致增加额外费用，**雇主**应有权根据第 20.2 款 [*付款和/或竣工时间延长的索赔*] 的规定，要求**承包商**支付合理增加的费用。

无论**承包商**是否参加，这些**竣工试验**应被视为是**承包商**在场时进行的，试验结果应认为准确，予以认可。

9.3 重新试验

如果**工程**或某**分项工程**未能通过**竣工试验**，第 7.5 款 [*缺陷和拒收*] 应适用。**工程师**或**承包商**可要求按相同的条款和条件，重新进行此项未通过的试验和相关**工程**的**竣工试验**。就本条而言，此类重复试验应视为**竣工试验**。

9.4 未能通过竣工试验

如果**工程**或某**分项工程**未能通过根据第 9.3 款 [*重新试验*] 的规定重新进行的**竣工试验**，**工程师**应有权：

(a) 下令根据第 9.3 款 [*重新试验*] 再次重复**竣工试验**；

(b) 如果此项试验未通过，使**雇主**实质上丧失了**工程**的整个利益时，拒收**工程**，在此情况下，**雇主**应采取与第 11.4 款 [*未能修补缺陷*]（d）段规定的相同的补救措施；

(c) 如果此项试验未通过，该**分项工程**不能用于合同规定的预期目的，拒收**分项工程**，在此情况下，**雇主**应采取与第 11.4 款 [*未能修补缺陷*]（c）段规定的相同的补救措施；（或）

(d) 如果**雇主**要求，签发**接收证书**。

在采用上述（d）段办法的情况下，**承包商**应继续履行合同规定的所有其他义务，**雇主**应有权分别根据第 20.2 款 [*付款和/或竣工时间延长的索赔*] 的规定，要求**承包商**付款或按照第 11.4 款 [*未能修补缺陷*]（b）段的规定，降低**合同价格**。该权利不应损害**雇主**根据本合同或其他规定可能享有的任何其他权利。

10 Employer's Taking Over

10.1 Taking Over the Works and Sections

Except as stated in Sub-Clause 9.4 [*Failure to Pass Tests on Completion*], Sub-Clause 10.2 [*Taking Over Parts*] and Sub-Clause 10.3 [*Interference with Tests on Completion*], the Works shall be taken over by the Employer when:

(a) the Works have been completed in accordance with the Contract, including the passing of the Tests on Completion and except as allowed in sub-paragraph (i) below;
(b) if applicable, the Engineer has given (or is deemed to have given) a Notice of No-objection to the as-built records submitted under Sub-Clause 4.4.2 [*As-Built Records*];
(c) if applicable, the Engineer has given (or is deemed to have given) a Notice of No-objection to the operation and maintenance manuals under Sub-Clause 4.4.3 [*Operation and Maintenance Manuals*];
(d) if applicable, the Contractor has carried out the training as described under Sub-Clause 4.5 [*Training*]; and
(e) a Taking-Over Certificate for the Works has been issued, or is deemed to have been issued in accordance with this Sub-Clause.

The Contractor may apply for a Taking-Over Certificate by giving a Notice to the Engineer not more than 14 days before the Works will, in the Contractor's opinion, be complete and ready for taking over. If the Works are divided into Sections, the Contractor may similarly apply for a Taking-Over Certificate for each Section.

If any Part of the Works is taken over under Sub-Clause 10.2 [*Taking Over Parts*], the remaining Works or Section shall not be taken over until the conditions described in sub-paragraphs (a) to (e) above (where applicable) have been fulfilled.

The Engineer shall, within 28 days after receiving the Contractor's Notice, either:

(i) issue the Taking-Over Certificate to the Contractor, stating the date on which the Works or Section were completed in accordance with the Contract, except for any minor outstanding work and defects (as listed in the Taking-Over Certificate) which will not substantially affect the safe use of the Works or Section for their intended purpose (either until or whilst this work is completed and these defects are remedied); or
(ii) reject the application by giving a Notice to the Contractor, with reasons. This Notice shall specify the work required to be done, the defects required to be remedied and/or the documents required to be submitted by the Contractor to enable the Taking-Over Certificate to be issued. The Contractor shall then complete this work, remedy such defects and/or submit such documents before giving a further Notice under this Sub-Clause.

If the Engineer does not issue the Taking-Over Certificate or reject the Contractor's application within this period of 28 days, and if the conditions described in sub-paragraphs (a) to (d) above (where applicable) have been fulfilled, the Works or Section shall be deemed to have been completed in accordance with the Contract on the fourteenth day after the Engineer receives the Contractor's Notice of application and the Taking-Over Certificate shall be deemed to have been issued.

10 雇主的接收

10.1 工程和分项工程的接收

除第9.4款［*未能通过竣工试验*］、第10.2款［*部分工程的接收*］和第10.3款［*对竣工试验的干扰*］中所述情况外，在下列情况下，工程应由雇主接收：

(a) 工程已按合同规定完成，包括通过**竣工试验**，但以下（i）段允许的情况除外；

(b) 如适用，**工程师**已根据**第4.4.2项**［*竣工记录*］的规定，对提交的竣工记录发出（或视为已发出）**不反对通知**；

(c) 如适用，**工程师**已根据**第4.4.3项**［*操作和维护手册*］的规定，对操作和维护手册发出（或视为已发出）**不反对通知**；

(d) 如适用，**承包商**已按照**第4.5款**［*培训*］的规定进行了培训；（以及）

(e) 已按照**本款**规定签发**工程接收证书**，或视为已签发。

承包商可在其认为工程将竣工并做好接收准备的日期前不少于14天，向**工程师**发出申请**接收证书**的**通知**。如工程分成若干**分项工程**，**承包商**可类似地为每个**分项工程**申请**接收证书**。

如果根据**第10.2款**［*部分工程的接收*］的规定接收工程的任何部分，则在满足上述（a）段至（e）段（如适用）所述的条件之前，不得接收还需修复的**工程**或**分项工程**。

工程师在收到**承包商**的通知后28天内，或者：

(i) 向**承包商**签发**接收证书**，注明**工程**或**分项工程**按照合同要求竣工的日期，任何对**工程**或**竣工时间**的延长预期安全使用目的没有实质影响（如**接收证书**中所列）的少量扫尾工作和缺陷（直到或当扫尾工作和缺陷修补完成时）除外；（或）

(ii) 拒绝申请，向**承包商**发出**通知**并说明理由。该**通知**应说明在能签发**接收证书**前**承包商**需做的工作，需修补的缺陷和／或需提交的文件。**承包商**应在再次根据**本款**发出申请**通知**前，完成此项工作，修补此类缺陷和／或提交此类文件。

如果工程师在28天期限内没有签发**接收证书**，或拒绝**承包商**的申请，以及上述（a）段至（d）段所述的条件（如适用）已得到满足，在**工程师**收到**承包商**申请的通知后的14天内，**工程**或**竣工时间**的延长应视为已按照**合同**规定完成，**接收证书**应视为已签发。

10.2
Taking Over Parts

The Engineer may, at the sole discretion of the Employer, issue a Taking-Over Certificate for any part of the Permanent Works.

The Employer shall not use any part of the Works (other than as a temporary measure, which is either stated in the Specification or with the prior agreement of the Contractor) unless and until the Engineer has issued a Taking-Over Certificate for this part. However, if the Employer does use any part of the Works before the Taking-Over Certificate is issued the Contractor shall give a Notice to the Engineer identifying such part and describing such use, and:

(a) that Part shall be deemed to have been taken over by the Employer as from the date on which it is used;
(b) the Contractor shall cease to be liable for the care of such Part as from this date, when responsibility shall pass to the Employer; and
(c) the Engineer shall immediately issue a Taking-Over Certificate for this Part, and any outstanding work to be completed (including Tests on Completion) and/or defects to be remedied shall be listed in this certificate.

After the Engineer has issued a Taking-Over Certificate for a Part, the Contractor shall be given the earliest opportunity to take such steps as may be necessary to carry out the outstanding work (including Tests on Completion) and/or remedial work for any defects listed in the certificate. The Contractor shall carry out these works as soon as practicable and, in any case, before the expiry date of the relevant DNP.

If the Contractor incurs Cost as a result of the Employer taking over and/or using a Part, the Contractor shall be entitled subject to Sub-Clause 20.2 [*Claims For Payment and/or EOT*] to payment of such Cost Plus Profit.

If the Engineer issues a Taking-Over Certificate for any part of the Works, or if the Employer is deemed to have taken over a Part under sub-paragraph (a) above, for any period of delay after the date under sub-paragraph (a) above, the Delay Damages for completion of the remainder of the Works shall be reduced. Similarly, the Delay Damages for the remainder of the Section (if any) in which this Part is included shall also be reduced. This reduction shall be calculated as the proportion which the value of the Part (except the value of any outstanding works and/or defects to be remedied) bears to the value of the Works or Section (as the case may be) as a whole. The Engineer shall proceed under Sub-Clause 3.7 [*Agreement or Determination*] to agree or determine this reduction (and for the purpose of Sub-Clause 3.7.3 [*Time limits*], the date the Engineer receives the Contractor's Notice under this Sub-Clause shall be the date of commencement of the time limit for agreement under Sub-Clause 3.7.3). The provisions of this paragraph shall only apply to the daily rate of Delay Damages, and shall not affect the maximum amount of these damages.

10.3
Interference with Tests on Completion

If the Contractor is prevented, for more than 14 days (either a continuous period, or multiple periods which total more than 14 days), from carrying out the Tests on Completion by the Employer's Personnel or by a cause for which the Employer is responsible:

(a) the Contractor shall give a Notice to the Engineer describing such prevention;
(b) the Employer shall be deemed to have taken over the Works or Section (as the case may be) on the date when the Tests on Completion would otherwise have been completed; and

10.2 部分工程的接收

在**雇主**完全自主决定情况下，**工程师**可签发永久**工程**任何**部分**的**接收证书**。

除非并直到**工程师**已签发任何部分**工程**的**接收证书**，**雇主**不得使用该部分**工程**（**规范要求**规定或**承包商**事先同意的临时措施除外）。但是，如果**雇主**在签发**接收证书**前确实使用了任何部分**工程**，**承包商**应向**工程师**发出**通知**，明确该部分**工程**并描述其用途，（以及）：

(a) 使用的**部分**应视为从开始使用的日期起已被**雇主**接收；

(b) **承包商**应从此日起不再承担该**部分**的照管责任，应转由**雇主**负责；（以及）

(c) **工程师**应立即签发该部分的**接收证书**。任何尚未完成的扫尾工作（包括**竣工试验**）和/或需修补的缺陷均应在证书中列出。

工程师签发部分**工程**的**接收证书**后，应尽早给予**承包商**机会采取可能必要的步骤，进行证书中列出的任何尚未完成的扫尾工作（包括**竣工试验**）和/或任何缺陷的修补工作。**承包商**应在切实可行的范围内，尽快开展这些工作。在任何情况下，均应在有关**缺陷通知期限**期满日期前进行。

如果**承包商**因**雇主**接收和/或使用**部分工程**，导致**承包商**增加**费用**，**承包商**应有权根据第20.2款[*付款和/或竣工时间延长的索赔*]的规定，要求获得此类**成本加利润**的支付。

如果**工程师**签发部分**工程**的**接收证书**，或如果**雇主**被视为已根据上述(a)段规定接收了**部分工程**，对上述(a)段规定日期后的任何延误期，工程剩余部分的竣工**误期损害赔偿费**应予减少。与此类似，包括该部分**分项工程**（如果有）的剩余部分的误期损害赔偿费也应减少。这些**误期损害赔偿费**的减少额，应按该**部分工程**的价值（但任何尚未完成的扫尾工作和/或需修补缺陷的价值除外），与整个**工程**或分项工程（视情况而定）价值的比例计算。**工程师**应按照第3.7款[*商定或确定*]的规定，对这些减少额（就第3.7.3项[*时限*]而言，**工程师**根据本款规定收到**承包商**通知的日期，应为根据第3.7.3项规定的商定时限的开始日期）进行商定或确定。本段的规定仅适用于**误期损害赔偿费**的每日费率，不应影响该损害赔偿费的最高限额。

10.3 对竣工试验的干扰

如果由于**雇主人员**或**雇主**应负责的原因妨碍**承包商**进行**竣工试验**达14天以上（连续一段时间，或总时间超过14天的多段时间）：

(a) **承包商**应向**工程师**发出**通知**，描述此类预防措施；

(b) **雇主**应被视为已在**竣工试验**原应完成的日期接收了**工程**或分项工程（视情况而定）；（以及）

(c) the Engineer shall immediately issue a Taking-Over Certificate for the Works or Section (as the case may be).

After the Engineer has issued this Taking-Over Certificate, the Contractor shall carry out the Tests on Completion as soon as practicable and, in any case, before the expiry date of the DNP. The Engineer shall give a Notice to the Contractor, of not less than 14 days, of the date after which the Contractor may carry out each of the Tests on Completion. Thereafter, Sub-Clause 9.1 [*Contractor's Obligations*] shall apply.

If the Contractor suffers delay and/or incurs Cost as a result of being prevented from carrying out the Tests on Completion, the Contractor shall be entitled subject to Sub-Clause 20.2 [*Claims For Payment and/or EOT*] to EOT and/or payment of such Cost Plus Profit.

10.4 Surfaces Requiring Reinstatement

Except as otherwise stated in the Taking-Over Certificate, a certificate for a Section or Part of the Works shall not be deemed to certify completion of any ground or other surfaces requiring reinstatement.

11 Defects after Taking Over

11.1 Completion of Outstanding Work and Remedying Defects

In order that the Works and Contractor's Documents, and each Section and/or Part, shall be in the condition required by the Contract (fair wear and tear excepted) by the expiry date of the relevant Defects Notification Period or as soon as practicable thereafter, the Contractor shall:

(a) complete any work which is outstanding on the relevant Date of Completion, within the time(s) stated in the Taking-Over Certificate or such other reasonable time as is instructed by the Engineer; and

(b) execute all work required to remedy defects or damage, of which a Notice is given to the Contractor by (or on behalf of) the Employer on or before the expiry date of the DNP for the Works or Section or Part (as the case may be).

If a defect appears (including if the Works fail to pass the Tests after Completion, if any) or damage occurs during the relevant DNP, a Notice shall be given to the Contractor accordingly, by (or on behalf of) the Employer. Promptly thereafter:

(i) the Contractor and the Employer's Personnel shall jointly inspect the defect or damage;
(ii) the Contractor shall then prepare and submit a proposal for necessary remedial work; and
(iii) the second, third and fourth paragraphs of Sub-Clause 7.5 [*Defects and Rejection*] shall apply.

11.2 Cost of Remedying Defects

All work under sub-paragraph (b) of Sub-Clause 11.1 [*Completion of Outstanding Work and Remedying Defects*] shall be executed at the risk and cost of the Contractor, if and to the extent that the work is attributable to:

(a) design (if any) of the Works for which the Contractor is responsible;
(b) Plant, Materials or workmanship not being in accordance with the Contract;

（c）工程师应立即签发工程或分项工程接收证书（视情况而定）。

工程师签发**接收证书**后，**承包商**应在切实可行的范围内，尽快进行**竣工试验**，在任何情况下，均应在有关**缺陷通知期限**期满日期前进行。**工程师**应向**承包商**发出不少于14天的**通知**，说明**承包商**可在该日期后进行的每项**竣工试验**。此后，第9.1款［*承包商的义务*］应适用。

如果**承包商**因无法进行**竣工试验**而遭受延误和／或增加**费用**，**承包商**应有权根据第20.2款［*付款和／或竣工时间延长的索赔*］的规定，要求获得**竣工时间的延长**和／或**成本加利润**的支付。

10.4
需要复原的地面

除**接收证书**中另有说明外，**竣工时间的延长**或部分工程的**接收证书**，不应视为任何需要复原的场地或其他地面已经完成的证明。

11 接收后的缺陷

11.1
完成扫尾工作和修补缺陷

为了使工程、**承包商文件**和每个分项工程和／或部分工程在相应**缺陷通知期限**期满日期或其后，尽快达到**合同要求**（合理的损耗除外），**承包商**应：

（a）在**接收证书**规定的时间内或**工程师**指示的其他合理时间内，完成在相关**竣工日期**时尚未完成的扫尾工作；（以及）

（b）在工程或分项工程或部分工程（视情况而定）的**缺陷通知期限**期满日期前，按照**雇主**（或其代表）向**承包商**发出**通知**的要求，完成修补缺陷或损害所需的所有工作。

如果在相关的**缺陷通知期限**出现缺陷（包括工程未能通过**竣工后试验**，如果有）或发生损害，**雇主**（或其代表）应相应地向**承包商**发出**通知**。此后立即：

（i）**承包商**和**雇主人员**应共同检查缺陷或损害；

（ii）随后，**承包商**应准备并提交必要修补工作的建议书；（以及）

（iii）第7.5款［*缺陷和拒收*］第2、第3和第4段应适用。

11.2
修补缺陷的费用

第11.1款［*完成扫尾工作和修补缺陷*］（b）段中提出的所有工作，在下列情况下，其实施中的风险和费用应由**承包商**承担：

（a）**承包商**负责的工程设计（如果有）；
（b）生产设备、材料或工艺不符合**合同要求**；

(c) improper operation or maintenance which was attributable to matters for which the Contractor is responsible (under Sub-Clauses 4.4.2 [*As-Built Records*], Sub-Clause 4.4.3 [*Operation and Maintenance Manuals*] and/or Sub-Clause 4.5 [*Training*] (where applicable) or otherwise); or

(d) failure by the Contractor to comply with any other obligation under the Contract.

If the Contractor considers that the work is attributable to any other cause, the Contractor shall promptly give a Notice to the Engineer and the Engineer shall proceed under Sub-Clause 3.7 [*Agreement or Determination*] to agree or determine the cause (and, for the purpose of Sub-Clause 3.7.3 [*Time limits*], the date of this Notice shall be the date of commencement of the time limit for agreement under Sub-Clause 3.7.3). If it is agreed or determined that the work is attributable to a cause other than those listed above, Sub-Clause 13.3.1 [*Variation by Instruction*] shall apply as if such work had been instructed by the Engineer.

11.3 Extension of Defects Notification Period

The Employer shall be entitled to an extension of the DNP for the Works, or a Section or a Part:

(a) if and to the extent that the Works, Section, Part or a major item of Plant (as the case may be, and after taking over) cannot be used for the intended purpose(s) by reason of a defect or damage which is attributable to any of the matters under sub-paragraphs (a) to (d) of Sub-Clause 11.2 [*Cost of Remedying Defects*]; and

(b) subject to Sub-Clause 20.2 [*Claims For Payment and/or EOT*].

However, a DNP shall not be extended by more than a period of two years after the expiry of the DNP stated in the Contract Data.

If delivery and/or erection of Plant and/or Materials was suspended under Sub-Clause 8.9 [*Employer's Suspension*] (other than where the cause of such suspension is the responsibility of the Contractor) or Sub-Clause 16.1 [*Suspension by Contractor*], the Contractor's obligations under this Clause shall not apply to any defects or damage occurring more than two years after the DNP for the Works, of which the Plant and/or Materials form part, would otherwise have expired.

11.4 Failure to Remedy Defects

If the remedying of any defect or damage under Sub-Clause 11.1 [*Completion of Outstanding Works and Remedying Defects*] is unduly delayed by the Contractor, a date may be fixed by (or on behalf of) the Employer, on or by which the defect or damage is to be remedied. A Notice of this fixed date shall be given to the Contractor by (or on behalf of) the Employer, which Notice shall allow the Contractor reasonable time (taking due regard of all relevant circumstances) to remedy the defect or damage.

If the Contractor fails to remedy the defect or damage by the date stated in this Notice and this remedial work was to be executed at the cost of the Contractor under Sub-Clause 11.2 [*Cost of Remedying Defects*], the Employer may (at the Employer's sole discretion):

(a) carry out the work or have the work carried out by others (including any retesting), in the manner required under the Contract and at the Contractor's cost, but the Contractor shall have no responsibility for this work. The Employer shall be entitled subject to Sub-Clause 20.2 [*Claims For Payment and/or EOT*] to payment by the Contractor of the costs reasonably incurred by the Employer in remedying the defect or damage;

(c) 由于**承包商**负责的事项（根据第 4.4.2 项［*竣工记录*］、第 4.4.3 项［*操作和维护手册*］和/或第 4.5 款［*培训*］（如适用）或其他规定）引起的不当操作或维护；（或）

(d) **承包商**未能履行**合同**规定的任何其他义务。

如果**承包商**认为此类工作由其他原因引起，**承包商**应立即向**工程师**发出**通知**，而**工程师**应根据第 3.7 款［*商定或确定*］的规定，商定或确定原因（以及，就第 3.7.3 项［*时限*］而言，该通知的日期应为第 3.7.3 项规定的商定时限的开始日期）。如果商定或确定工作是由上述以外的原因引起的，则第 13.3.1 项［*指示变更*］应适用，如同该工作是由**工程师**指示的。

11.3 缺陷通知期限的延长

雇主有权对**工程**或某一分项**工程**或某部分**工程**的**缺陷通知期限**提出延长：

(a) 如果由于第 11.2 款［*修补缺陷的费用*］(a) 段至 (b) 段所述的任何事项造成的某项缺陷或损害达到使**工程**、分项**工程**、部分**工程**或某项主要**生产设备**（视情况而定，并在接收以后）不能按预期目的使用的程度；（以及）

(b) 按照第 20.2 款［*付款和/或竣工时间延长的索赔*］的规定办理。

但是，根据**合同数据**中的规定，**缺陷通知期限**期满后的延长不得超过两年。

当**生产设备**和/或**材料**的交付和/或安装，已根据第 8.9 款［*雇主暂停*］（**承包商**负责暂停的情况除外）或第 16.1 款［*由承包商暂停*］的规定暂停进行时，对于**生产设备**和/或**材料**构成的部分**工程**的**缺陷通知期限**原期满日期两年后发生的任何缺陷或损害，本**条**规定的**承包商**各项义务应不适用。

11.4 未能修补缺陷

如果**承包商**不适当地延误了根据第 11.1 款［*完成收尾工作和修补缺陷*］中所述的修补任何缺陷或损害，**雇主**（或其代表）可确定一个日期，要求在该日期或不迟于该日期修补好缺陷和损害。**雇主**（或其代表）应将该固定日期向**承包商**发出**通知**，该通知应允许**承包商**有合理的时间（适当考虑所有相关情况）修补缺陷和损害。

如果**承包商**未能在该**通知**中规定的日期修补好缺陷或损害，并且根据第 11.2 款［*修补缺陷的费用*］的规定，此项修补工作应由**承包商**承担费用，**雇主**可以（自行决定）：

(a) 按合同要求的方式或由他人进行此项工作（包括任何重新试验），由**承包商**承担费用，但**承包商**对此项工作将不再负责任。**雇主**应有权按照第 20.2 款［*付款和/或竣工时间延长的索赔*］的规定，要求**承包商**支付由**雇主**修补缺陷或损害而合理发生的费用；

(b) accept the damaged or defective work, in which case the Employer shall be entitled subject to Sub-Clause 20.2 [*Claims For Payment and/or EOT*] to a reduction in the Contract Price. The reduction shall be in full satisfaction of this failure only and shall be in the amount as shall be appropriate to cover the reduced value to the Employer as a result of this failure;

(c) require the Engineer to treat any part of the Works which cannot be used for its intended purpose(s) under the Contract by reason of this failure as an omission, as if such omission had been instructed under Sub-Clause 13.3.1 [*Variation by Instruction*]; or

(d) terminate the Contract as a whole with immediate effect (and Sub-Clause 15.2 [*Termination for Contractor's Default*] shall not apply) if the defect or damage deprives the Employer of substantially the whole benefit of the Works. The Employer shall then be entitled subject to Sub-Clause 20.2 [*Claims For Payment and/or EOT*] to recover from the Contractor all sums paid for the Works, plus financing charges and any costs incurred in dismantling the same, clearing the Site and returning Plant and Materials to the Contractor.

The exercise of discretion by the Employer under sub-paragraph (c) or (d) above shall be without prejudice to any other rights the Employer may have, under the Contract or otherwise.

11.5 Remedying of Defective Work off Site

If, during the DNP, the Contractor considers that any defect or damage in any Plant cannot be remedied expeditiously on the Site the Contractor shall give a Notice, with reasons, to the Employer requesting consent to remove the defective or damaged Plant off the Site for the purposes of repair. This Notice shall clearly identify each item of defective or damaged Plant, and shall give details of:

(a) the defect or damage to be repaired;
(b) the place to which defective or damaged Plant is to be taken for repair;
(c) the transportation to be used (and insurance cover for such transportation);
(d) the proposed inspections and testing off the Site;
(e) the planned duration required before the repaired Plant shall be returned to the Site; and
(f) the planned duration for reinstallation and retesting of the repaired Plant (under Sub-Clause 7.4 [*Testing by the Contractor*] and/or Clause 9 [*Tests on Completion*] if applicable).

The Contractor shall also provide any further details that the Employer may reasonably require.

When the Employer gives consent (which consent shall not relieve the Contractor from any obligation or responsibility under this Clause), the Contractor may remove from the Site such items of Plant as are defective or damaged. As a condition of this consent, the Employer may require the Contractor to increase the amount of the Performance Security by the full replacement cost of the defective or damaged Plant.

11.6 Further Tests after Remedying Defects

Within 7 days of completion of the work of remedying of any defect or damage, the Contractor shall give a Notice to the Engineer describing the remedied Works, Section, Part and/or Plant and the proposed repeated tests (under Clause 9 [*Tests on Completion*]). Within 7 days after receiving this Notice, the Engineer shall give a Notice to the Contractor either:

(a) agreeing with such proposed testing; or
(b) instructing the repeated tests that are necessary to demonstrate that the remedied Works, Section, Part and/or Plant comply with the Contract.

（b）接受损害或有缺陷的工作，在这种情况下，**雇主**有权根据**第 20.2 款**［*付款和／或竣工时间延长的索赔*］的规定，要求降低**合同价格**。扣减额应仅限于完全满足未能完成的工作，并应以适当的金额弥补因此项疏忽给**雇主**造成的扣减价值；

（c）要求**工程师**将因该疏忽而不能用于**合同**规定的预期目的的**工程**任何部分视为删减项目，如同此类删减是根据**第 13.3.1 项**［*指示变更*］指示的；（或）

（d）如果缺陷或损害使**雇主**实质上丧失了**工程**的整个利益，则立即终止整个合同（**第 15.2 款**［*因承包商违约的终止*］不适用）。按照**第 20.2 款**［*付款和／或竣工时间延长的索赔*］的规定，**雇主**还应有权向**承包商**收回对**工程**的全部支出总额，加上融资费用和拆除**工程**、清理**现场**，以及将生产**设备**和材料退还给**承包商**所支付的任何费用。

雇主根据上述（c）或（d）段的规定行使酌处权，不得损害**雇主**根据**合同**或其他规定可能享有的任何其他权利。

11.5 现场外缺陷工程的修补

如果在**缺陷通知期限**期间，**承包商**认为任何**生产设备**的任何缺陷或损害在**现场**无法迅速修复，**承包商**应向**雇主**发出**通知**，说明理由，要求**雇主**同意将此类有缺陷或损害的**生产设备**移出**现场**进行修复。该**通知**应明确指出每项有缺陷或损害的**生产设备**，并应详细说明：

（a）需要修复的缺陷或损害；
（b）有缺陷或损害的**生产设备**将被送修的地点；
（c）要使用的运输方式（以及此类运输的保险范围）；
（d）建议在**现场**外进行的检测和试验；
（e）修复后的**生产设备**返回**现场**前所需的计划时间；（以及）
（f）修复后的**生产设备**重新安装和重新试验的计划时间（根据**第 7.4 款**［*由承包商试验*］和／或**第 9 条**［*竣工试验*］如适用）。

承包商还应提供**雇主**可能合理要求的任何进一步细节。

雇主同意时（该同意不应免除本**条**规定的**承包商**的任何义务和责任），**承包商**可将此类有缺陷或损害的各项**生产设备**移出**现场**进行修复。作为该同意的一个条件，**雇主**可要求**承包商**按有缺陷或损害的**生产设备**的全部重置成本，增加履约担保的金额。

11.6 修补缺陷后进一步试验

在完成任何缺陷或损害的修补工作后 7 天内，**承包商**应向**工程师**发出**通知**，说明修补的**工程**、分项**工程**、部分**工程**和／或**生产设备**以及建议的重新试验（根据**第 9 条**［*竣工试验*］的规定）。在收到此**通知**后 7 天内，**工程师**应向**承包商**发出**通知**，或者：

（a）同意所建议的试验；（或）
（b）指示进行必要的重新试验，以证明修补的**工程**、分项**工程**、部分**工程**和／或**生产设备**符合**合同**要求。

If the Contractor fails to give such a Notice within the 7 days, the Engineer may give a Notice to the Contractor, within 14 days after the defect or damage is remedied, instructing the repeated tests that are necessary to demonstrate that the remedied Works, Section, Part and/or Plant comply with the Contract.

All repeated tests under this Sub-Clause shall be carried out in accordance with the terms applicable to the previous tests, except that they shall be carried out at the risk and cost of the Party liable, under Sub-Clause 11.2 [*Cost of Remedying Defects*], for the cost of the remedial work.

11.7 Right of Access after Taking Over

Until the date 28 days after issue of the Performance Certificate, the Contractor shall have the right of access to the Works as is reasonably required in order to comply with this Clause, except as may be inconsistent with the Employer's reasonable security restrictions.

Whenever the Contractor intends to access any part of the Works during the relevant DNP:

(a) the Contractor shall request access by giving a Notice to the Employer, describing the parts of the Works to be accessed, the reasons for such access, and the Contractor's preferred date for access. This Notice shall be given in reasonable time in advance of the preferred date for access, taking due regard of all relevant circumstances including the Employer's security restrictions; and

(b) within 7 days after receiving the Contractor's Notice, the Employer shall give a Notice to the Contractor either:

   (i) stating the Employer's consent to the Contractor's request; or
   (ii) proposing reasonable alternative date(s), with reasons. If the Employer fails to give this Notice within the 7 days, the Employer shall be deemed to have given consent to the Contractor's access on the preferred date stated in the Contractor's Notice.

If the Contractor incurs additional Cost as a result of any unreasonable delay by the Employer in permitting access to the Works by the Contractor, the Contractor shall be entitled subject to Sub-Clause 20.2 [*Claims For Payment and/or EOT*] to payment of any such Cost Plus Profit.

11.8 Contractor to Search

The Contractor shall, if instructed by the Engineer, search for the cause of any defect, under the direction of the Engineer. The Contractor shall carry out the search on the date(s) stated in the Engineer's instruction or other date(s) agreed with the Engineer.

Unless the defect is to be remedied at the cost of the Contractor under Sub-Clause 11.2 [*Cost of Remedying Defects*], the Contractor shall be entitled subject to Sub-Clause 20.2 [*Claims For Payment and/or EOT*] to payment of the Cost Plus Profit of the search.

If the Contractor fails to carry out the search in accordance with this Sub-Clause, the search may be carried out by the Employer's Personnel. The Contractor shall be given a Notice of the date when such a search will be carried out and the Contractor may attend at the Contractor's own cost. If the defect is to be remedied at the cost of the Contractor under Sub-Clause 11.2 [*Cost of Remedying Defects*], the Employer shall be entitled subject to Sub-Clause 20.2 [*Claims For Payment and/or EOT*] to payment by the Contractor of the costs of the search reasonably incurred by the Employer.

如果**承包商**未能在 7 天内发出此类**通知**，**工程师**可在缺陷或损害修补后 14 天内向**承包商**发出**通知**，指示进行必要的重新试验，以证明修补的**工程、分项工程、部分工程**和／或**生产设备**符合**合同**要求。

本**款**规定的所有重复试验，应根据适用于前述试验的条款实施，由根据第 11.2 款［*修补缺陷的费用*］的规定，对负责修补工作费用的一方承担风险和费用，而实施重复试验的情况除外。

11.7
接收后的进入权

直至签发**履约证书**后 28 天的日期，**承包商**应有为遵照本条要求而合理需要的**工程**进入权，但不符合**雇主**的合理安全限制的情况除外。

当**承包商**想在相关**缺陷通知期限**期间进入**工程**的任何部分时：

（a） **承包商**应向**雇主**发出**通知**以请求进入，并说明要进入的**工程**部分、进入原因以及**承包商**的首选进入日期。该**通知**应在首选日期之前的合理时间内发出，并适当考虑所有相关情况，包括**雇主**的安全限制情况；（以及）

（b） **雇主**在收到**承包商**的通知后 7 天内，应向**承包商**发出**通知**：

  （i） 说明**雇主**同意**承包商**的要求；（或）
  （ii） 提出合理的备选日期，并说明理由。如果**雇主**未能在 7 天内发出此**通知**，**雇主**应被视为同意**承包商**在**通知**中规定的首选日期进入。

如果由于**雇主**不合理地延误对**承包商**进入工程的许可，而使**承包商**招致额外**费用**，**承包商**应有权根据第 20.2 款［*付款和／或竣工时间延长的索赔*］的规定，要求获得任何此类**成本加利润**的支付。

11.8
承包商调查

如果**工程师**指示调查任何缺陷的原因，**承包商**应在工程师指示中规定的日期或与**工程师**商定的其他日期进行调查。

除非根据第 11.2 款［*修补缺陷的费用*］的规定，应由**承包商**承担修补费用的情况，否则**承包商**应有权根据第 20.2 款［*付款和／或竣工时间延长的索赔*］的规定，要求获得调查的**成本加利润**的支付。

如果**承包商**未能按照本**款**的规定进行调查，调查可以由**雇主**人员进行。应**通知承包商**进行这类调查的日期，**承包商**可自费参加。如果根据第 11.2 款［*修补缺陷的费用*］的规定，应由**承包商**承担修补缺陷的费用，**雇主**应有权根据第 20.2 款［*付款和／或竣工时间延长的索赔*］的规定，由**承包商**支付**雇主**合理发生的调查费用。

11.9
Performance Certificate

Performance of the Contractor's obligations under the Contract shall not be considered to have been completed until the Engineer has issued the Performance Certificate to the Contractor, stating the date on which the Contractor fulfilled the Contractor's obligations under the Contract.

The Engineer shall issue the Performance Certificate to the Contractor (with a copy to the Employer and to the DAAB) within 28 days after the latest of the expiry dates of the Defects Notification Periods, or as soon thereafter as the Contractor has:

(a) supplied all the Contractor's Documents; and
(b) completed and tested all the Works (including remedying any defects) in accordance with the Contract.

If the Engineer fails to issue the Performance Certificate within this period of 28 days, the Performance Certificate shall be deemed to have been issued on the date 28 days after the date on which it should have been issued, as required by this Sub-Clause.

Only the Performance Certificate shall be deemed to constitute acceptance of the Works.

11.10
Unfulfilled Obligations

After the issue of the Performance Certificate, each Party shall remain liable for the fulfilment of any obligation which remains unperformed at that time. For the purposes of determining the nature and extent of unperformed obligations, the Contract shall be deemed to remain in force.

However in relation to Plant, the Contractor shall not be liable for any defects or damage occurring more than two years after expiry of the DNP for the Plant except if prohibited by law or in any case of fraud, gross negligence, deliberate default or reckless misconduct.

11.11
Clearance of Site

Promptly after the issue of the Performance Certificate, the Contractor shall:

(a) remove any remaining Contractor's Equipment, surplus material, wreckage, rubbish and Temporary Works from the Site;
(b) reinstate all parts of the Site which were affected by the Contractor's activities during the execution of the Works and are not occupied by the Permanent Works; and
(c) leave the Site and the Works in the condition stated in the Specification (if not stated, in a clean and safe condition).

If the Contractor fails to comply with sub-paragraphs (a), (b) and/or (c) above within 28 days after the issue of the Performance Certificate, the Employer may sell (to the extent permitted by applicable Laws) or otherwise dispose of any remaining items and/or may reinstate and clean the Site (as may be necessary) at the Contractor's cost.

The Employer shall be entitled subject to Sub-Clause 20.2 [*Claims For Payment and/or EOT*] to payment by the Contractor of the costs reasonably incurred in connection with, or attributable to, such sale or disposal and reinstating and/or cleaning the Site, less an amount equal to the moneys from the sale (if any).

| 11.9 履约证书 | 直到**工程师**向**承包商**签发**履约证书**，注明**承包商**完成合同规定的各项义务的日期后，才应认为**承包商**的义务已经完成。

工程师应在最后一个**缺陷通知期限**期满日期后 28 天内，或在**承包商**有下列情况时签发（抄送**雇主**和**争端避免／裁决委员会**）**履约证书**：

（a） 提供所有**承包商文件**；（以及）
（b） 根据合同要求完成所有**工程**的施工和试验（包括修补任何缺陷）。

如果**工程师**未能在这 28 天的期限内签发**履约证书**，**履约证书**应被视为已根据**本款**的要求，在应签发日期后的 28 天的日期签发。

只有**履约证书**才应被视为构成对**工程**的认可。 |

| 11.10 未履行的义务 | 签发**履约证书**后，每一方仍应负责完成当时尚未履行的任何义务。为了确定这些未履行义务的性质和范围，本合同应被视为仍然有效。

但是，对于**生产设备**，除非法律禁止，或在任何欺诈、重大过失、故意违约或轻率不当行为情况下，**承包商**对**生产设备缺陷通知期限**期满后两年以上发生的任何缺陷和损害不承担责任。 |

| 11.11 现场清理 | 签发**履约证书**后，**承包商**应立即：

（a） 从**现场**撤走任何剩余的**承包商设备**、多余材料、残余物、垃圾和**临时工程**等；
（b） 恢复在施工期间受**承包商**活动影响和不被**永久工程**占用的**现场**所有部分；（以及）
（c） 让**现场**和**工程**保持在**规范要求**规定的状态（如未规定，保持在整洁和安全的状态下）。

如果**承包商**未能在**履约证书**签发后 28 天内，遵守上述（a）（b）和／或（c）段的规定，**雇主**可（在适用**法律**允许的范围内）出售或另行处理任何这些剩余物品，和／或可恢复和清理**现场**（视需要而定），费用由**承包商**承担。

雇主应有权根据第 20.2 款 *[付款和／或竣工时间延长的索赔]* 的规定，要求**承包商**支付与此类有关出售、处理、恢复和／或清理**现场**所发生的合理费用，减去销售所得的金额（如果有）。 |

12 Measurement and Valuation

12.1
Works to be Measured

The Works shall be measured, and valued for payment, in accordance with this Clause.

Whenever the Engineer requires any part of the Works to be measured on Site, he/she shall give a Notice to the Contractor of not less than 7 days, of the part to be measured and the date on which and place on Site at which the measurement will be made. Unless otherwise agreed with the Contractor, the measurement on Site shall be made on this date and the Contractor's Representative shall:

(a) either attend or send another qualified representative to assist the Engineer and to endeavour to reach agreement of the measurement, and

(b) supply any particulars requested by the Engineer.

If the Contractor fails to attend or send a representative at the time and place stated in the Engineer's Notice (or otherwise agreed with the Contractor), the measurement made by (or on behalf of) the Engineer shall be deemed to have been made in the Contractor's presence and the Contractor shall be deemed to have accepted the measurement as accurate.

Any part of the Permanent Works that is to be measured from records shall be identified in the Specification and, except as otherwise stated in the Contract, such records shall be prepared by the Engineer. Whenever the Engineer has prepared the records for such a part, he/she shall give a Notice to the Contractor of not less than 7 days, stating the date on which and place at which the Contractor's Representative shall attend to examine and agree the records with the Engineer. If the Contractor fails to attend or send a representative at the time and place stated in the Engineer's Notice (or otherwise agreed with the Contractor), the Contractor shall be deemed to have accepted the records as accurate.

If, for any part of the Works, the Contractor attends the measurement on Site or examines the measurement records (as the case may be) but the Engineer and the Contractor are unable to agree the measurement, then the Contractor shall give a Notice to the Engineer setting out the reasons why the Contractor considers the measurement on Site or records are inaccurate. If the Contractor does not give such a Notice to the Engineer within 14 days after attending the measurement on Site or examining the measurement records, the Contractor shall be deemed to have accepted the measurement as accurate.

After receiving a Contractor's Notice under this Sub-Clause, unless at that time such measurement is already subject to the last paragraph of Sub-Clause 13.3.1 [*Variation by Instruction*], the Engineer shall:

- proceed under Sub-Clause 3.7 [*Agreement or Determination*] to agree or determine the measurement; and
- for the purpose of Sub-Clause 3.7.3 [*Time limits*], the date on which the Engineer receives the Contractor's Notice shall be the date of commencement of the time limit for agreement under Sub-Clause 3.7.3.

Until such time as the measurement is agreed or determined, the Engineer shall assess a provisional measurement for the purposes of Interim Payment Certificates.

12 测量和估价

12.1 需测量的工程

为了付款，应按照本**条**规定对**工程**进行测量和估价。

当**工程师**要求在**现场**测量**工程**任何部分时，应提前 7 天向**承包商**发出**通知**，告知将测量的部分以及测量的日期和地点。除非与**承包商**另有商定，**现场**测量应在此日期进行。**承包商代表**应：

（a）亲自或另派合格代表，协助**工程师**进行测量并努力达成测量的一致意见；（以及）

（b）提供**工程师**要求的任何具体资料。

如果**承包商**未能或派代表在**工程师通知**中规定（或与**承包商**另有商定）的时间和地点到场，**工程师**（或其代表）所做的测量应被视为**承包商**在场的情况下进行的，并且应认为**承包商**已认可该测量为准确的。

除**合同**另有规定外，凡需根据记录进行测量的**永久工程**的任何部分，应在**规范要求**中确定，此类记录应由**工程师**准备。当**工程师**为该部分准备好记录时，其应向**承包商**提前 7 天发出**通知**，说明**承包商代表**应按日期和地点到场与**工程师**对记录进行检查和商定，如**承包商**或所派代表未能在**工程师通知**中规定（或与**承包商**另有商定）的时间和地点到场，则应认为**承包商**已认可该记录为准确的。

如果对于**工程**的任何部分，**承包商**到**场**测量或检查测量记录（视情况而定），但**工程师**和**承包商**不能就测量达成一致，然后，**承包商**应向**工程师**发出**通知**，说明**承包商**认为**现场**测量或记录不准确的原因。如果**承包商**在**现场**测量或检查测量记录后 14 天内，没有向**工程师**发出此类**通知**，应视为**承包商**接受了测量，认为是准确的。

在收到**承包商**根据本款发出的**通知**后，除非当时此类测量已符合第 13.3.1 项 [*指示变更*] 最后一段的规定，否则**工程师**应：

- 根据第 3.7 款 [*商定或确定*] 的规定，继续商定或确定测量；（以及）

- 就第 3.7.3 项 [*时限*] 而言，**工程师**收到**承包商通知**的日期应为第 3.7.3 项规定的商定时限的开始日期。

工程师应在商定或确定测量之前，评估用于**期中付款证书**的临时测量。

12.2 Method of Measurement

The method of measurement shall be as stated in the Contract Data or, if not so stated, that which shall be in accordance with the Bill of Quantities or other applicable Schedule(s).

Except as otherwise stated in the Contract, measurement shall be made of the net actual quantity of each item of the Permanent Works and no allowance shall be made for bulking, shrinkage or waste.

12.3 Valuation of the Works

Except as otherwise stated in the Contract, the Engineer shall value each item of work by applying the measurement agreed or determined in accordance with Sub-Clauses 12.1 [*Works to be Measured*] and 12.2 [*Method of Measurement*], and the appropriate rate or price for the item.

For each item of work, the appropriate rate or price for the item shall be the rate or price specified for such item in the Bill of Quantities or other Schedule or, if there is no such an item, specified for similar work.

Any item of work which is identified in the Bill of Quantities or other Schedule, but for which no rate or price is specified, shall be deemed to be included in other rates and prices in the Bill of Quantities or other Schedule(s).

A new rate or price shall be appropriate for an item of work if:

(a) the item is not identified in, and no rate or price for this item is specified in, the Bill of Quantities or other Schedule and no specified rate or price is appropriate because the item of work is not of similar character, or is not executed under similar conditions, as any item in the Contract;

(b) (i) the measured quantity of the item is changed by more than 10% from the quantity of this item in the Bill of Quantities or other Schedule,
 (ii) this change in quantity multiplied by the rate or price specified in the Bill of Quantities or other Schedule for this item exceeds 0.01% of the Accepted Contract Amount,
 (iii) this change in quantity directly changes the Cost per unit quantity of this item by more than 1%, and
 (iv) this item is not specified in the Bill of Quantities or other Schedule as a "fixed rate item", "fixed charge" or similar term referring to a rate or price which is not subject to adjustment for any change in quantity; and/or

(c) the work is instructed under Clause 13 [*Variations and Adjustments*] and sub-paragraph (a) or (b) above applies.

Each new rate or price shall be derived from any relevant rates or prices specified in the Bill of Quantities or other Schedule, with reasonable adjustments to take account of the matters described in sub-paragraph (a), (b) and/or (c), as applicable. If no specified rates or prices are relevant for the derivation of a new rate or price, it shall be derived from the reasonable Cost of executing the work, together with the applicable percentage for profit stated in the Contract Data (if not stated, five percent (5%)), taking account of any other relevant matters.

12.2 测量方法

测量方法应如**合同数据**中规定的,如未这样规定,应按照**工程量清单**或其他适用**资料表**中规定的方法。

除**合同**另有规定外,应测量**永久工程**各项内容的实际净数量,不应允许留有膨胀、收缩或浪费的数量。

12.3 工程的估价

除合同另有规定外,**工程师**应按照第 12.1 款[*需测量的工程*]和 12.2 款[*测量方法*]商定或确定的测量方法和适宜的费率和价格,对各项工作内容进行估价。

各项工作内容的适宜费率或价格,应为**工程量清单**或其他**资料表**对此类工作内容规定的费率或价格,如**工程量清单**或其他**资料表**中无某项内容,应取类似工作的费率或价格。

在**工程量清单**或其他**资料表**中确定的任何工作事项,但未规定费率或价格的,应视为包含在**工程量清单**或其他**资料表**中的其他费率和价格中。

在以下情况下,宜对有关工作内容采用新的费率或价格:

(a) 该项工作在**工程量清单**或其他**资料表**中没有确定,也没有在**工程量清单**或其他**资料表**中规定该项工作的费率或价格,由于工作性质不同,或在与**合同**中任何工作不同的条件下实施,未规定适宜的费率或价格;

(b) (i) 该项工作测量出的数量变化超过**工程量清单**或其他**资料表**中所列数量的 10% 以上;

  (ii) 此数量变化与该项工作在**工程量清单**或其他**资料表**中规定的费率或价格的乘积,超过**中标合同金额**的 0.01%;

  (iii) 此数量变化直接改变该项工作的单位成本超过 1%;(以及)

  (iv) **工程量清单**或其他**资料表**中没有规定该项工作为"固定费率项目""固定费用"或类似术语,指的是不因数量变化而调整的费率或价格;(和/或)

(c) 根据**第 13 条**[*变更和调整*]的规定指示工作,并且上述(a)或(b)段适用。

新的费率或价格应考虑(a)(b)和/或(c)段中描述的有关事项对**工程量清单**或其他**资料表**中相关费率或价格加以合理调整后得出。如果没有规定的费率或价格可供推算新的费率或价格,应根据实施该工作的合理**成本**,连同**合同数据**中规定的适用利润百分比(如未规定,则为 5%),并考虑其他相关事项后得出。

If, for any item of work, the Engineer and the Contractor are unable to agree the appropriate rate or price, then the Contractor shall give a Notice to the Engineer setting out the reasons why the Contractor disagrees. After receiving a Contractor's Notice under this Sub-Clause, unless at that time such rate or price is already subject to the last paragraph of Sub-Clause 13.3.1 [*Variation by Instruction*], the Engineer shall:

- proceed under Sub-Clause 3.7 [*Agreement or Determination*] to agree or determine the appropriate rate or price; and
- for the purpose of Sub-Clause 3.7.3 [*Time limits*], the date on which the Engineer receives the Contractor's Notice shall be the date of commencement of the time limit for agreement under Sub-Clause 3.7.3.

Until such time as an appropriate rate or price is agreed or determined, the Engineer shall assess a provisional rate or price for the purposes of Interim Payment Certificates.

12.4 Omissions

Whenever the omission of any work forms part (or all) of a Variation;

(a) the value of which has not otherwise been agreed;
(b) the Contractor will incur (or has incurred) cost which, if the work had not been omitted, would have been deemed to be covered by a sum forming part of the Accepted Contract Amount;
(c) the omission of the work will result (or has resulted) in this sum not forming part of the Contract Price; and
(d) this cost is not deemed to be included in the valuation of any substituted work;

then the Contractor shall, in the Contractor's proposal under sub-paragraph (c) of Sub-Clause 13.3.1 [*Variation by Instruction*], give details to the Engineer accordingly, with detailed supporting particulars.

13 Variations and Adjustments

13.1 Right to Vary

Variations may be initiated by the Engineer under Sub-Clause 13.3 [*Variation Procedure*] at any time before the issue of the Taking-Over Certificate for the Works.

Other than as stated under Sub-Clause 11.4 [*Failure to Remedy Defects*], a Variation shall not comprise the omission of any work which is to be carried out by the Employer or by others unless otherwise agreed by the Parties.

The Contractor shall be bound by each Variation instructed under Sub-Clause 13.3.1 [*Variation by Instruction*], and shall execute the Variation with due expedition and without delay, unless the Contractor promptly gives a Notice to the Engineer stating (with detailed supporting particulars) that:

(a) the varied work was Unforeseeable having regard to the scope and nature of the Works described in the Specification;
(b) the Contractor cannot readily obtain the Goods required for the Variation; or
(c) it will adversely affect the Contractor's ability to comply with Sub-Clause 4.8 [*Health and Safety Obligations*] and/or Sub-Clause 4.18 [*Protection of the Environment*].

如果**工程师**和**承包商**无法就任何工作事项商定适当的费率或价格，**承包商**应向**工程师**发出**通知**，说明**承包商**不同意的原因。在收到**承包商**根据**本款**规定发出的**通知**后，除非当时该费率或价格已受**第 13.3.1 项**[*指示变更*]最后一段的约束，**工程师**应：

- 根据**第 3.7 款**[*商定或确定*]的规定，商定或确定适当的费率或价格；（以及）
- 就**第 3.7.3 项**[*时限*]而言，**工程师**收到**承包商通知**的日期应为**第 3.7.3 项**规定的商定期限的开始日期。

工程师应在商定或确定适宜费率或价格前，评估用于**期中付款证书**的临时费率或价格。

12.4 删减

当任何工作的删减构成一项**变更**的一部分（或全部）：

(a) 对其价值尚未达成一致；
(b) 如该工作未被删减，**承包商**将（或已）招致的费用，本应包含在**中标合同金额**的某部分款额中；
(c) 删减该工作将（或已）导致此项款额不构成**合同价格**的一部分；（以及）
(d) 此项费用不被视为要包括在任何替代工作的估价中。

承包商应在其根据**第 13.3.1 项**[*指示变更*](c)段提交的建议书中，向**工程师**提供相应的详细资料并附详细证明资料。

13 变更和调整

13.1 变更权

在签发**工程接收证书**前的任何时间，**工程师**可根据**第 13.3 款**[*变更程序*]的规定提出**变更**。

除**第 11.4 款**[*未能修补缺陷*]规定的情况外，除非双方另有商定，**变更**不应包括**雇主**或其他方将要进行的任何工作的删减。

承包商应遵守根据**第 13.3.1 项**[*指示变更*]的规定，指示每项**变更**，并应尽快毫不延误地执行**变更**，除非**承包商**迅速向**工程师**发出**通知**，说明（附详细证明资料）：

(a) 考虑到**规范要求**中所述**工程**的范围和性质，被变更的工作是**不可预见的**；
(b) **承包商**难以取得**变更**所需的**货物**；（或）
(c) 这将对**承包商**遵守**第 4.8 款**[*健康和安全义务*]和/或**第 4.18 款**[*环境保护*]的能力产生不利影响。

133

Promptly after receiving this Notice, the Engineer shall respond by giving a Notice to the Contractor cancelling, confirming or varying the instruction. Any instruction so confirmed or varied shall be taken as an instruction under Sub-Clause 13.3.1 [*Variation by instruction*].

Each Variation may include:

(i) changes to the quantities of any item of work included in the Contract (however, such changes do not necessarily constitute a Variation);
(ii) changes to the quality and other characteristics of any item of work;
(iii) changes to the levels, positions and/or dimensions of any part of the Works;
(iv) the omission of any work, unless it is to be carried out by others without the agreement of the Parties;
(v) any additional work, Plant, Materials or services necessary for the Permanent Works, including any associated Tests on Completion, boreholes and other testing and exploratory work; or
(vi) changes to the sequence or timing of the execution of the Works.

The Contractor shall not make any alteration to and/or modification of the Permanent Works, unless and until the Engineer instructs a Variation under Sub-Clause 13.3.1 [*Variation by Instruction*].

13.2 Value Engineering

The Contractor may, at any time, submit to the Engineer a written proposal which (in the Contractor's opinion) will, if adopted:

(a) accelerate completion;
(b) reduce the cost to the Employer of executing, maintaining or operating the Works;
(c) improve the efficiency or value to the Employer of the completed Works; or
(d) otherwise be of benefit to the Employer.

The proposal shall be prepared at the cost of the Contractor and shall include the details as stated in sub-paragraphs (a) to (c) of Sub-Clause 13.3.1 [*Variation by Instruction*].

The Engineer shall, as soon as practicable after receiving such a proposal, respond by giving a Notice to the Contractor stating his/her consent or otherwise. The Engineer's consent or otherwise shall be at the sole discretion of the Employer. The Contractor shall not delay any work while awaiting a response.

If the Engineer gives his/her consent to the proposal, with or without comments, the Engineer shall then instruct a Variation. Thereafter, the Contractor shall submit any further particulars that the Engineer may reasonably require, and the last paragraph of Sub-Clause 13.3.1 [*Variation by Instruction*] shall apply which shall include consideration by the Engineer of the sharing (if any) of the benefit, costs and/or delay between the Parties stated in the Particular Conditions.

If a proposal under this Sub-Clause, to which the Engineer gives his/her consent, includes a change in the design of part of the Permanent Works, then unless otherwise agreed by both Parties:

(i) the Contractor shall design this part at his/her cost; and
(ii) sub-paragraphs (a) to (h) of Sub-Clause 4.1 [*Contractor's General Obligations*] shall apply.

工程师接到该**通知**后，应立即向**承包商**发出**通知**，取消、确认或更改该指示。任何经这样确认或更改的指示，应视为根据**第 13.3.1 项**[*指示变更*]的规定做出的。

每项**变更**可包括：

（i） 合同中包括的任何工作内容的数量的改变（但此类改变不一定构成**变更**）；
（ii） 任何工作内容的质量或其他特性的改变；
（iii） 任何部分**工程**的标高、位置和 / 或尺寸的改变；
（iv） 任何工作的删减，但要交他人未经双**方**同意实施的工作除外；
（v） **永久工程**所需的任何附加工作、**生产设备**、**材料**或服务，包括任何有关的**竣工试验**、钻孔和其他试验和勘探工作；（或）
（vi） 实施**工程**的顺序或时间安排的改变。

除非并直到**工程师**根据**第 13.3.1 项**[*指示变更*]的规定指示了**变更**，**承包商**不得对**永久工程**做任何改变和 / 或修改。

13.2 价值工程

承包商可随时向**工程师**提交书面建议，该建议（**承包商**认为）采纳后将：

（a） 加快竣工；
（b） 降低**雇主**的**工程**施工、维护或运行的费用；
（c） 提高**雇主**竣工**工程**的效率或价值；（或）
（d） 给**雇主**带来其他利益的建议。

此类建议书应由**承包商**自费编制，并应包括**第 13.3.1 项**[*指示变更*]（a）至（c）段所列的详细内容。

工程师在收到此类建议书后，应在切实可行的范围内尽快做出回应，向**承包商**发出**通知**说明其同意或其他意见。**工程师**的同意或其他意见应由**雇主**自主决定。**承包商**在等待答复期间不得延误任何工作。

如果**工程师**同意该建议书，无论是否有意见，**工程师**应指示**变更**。此后，**承包商**应提交**工程师**可能合理要求的任何进一步的资料，**第 13.3.1 项**[*指示变更*]最后一段应适用，其中应包括**工程师**考虑对**专用条件**中规定的双方利益、费用和 / 或延误的分担（如果有）。

如果**工程师**同意根据本款提交的建议书中包括部分**永久工程**设计的改变，则除非经双方同意：

（i） **承包商**应自费设计这一部分；（以及）
（ii） **第 4.1 款**[*承包商的一般义务*]中的（a）至（h）段应适用。

13.3 Variation Procedure

Subject to Sub-Clause 13.1 [*Right to Vary*], Variations shall be initiated by the Engineer in accordance with either of the following procedures:

13.3.1 Variation by Instruction

The Engineer may instruct a Variation by giving a Notice (describing the required change and stating any requirements for the recording of Costs) to the Contractor in accordance with Sub-Clause 3.5 [*Engineer's Instructions*].

The Contractor shall proceed with execution of the Variation and shall within 28 days (or other period proposed by the Contractor and agreed by the Engineer) of receiving the Engineer's instruction, submit to the Engineer detailed particulars including:

(a) a description of the varied work performed or to be performed, including details of the resources and methods adopted or to be adopted by the Contractor;
(b) a programme for its execution and the Contractor's proposal for any necessary modifications (if any) to the Programme according to Sub-Clause 8.3 [*Programme*] and to the Time for Completion; and
(c) the Contractor's proposal for adjustment to the Contract Price by valuing the Variation in accordance with Clause 12 [*Measurement and Valuation*], with supporting particulars (which shall include identification of any estimated quantities and, if the Contractor incurs or will incur Cost as a result of any necessary modification to the Time for Completion, shall show the additional payment (if any) to which the Contractor considers that the Contractor is entitled). If the Parties have agreed to the omission of any work which is to be carried out by others, the Contractor's proposal may also include the amount of any loss of profit and other losses and damages suffered (or to be suffered) by the Contractor as a result of the omission.

Thereafter, the Contractor shall submit any further particulars that the Engineer may reasonably require.

The Engineer shall then proceed under Sub-Clause 3.7 [*Agreement or Determination*] to agree or determine:

(i) EOT, if any; and/or
(ii) the adjustment to the Contract Price (including valuation of the Variation in accordance with Clause 12 [*Measurement and Valuation*] using measured quantities of the varied work)

(and, for the purpose of Sub-Clause 3.7.3 [*Time limits*], the date the Engineer receives the Contractor's submission (including any requested further particulars) shall be the date of commencement of the time limit for agreement under Sub-Clause 3.7.3). The Contractor shall be entitled to such EOT and/or adjustment to the Contract Price, without any requirement to comply with Sub-Clause 20.2 [*Claims For Payment and/or EOT*].

13.3.2 Variation by Request for Proposal

The Engineer may request a proposal, before instructing a Variation, by giving a Notice (describing the proposed change) to the Contractor.

The Contractor shall respond to this Notice as soon as practicable, by either:

(a) submitting a proposal, which shall include the matters as described in sub-paragraphs (a) to (c) of Sub-Clause 13.3.1 [*Variation by Instruction*]; or

13.3 变更程序

根据第 13.1 款 [*变更权*] 的规定，工程师应按照下列任一项程序提出变更：

13.3.1 指示变更

工程师可根据第 3.5 款 [*工程师的指示*] 的规定，向**承包商**发出**通知**（说明所需的改变和记录**成本**的任何要求）指示**变更**。

承包商应继续执行**变更**，并应在收到**工程师**指示后 28 天内（或**承包商**建议并经**工程师**商定的其他期限）向**工程师**提交详细资料，包括：

(a) 对已执行或将要执行的各种工作的说明，包括**承包商**已采用或将要采用的资源和方法的详细资料；

(b) 执行进度计划和根据第 8.3 款 [*进度计划*] 和**竣工时间**的要求，**承包商**对**进度计划**做出必要修改（如果有）的建议书；（以及）

(c) **承包商**根据第 12 条 [*测量和估价*] 的规定对**变更**进行估价的**合同价格**调整建议书，并附详细资料 [其中应包括确定任何估计数量，如果**承包商**对**竣工时间**进行任何必要的修改而产生或将产生**费用**，应表明**承包商**认为其有权获得的额外付款（如果有）]。如果双方同意删减由其他人实施的任何工作，**承包商**建议书也可包括由于该项删减工作，使**承包商**遭受（或将遭受）的任何利润损失和其他损失以及损害的金额。

随后，**承包商**应提交**工程师**可能合理要求的任何进一步详细资料。

然后**工程师**应按照第 3.7 款 [*商定或确定*] 的规定，商定或确定：

(i) **竣工时间**的延长，如果有；（和/或）
(ii) **合同价格**的调整（包括根据第 12 条 [*测量和估价*] 的规定，使用不相同工作的测量数量对**变更**进行估价）。

[以及，就 3.7.3 项 [*时限*] 而言，工程师收到**承包商**提交的文件（包括所要求的任何进一步资料）的日期应为根据第 3.7.3 项规定的商定时限的开始日期]。**承包商**应有权获得此类**竣工时间**的延长和/或**合同价格**的调整，不需要遵守第 20.2 款 [*付款和/或竣工时间延长的索赔*] 的任何要求。

13.3.2 建议书要求的变更

工程师在发出**变更**指示前向**承包商**发出**通知**（说明对建议的改变），要求**承包商**提出一份建议书。

承包商应在切实可行的范围内尽快做出下述任一回应：

(a) 提交一份建议书，其中应包括第 13.3.1 项 [*指示变更*] (a) 至 (c) 段中所述的事项；（或）

(b) giving reasons why the Contractor cannot comply (if this is the case), by reference to the matters described in sub-paragraphs (a) to (c) of Sub-Clause 13.1 [*Right to Vary*].

If the Contractor submits a proposal, the Engineer shall, as soon as practicable after receiving it, respond by giving a Notice to the Contractor stating his/her consent or otherwise. The Contractor shall not delay any work whilst awaiting a response.

If the Engineer gives consent to the proposal, with or without comments, the Engineer shall then instruct the Variation. Thereafter, the Contractor shall submit any further particulars that the Engineer may reasonably require and the last paragraph of Sub-Clause 13.3.1 [*Variation by Instruction*] shall apply.

If the Engineer does not give consent to the proposal, with or without comments, and if the Contractor has incurred Cost as a result of submitting it, the Contractor shall be entitled subject to Sub-Clause 20.2 [*Claims For Payment and/or EOT*] to payment of such Cost.

13.4 Provisional Sums

Each Provisional Sum shall only be used, in whole or in part, in accordance with the Engineer's instructions, and the Contract Price shall be adjusted accordingly. The total sum paid to the Contractor shall include only such amounts for the work, supplies or services to which the Provisional Sum relates, as the Engineer shall have instructed.

For each Provisional Sum, the Engineer may instruct:

(a) work to be executed (including Plant, Materials or services to be supplied) by the Contractor, and for which adjustments to the Contract Price shall be agreed or determined under Sub-Clause 13.3.1 [*Variation by Instruction*]; and/or

(b) Plant, Materials, works or services to be purchased by the Contractor from a nominated Subcontractor (as defined in Sub-Clause 5.2 [*Nominated Subcontractors*]) or otherwise; and for which there shall be included in the Contract Price:

  (i) the actual amounts paid (or due to be paid) by the Contractor; and
  (ii) a sum for overhead charges and profit, calculated as a percentage of these actual amounts by applying the relevant percentage rate (if any) stated in the applicable Schedule. If there is no such rate, the percentage rate stated in the Contract Data shall be applied.

If the Engineer instructs the Contractor under sub-paragraph (a) and/or (b) above, this instruction may include a requirement for the Contractor to submit quotations from the Contractor's suppliers and/or subcontractors for all (or some) of the items of the work to be executed or Plant, Materials, works or services to be purchased. Thereafter, the Engineer may respond by giving a Notice either instructing the Contractor to accept one of these quotations (but such an instruction shall not be taken as an instruction under Sub-Clause 5.2 [*Nominated Subcontractors*]) or revoking the instruction. If the Engineer does not so respond within 7 days of receiving the quotations, the Contractor shall be entitled to accept any of these quotations at the Contractor's discretion.

Each Statement that includes a Provisional Sum shall also include all applicable invoices, vouchers and accounts or receipts in substantiation of the Provisional Sum.

（b）参照第 13.1 款 [变更权] (a) 至 (c) 段所述事项，说明**承包商**不能遵守的理由（如果情况如此）。

如果**承包商**提交了建议书，**工程师**在收到建议书后应尽快做出回应，向**承包商**发出**通知**说明其同意或其他意见。**承包商**在等待答复期间不得延误任何工作。

如果**工程师**同意该建议书，无论是否有意见，**工程师**应指示**变更**。此后，**承包商**应提交工程师可能合理要求的任何进一步的资料，以及**第 13.3.1 项 [指示变更]** 最后一段应适用。

如果**工程师**不同意该建议书，无论是否有意见，如果**承包商**因提交建议书而产生**费用**，**承包商**应有权根据第 20.2 款 [*付款和/或竣工时间延长的索赔*] 的规定，要求获得此类费用的支付。

13.4 暂列金额

每笔**暂列金额**只应按照**工程师**的指示全部或部分地使用，**合同价格**应相应进行调整。付给**承包商**的总金额只应包括**工程师**已指示的、与**暂列金额**有关的工作、供货或服务的应付款项。

对于每笔**暂列金额**，**工程师**可指示用于下列支付：

（a）要由**承包商**实施的工作（包括要提供的**生产设备、材料**或服务），和**合同价格**的调整应根据**第 13.3.1 项 [*指示变更*]** 的规定进行商定或确定；（和/或）

（b）应包括在**合同价格**中的，要由**承包商**从指定**分包商**（按第 5.2 款 [*指定分包商*] 的定义）或其他单位购买的**生产设备、材料**或服务，所需的下列费用：

（i）**承包商**已付（或应付）的实际金额；（以及）
（ii）以相应**资料表**规定的有关百分率（如果有）计算的这些实际金额的一个百分比，作为管理费和利润的金额。如无此类百分率，应采用**合同数据**中的百分率。

如果**工程师**根据上述（a）和/或（b）段指示**承包商**，则该指示可包括要求**承包商**提交其供应商和/或分包商对所有（或部分）将要实施的工作事项或拟采购的**生产设备、材料**、工作或服务的报价。此后，**工程师**可发出**通知**，指示**承包商**接受其中一个报价（但此指示不得作为第 5.2 款 [*指定分包商*] 规定的指示）或撤销该指示。如果**工程师**在收到报价后 7 天内没有做出回应，**承包商**有权自行决定接受任何此类报价。

暂列金额的每份**报表**还应包括**暂列金额**的所有适用发票、凭证以及账单或收据等证明。

13.5 Daywork

If a Daywork Schedule is not included in the Contract, this Sub-Clause shall not apply.

For work of a minor or incidental nature, the Engineer may instruct that a Variation shall be executed on a daywork basis. The work shall then be valued in accordance with the Daywork Schedule, and the following procedure shall apply.

Before ordering Goods for such work (other than any Goods priced in the Daywork Schedule), the Contractor shall submit one or more quotations from the Contractor's suppliers and/or subcontractors to the Engineer. Thereafter, the Engineer may instruct the Contractor to accept one of these quotations (but such an instruction shall not be taken as an instruction under Sub-Clause 5.2 [*Nominated Subcontractors*]). If the Engineer does not so instruct the Contractor within 7 days of receiving the quotations, the Contractor shall be entitled to accept any of these quotations at the Contractor's discretion.

Except for any items for which the Daywork Schedule specifies that payment is not due, the Contractor shall deliver each day to the Engineer accurate statements in duplicate (and one electronic copy), which shall include records (as described under Sub-Clause 6.10 [*Contractor's Records*]) of the resources used in executing the previous day's work.

One copy of each statement shall, if correct and agreed, be signed by the Engineer and promptly returned to the Contractor. If not correct or agreed, the Engineer shall proceed under Sub-Clause 3.7 [*Agreement or Determination*] to agree or determine the resources (and for the purpose of Sub-Clause 3.7.3 [*Time limits*], the date the works which are the subject of the Variation under this Sub-Clause are completed by the Contractor shall be the date of commencement of the time limit for agreement under Sub-Clause 3.7.3).

In the next Statement, the Contractor shall then submit priced statements of the agreed or determined resources to the Engineer, together with all applicable invoices, vouchers and accounts or receipts in substantiation of any Goods used in the daywork (other than Goods priced in the Daywork Schedule).

Unless otherwise stated in the Daywork Schedule, the rates and prices in the Daywork Schedule shall be deemed to include taxes, overheads and profit.

13.6 Adjustments for Changes in Laws

Subject to the following provisions of this Sub-Clause, the Contract Price shall be adjusted to take account of any increase or decrease in Cost resulting from a change in:

(a) the Laws of the Country (including the introduction of new Laws and the repeal or modification of existing Laws);
(b) the judicial or official governmental interpretation or implementation of the Laws referred to in sub-paragraph (a) above;
(c) any permit, permission, license or approval obtained by the Employer or the Contractor under sub-paragraph (a) or (b), respectively, of Sub-Clause 1.13 [*Compliance with Laws*]; or
(d) the requirements for any permit, permission, licence and/or approval to be obtained by the Contractor under sub-paragraph (b) of Sub-Clause 1.13 [*Compliance with Laws*],

13.5 计日工作

如果合同中未包括计日工作计划表，则本款不适用。

对于一些小的或附带性的工作，**工程师**可指示按计日工作实施**变更**。这时，工作应按照**计日工作计划表**进行估价，并应适用下述程序。

在为此类工作订购**货物**前（在**计日工作计划表**中标价的任何**货物**除外），**承包商**应向**工程师**提交一份或多份**承包商**的供应商和/或分包商的报价单。此后，**工程师**可指示**承包商**接受其中一个报价（但此指示不得作为第 5.2 款［*指定分包商*］规定的指示）。如果**工程师**在收到报价后 7 天内没有指示**承包商**，**承包商**有权自行决定接受任何此类报价。

除**计日工作计划表**中规定不应支付的任何事项外，**承包商**应向**工程师**提交每日的精确报表，一式两份（和一份电子副本），报表应包括前一日工作中使用的各项资源的记录（如第 6.10 款［*承包商的记录*］中所述）。

如果正确并经同意，每份报表应由**工程师**签署并迅速退回**承包商**一份。如果不正确或不同意，**工程师**应根据第 3.7 款［*商定或确定*］的规定对资源进行商定或确定（就第 3.7.3 项［*时限*］而言，**承包商**完成本款规定的**变更**工作的日期，应为第 3.7.3 项规定的商定时限的开始日期）。

在下一份**报表**中，**承包商**应向**工程师**提交商定或确定的资源的估价报表，连同所有适用发票、凭证以及账单或收据，以证明计日工作中使用的任何**货物**（在**计日工作计划表**中标价的**货物**除外）。

除非**计日工作计划表**中另有规定，否则**计日工作计划表**中的费率和价格应视为包括税金、管理费和利润。

13.6 因法律改变的调整

根据本**款**的下列规定，**合同价格**应考虑因以下改变导致的任何费用增减进行调整：

(a) **工程所在国**的**法律**有改变（包括实施新的**法律**，废除或修改现有**法律**）；
(b) 上述（a）段所指**法律**的司法或政府官方解释或实施；
(c) **雇主**或**承包商**分别根据第 1.13 款［*遵守法律*］(a) 段或 (b) 段获得的任何许可证、准许、执照或批准；（或）
(d) **承包商**根据第 1.13 款［*遵守法律*］(b) 段的规定获得的任何许可证、准许、执照或批准的要求。

made and/or officially published after the Base Date, which affect the Contractor in the performance of obligations under the Contract. In this Sub-Clause "change in Laws" means any of the changes under sub-paragraphs (a), (b), (c) and/or (d) above.

If the Contractor suffers delay and/or incurs an increase in Cost as a result of any change in Laws, the Contractor shall be entitled subject to Sub-Clause 20.2 [*Claims For Payment and/or EOT*] to EOT and/or payment of such Cost.

If there is a decrease in Cost as a result of any change in Laws, the Employer shall be entitled subject to Sub-Clause 20.2 [*Claims For Payment and/or EOT*] to a reduction in the Contract Price.

If any adjustment to the execution of the Works becomes necessary as a result of any change in Laws:

(i) the Contractor shall promptly give a Notice to the Engineer, or
(ii) the Engineer shall promptly give a Notice to the Contractor
(with detailed supporting particulars).

Thereafter, the Engineer shall either instruct a Variation under Sub-Clause 13.3.1 [*Variation by Instruction*] or request a proposal under Sub-Clause 13.3.2 [*Variation by Request for Proposal*].

13.7 Adjustments for Changes in Cost

If Schedule(s) of cost indexation are not included in the Contract, this Sub-Clause shall not apply.

The amounts payable to the Contractor shall be adjusted for rises or falls in the cost of labour, Goods and other inputs to the Works, by the addition or deduction of the amounts calculated in accordance with the Schedule(s) of cost indexation.

To the extent that full compensation for any rise or fall in Costs is not covered by this Sub-Clause or other Clauses of these Conditions, the Accepted Contract Amount shall be deemed to have included amounts to cover the contingency of other rises and falls in costs.

The adjustment to be applied to the amount otherwise payable to the Contractor, as certified in Payment Certificates, shall be calculated for each of the currencies in which the Contract Price is payable. No adjustment shall be applied to work valued on the basis of Cost or current prices.

Until such time as each current cost index is available, the Engineer shall use a provisional index for the issue of Interim Payment Certificates. When a current cost index is available, the adjustment shall be recalculated accordingly.

If the Contractor fails to complete the Works within the Time for Completion, adjustment of prices thereafter shall be made using either:

(a) each index or price applicable on the date 49 days before the expiry of the Time for Completion of the Works; or
(b) the current index or price

whichever is more favourable to the Employer.

在**基准日期**后制定和／或正式公布的法律改变，影响了**承包商**履行合同规定的义务。在本款中，"**法律改变**"是指上述（a）（b）（c）和／或（d）段所述的任何改变。

如果**承包商**因任何**法律**的改变，已遭受延误和／或已招致增加**费用**，**承包商**应有权根据第 20.2 款［*付款和／或竣工时间延长的索赔*］的规定，要求获得**竣工时间的延长**和／或此类**费用**的支付。

如果由于**法律**的任何改变而导致**费用**减少，**雇主**有权根据第 20.2 款［*付款和／或竣工时间延长的索赔*］的规定，要求降低合同价格。

如果由于**法律**的任何改变而需要对**工程**的实施进行任何调整：

(i) **承包商**应立即向**工程师**发出**通知**；（或）
(ii) **工程师**应立即向**承包商**发出**通知**（附详细证明材料）。

此后，**工程师**应根据第 13.3.1 项［*指示变更*］的规定指示**变更**，或根据第 13.3.2 项［*建议书要求的变更*］的规定要求提交建议书。

13.7
因成本改变的调整

如果合同中没有包括成本指数表，本款应不适用。

可付给**承包商**的款额，应就**工程**所用的劳动力、**货物**和其他投入的成本的涨落，按成本指数**资料表**计算的增减额进行调整。

在本**条款**或本**条件**的其他条款对**成本**的任何涨落不能完全补偿的情况下，**中标合同金额**应视为已包括其他成本涨落的应急费用。

在**付款证书**中确认的，付给**承包商**的其他应付款要做的调整，应按**合同价格**应付每种货币计算。对于根据**成本**或现行价格进行估价的工作，不予调整。

在获得每种现行成本指数前，**工程师**应使用一个临时指数，用以签发**期中付款证书**。当得到现行成本指数时，应据此重新计算调整。

如果**承包商**未能在**竣工时间**内完成**工程**，其后应利用下列任一方法调整价格：

(a) 适用于**工程竣工时间**期满前第 49 天的各指数或价格；（或）

(b) 现行指数或价格。

取两者中对**雇主**更有利的，对价格做出调整。

14 Contract Price and Payment

14.1
The Contract Price

Unless otherwise stated in the Particular Conditions:

(a) the Contract Price shall be the value of the Works in accordance with Sub-Clause 12.3 [*Valuation of the Works*] and be subject to adjustments, additions (including Cost or Cost Plus Profit to which the Contractor is entitled under these Conditions) and/or deductions in accordance with the Contract;

(b) the Contractor shall pay all taxes, duties and fees required to be paid by the Contractor under the Contract, and the Contract Price shall not be adjusted for any of these costs except as stated in Sub-Clause 13.6 [*Adjustments for Changes in Laws*];

(c) any quantities which may be set out in the Bill of Quantities or other Schedule(s) are estimated quantities and are not to be taken as the actual and correct quantities:

   (i) of the Works which the Contractor is required to execute, or
   (ii) for the purposes of Clause 12 [*Measurement and Valuation*]; and

(d) the Contractor shall submit to the Engineer, within 28 days after the Commencement Date, a proposed breakdown of each lump sum price (if any) in the Schedules. The Engineer may take account of the breakdown when preparing Payment Certificates, but shall not be bound by it.

14.2
Advance Payment

If no amount of advance payment is stated in the Contract Data, this Sub-Clause shall not apply.

After receiving the Advance Payment Certificate, the Employer shall make an advance payment, as an interest-free loan for mobilisation (and design, if any). The amount of the advance payment and the currencies in which it is to be paid shall be as stated in the Contract Data.

14.2.1
Advance Payment Guarantee

The Contractor shall obtain (at the Contractor's cost) an Advance Payment Guarantee in amounts and currencies equal to the advance payment, and shall submit it to the Employer with a copy to the Engineer. This guarantee shall be issued by an entity and from within a country (or other jurisdiction) to which the Employer gives consent, and shall be based on the sample form included in the Tender documents or on another form agreed by the Employer (but such consent and/or agreement shall not relieve the Contractor from any obligation under this Sub-Clause).

The Contractor shall ensure that the Advance Payment Guarantee is valid and enforceable until the advance payment has been repaid, but its amount may be progressively reduced by the amount repaid by the Contractor as stated in the Payment Certificates.

If the terms of the Advance Payment Guarantee specify its expiry date, and the advance payment has not been repaid by the date 28 days before the expiry date:

(a) the Contractor shall extend the validity of this guarantee until the advance payment has been repaid;

14 合同价格和付款

14.1 合同价格

除非**专用条件**中另有规定：

(a) **合同价格**应为第 12.3 款［**工程的估价**］规定的工程价值，并可根据合同进行调整、增加（包括**承包商**根据本**条件**的规定有权获得的**成本**或**成本加利润**）和／或扣减；

(b) **承包商**应支付根据合同要求应由其支付的各项税金、关税和费用。除第 13.6 款［*因法律改变的调整*］说明的情况外，**合同价格**不应因任何这些费用进行调整；

(c) **工程量清单**或其他**资料表**中可能列出的任何数量都是估计数量，不能作为实际和正确的数量：

(i) 要求**承包商**实施的工程的；（或）
(ii) 用于第 12 条［*测量和估价*］的；（以及）

(d) **承包商**在开工日期后 28 天内，应向**工程师**提交**资料表**中所列每项总价（如果有）的建议分类细目。**工程师**编制**付款证书**时可以考虑此类分类细目，但不受其约束。

14.2 预付款

如果**合同数据**中没有规定预付款金额，本**款**应不适用。

在收到**预付款证书**后，**雇主**应支付一笔预付款，作为用于动员的无息贷款（以及设计，如果有）。预付款的金额和支付货币应按**合同数据**中的规定。

14.2.1 预付款保函

承包商应获得（由**承包商**承担费用）金额和货币种类与预付款一致的**预付款保函**，并应将其提交给**雇主**，并抄送一份副本给**工程师**。该保函应由**雇主**同意的国家（或其他司法管辖区）的实体，以**专用条件**所附的格式或**雇主**同意的其他格式出具（但此类同意和／或商定不得免除**承包商**根据本**款**规定的任何义务）。⊖

在还清预付款前，**承包商**应确保**预付款保函**一直有效并可执行，但其总额可根据**付款证书**规定的**承包商**付还的金额逐渐减少。

如果**预付款保函**条款中规定了期满日期，而在期满日期前 28 天预付款尚未还清时：

(a) **承包商**应将保函有效期延至预付款还清为止；

⊖ 此处中文按照勘误表改正后的英文翻译。——译者注

(b) the Contractor shall immediately submit evidence of this extension to the Employer, with a copy to the Engineer; and

(c) if the Employer does not receive this evidence 7 days before the expiry date of this guarantee, the Employer shall be entitled to claim under the guarantee the amount of advance payment which has not been repaid.

When submitting the Advance Payment Guarantee, the Contractor shall include an application (in the form of a Statement) for the advance payment.

14.2.2 Advance Payment Certificate

The Engineer shall issue an Advance Payment Certificate for the advance payment within 14 days after:

(a) the Employer has received both the Performance Security and the Advance Payment Guarantee, in the form and issued by an entity in accordance with Sub-Clause 4.2.1 [*Contractor's Obligations*] and Sub-Clause 14.2.1 [*Advance Payment Guarantee*] respectively; and

(b) the Engineer has received a copy of the Contractor's application for the advance payment under Sub-Clause 14.2.1 [*Advance Payment Guarantee*].

14.2.3 Repayment of Advance Payment

The advance payment shall be repaid through percentage deductions in Payment Certificates. Unless other percentages are stated in the Contract Data:

(a) deductions shall commence in the IPC in which the total of all certified interim payments in the same currency as the advance payment (excluding the advance payment and deductions and release of retention moneys) exceeds ten percent (10%) of the portion of the Accepted Contract Amount payable in that currency less Provisional Sums; and

(b) deductions shall be made at the amortisation rate of one quarter (25%) of the amount of each IPC (excluding the advance payment and deductions and release of retention moneys) in the currencies and proportions of the advance payment, until such time as the advance payment has been repaid.

If the advance payment has not been repaid before the issue of the Taking-Over Certificate for the Works, or before termination under Clause 15 [*Termination by Employer*], Clause 16 [*Suspension and Termination by Contractor*] or Clause 18 [*Exceptional Events*] (as the case may be), the whole of the balance then outstanding shall immediately become due and payable by the Contractor to the Employer.

14.3 Application for Interim Payment

The Contractor shall submit a Statement to the Engineer after the end of the period of payment stated in the Contract Data (if not stated, after the end of each month). Each Statement shall:

(a) be in a form acceptable to the Engineer;

(b) be submitted in one paper-original, one electronic copy and additional paper copies (if any) as stated in the Contract Data; and

(c) show in detail the amounts to which the Contractor considers that the Contractor is entitled, with supporting documents which shall include sufficient detail for the Engineer to investigate these amounts together with the relevant report on progress in accordance with Sub-Clause 4.20 [*Progress Reports*].

The Statement shall include the following items, as applicable, which shall be expressed in the various currencies in which the Contract Price is payable, in the sequence listed:

（b）**承包商**应立即向**雇主**提交延期证据，并向**工程师**提交一份副本；（以及）

（c）如果**雇主**在该保函到期日前 7 天没有收到证据，**雇主**有权根据保函要求支付尚未偿还的预付款。

承包商在提交**预付款保函**时，应包括一份预付款申请书（以**报表**的格式）。

14.2.2
预付款证书

工程师应在下列情况发生后 14 天内签发**预付款证书**，用于支付预付款：

（a）**雇主**已收到**履约保函**和**预付款保函**，其格式和签发的实体均符合第 4.2.1 项［*承包商的义务*］和第 14.2.1 项［*预付款保函*］的规定；（以及）

（b）**工程师**已收到第 14.2.1 项［*预付款保函*］规定的**承包商**的预付款申请副本。

14.2.3
预付款付还

预付款应通过**付款证书**中按百分比扣减的方式付还。除非**合同数据**中规定了其他百分比：

（a）扣减应从确认的、与预付款相同货币的期中付款（不包括预付款、扣减额和保留金的发放）累计额超过以该货币支付的**中标合同金额**，减去**暂列金额**后余额的百分之十（10%）时的**期中付款证书**开始；（以及）

（b）扣减应按每次**期中付款证书**中金额（不包括预付款、扣减额和保留金的发放）的四分之一（25%）的摊还比率，并按预付款的货币和比例计算，直到预付款还清时为止。

如果在签发**工程接收证书**前，或根据第 15 条［*由雇主终止*］、第 16 条［*由承包商暂停和终止*］或第 18 条［*例外事件*］（视情况而定）的规定终止前，预付款尚未还清，则全部余额应立即成为**承包商**对**雇主**的到期应付款。

14.3
期中付款的申请

承包商应在每个付款期结束后，按**合同数据**中的规定（如未规定，按每个月结束后）向**工程师**提交**报表**。每份**报表**应：

（a）采用**工程师**可接受的格式；
（b）按**合同数据**中的规定，提交一份纸质原件、一份电子副本和附加纸质副本（如果有）；（以及）
（c）根据第 4.20 款［*进度报告*］的规定，详细说明**承包商**认为其有权得到的款额，并附上证明文件，其中应包括足够的详细资料，以便**工程师**调查这些款额以及相关的进度报告。

适用时，该**报表**应包括下列项目，以**合同价格**应付的各种货币表示，并按下列顺序排列：

(i) the estimated contract value of the Works executed, and the Contractor's Documents produced, up to the end of the period of payment (including Variations but excluding items described in sub-paragraphs (ii) to (x) below);

(ii) any amounts to be added and/or deducted for changes in Laws under Sub-Clause 13.6 [*Adjustments for Changes in Laws*], and for changes in Cost under Sub-Clause 13.7 [*Adjustments for Changes in Cost*];

(iii) any amount to be deducted for retention, calculated by applying the percentage of retention stated in the Contract Data to the total of the amounts under sub-paragraphs (i), (ii) and (vi) of this Sub-Clause, until the amount so retained by the Employer reaches the limit of Retention Money (if any) stated in the Contract Data;

(iv) any amounts to be added and/or deducted for the advance payment and repayments under Sub-Clause 14.2 [*Advance Payment*];

(v) any amounts to be added and/or deducted for Plant and Materials under Sub-Clause 14.5 [*Plant and Materials intended for the Works*];

(vi) any other additions and/or deductions which have become due under the Contract or otherwise, including those under Sub-Clause 3.7 [*Agreement or Determination*];

(vii) any amounts to be added for Provisional Sums under Sub-Clause 13.4 [*Provisional Sums*];

(viii) any amount to be added for release of Retention Money under Sub-Clause 14.9 [*Release of Retention Money*];

(ix) any amount to be deducted for the Contractor's use of utilities provided by the Employer under Sub-Clause 4.19 [*Temporary Utilities*]; and

(x) the deduction of amounts certified in all previous Payment Certificates.

14.4 Schedule of Payments

If the Contract includes a Schedule of Payments specifying the instalments in which the Contract Price will be paid then, unless otherwise stated in this Schedule:

(a) the instalments quoted in the Schedule of Payments shall be treated as the estimated contract values for the purposes of sub-paragraph (i) of Sub-Clause 14.3 [*Application for Interim Payment*];

(b) Sub-Clause 14.5 [*Plant and Materials intended for the Works*] shall not apply; and

(c) if:

(i) these instalments are not defined by reference to the actual progress achieved in the execution of the Works; and

(ii) actual progress is found by the Engineer to differ from that on which the Schedule of Payments was based,

then the Engineer may proceed under Sub-Clause 3.7 [*Agreement or Determination*] to agree or determine revised instalments (and for the purpose of Sub-Clause 3.7.3 [*Time limits*] the date when the difference under sub-paragraph (ii) above was found by the Engineer shall be the date of commencement of the time limit for agreement under Sub-Clause 3.7.3). Such revised instalments shall take account of the extent to which progress differs from that on which the Schedule of Payments was based.

If the Contract does not include a Schedule of Payments, the Contractor shall submit non-binding estimates of the payments which the Contractor expects to become due during each period of 3 months. The first estimate shall be submitted within 42 days after the Commencement Date. Revised estimates shall be submitted at intervals of 3 months, until the issue of the Taking-Over Certificate for the Works.

（i）截至付款期结束已实施的**工程**和已提出的**承包商文件**的估算合同价值〔包括各项**变更**，但不包括以下（ii）至（x）段所列事项〕；

（ii）按照**第 13.6 款**〔*因法律改变的调整*〕和**第 13.7 款**〔*因成本改变的调整*〕的规定，由于**法律**改变和**成本**改变，应增减的任何款额；

（iii）至**雇主**提取的保留金额达到**合同数据**中规定的**保留金**限额（如果有）前，用**合同数据**中规定的保留金百分比乘以根据本款（i）（ii）和（vi）段所述的款项总额计算的应扣减的任何保留金额；

（iv）按照**第 14.2 款**〔*预付款*〕的规定，因预付款的支付和付还，应增加和/或扣减的任何款额；

（v）按照**第 14.5 款**〔*拟用于工程的生产设备和材料*〕的规定，为**生产设备**和**材料**应增加和/或扣减的任何款额；

（vi）根据合同或包括**第 3.7 款**〔*商定或确定*〕规定等其他理由，应付的任何其他增加和/或扣减额；

（vii）根据**第 13.4 款**〔*暂列金额*〕的规定，为**暂列金额**增加的任何款额；

（viii）根据**第 14.9 款**〔*保留金的发放*〕的规定，为发放**保留金**增加的任何款额；

（ix）**承包商**使用**雇主**根据**第 4.19 款**〔*临时公用设施*〕的规定，提供的公共设施扣除的任何款额；（以及）

（x）所有以前**付款证书**中确认的扣减额。

14.4 付款计划表

如果合同包括一份**付款计划表**，其中规定了**合同价格**的分期付款，除非该表中另有规定：

（a）该**付款计划表**所述的分期付款，应视为**第 14.3 款**〔*期中付款的申请*〕（i）段需要的估算合同价值；

（b）**第 14.5 款**〔*拟用于工程的生产设备和材料*〕的规定应不适用；（以及）

（c）如果：

（i）分期付款额不是参照**工程**实施达到的实际进度确定；（以及）

（ii）**工程师**发现实际进度与**付款计划表**依据的进度不同。

工程师可按照**第 3.7 款**〔*商定或确定*〕的要求进行商定或确定，修改分期付款额〔就**第 3.7.3 项**〔*时限*〕而言，**工程师**发现上述（ii）段所述不同的日期，应为根据**第 3.7.3 项**商定的时限的开始日期〕，这种修改的分期付款应考虑实际进度与该分期**付款计划表**所依据的进度的不同程度。

如果合同未包括**付款计划表**，**承包商**应每 3 个月提交其预计应付的无约束性估算付款额。第一次估算应在**开工日期**后 42 天内提交。直到签发**工程接收证书**前，应按每 3 个月提交一次修正的估算。

14.5 Plant and Materials intended for the Works

If no Plant and/or Materials are listed in the Contract Data for payment when shipped and/or payment when delivered, this Sub-Clause shall not apply.

The Contractor shall include, under sub-paragraph (v) of Sub-Clause 14.3 [*Application for Interim Payment*]:

- an amount to be added for Plant and Materials which have been shipped or delivered (as the case may be) to the Site for incorporation in the Permanent Works; and
- an amount to be deducted when the contract value of such Plant and Materials is included as part of the Permanent Works under sub-paragraph (i) of Sub-Clause 14.3 [*Application for Interim Payment*].

The Engineer shall proceed under Sub-Clause 3.7 [*Agreement or Determination*] to agree or determine each amount to be added for Plant and Materials if the following conditions are fulfilled (and for the purpose of Sub-Clause 3.7.3 [*Time limits*] the date these conditions are fulfilled shall be the date of commencement of the time limit for agreement under Sub-Clause 3.7.3):

(a) the Contractor has:

(i) kept satisfactory records (including the orders, receipts, Costs and use of Plant and Materials) which are available for inspection by the Engineer;

(ii) submitted evidence demonstrating that the Plant and Materials comply with the Contract (which may include test certificates under Sub-Clause 7.4 [*Testing by the Contractor*] and/or compliance verification documentation under Sub-Clause 4.9.2 [*Compliance Verification System*]) to the Engineer; and

(iii) submitted a statement of the Cost of acquiring and shipping or delivering (as the case may be) the Plant and Materials to the Site, supported by satisfactory evidence;

and either:

(b) the relevant Plant and Materials:

(i) are those listed in the Contract Data for payment when shipped;
(ii) have been shipped to the Country, en route to the Site, in accordance with the Contract; and
(iii) are described in a clean shipped bill of lading or other evidence of shipment, which has been submitted to the Engineer together with:

- evidence of payment of freight and insurance;
- any other documents reasonably required by the Engineer; and
- a written undertaking by the Contractor that the Contractor will deliver to the Employer (prior to submitting the next Statement) a bank guarantee in a form and issued by an entity to which the Employer gives consent (but such consent shall not relieve the Contractor from any obligation in the following provisions of this sub-paragraph), in amounts and currencies equal to the amount due under this Sub-Clause. This guarantee shall be in a similar form to the form described in Sub-Clause 14.2.1 [*Advance Payment Guarantee*] and shall be valid until the Plant and Materials are properly stored on Site and protected against loss, damage or deterioration;

| 14.5 拟用于工程的生产设备和材料 | 如果**合同数据**中没有列出**生产设备**和/或**材料**装运时和/或交付时的付款，**本款**应不适用。

根据**第 14.3 款**［*期中付款的申请*］（v）段的规定，**承包商**应包括：

- 已装运或交付（视情况而定）到**现场**用于**永久工程**的**生产设备**和**材料**所需增加的金额；（以及）

- 当此类**生产设备**和**材料**的合同价值已根据**第 14.3 款**［*期中付款的申请*］（i）段规定，作为**永久工程**一部分包含在内时的减少额。

如满足以下条件，**工程师**应根据**第 3.7 款**［*商定或确定*］的规定，商定或确定**生产设备**和**材料**的各项增加金额（就**第 3.7.3 项**［*时限*］而言，这些条件的满足日期应为根据**第 3.7.3 段**规定的商定的时限的开始日期）：

（a） **承包商**已：

（i） 保存了符合要求、可供**工程师**检验的（包括**生产设备**和**材料**的订单、收据、**费用**和使用的）记录；

（ii） 向**工程师**提交了证明**生产设备**和**材料**符合**合同**要求的证据（可能包括**第 7.4 款**［*由承包商试验*］规定的试验证书和/或**第 4.9.2 项**［*合规验证体系*］规定的合规审核文件）；（以及）

（iii） 提交了购买**生产设备**和**材料**并将其装运或交付至**现场**的**费用**报表（视情况而定），并附有符合要求的证据。

以及，或：

（b） 有关**生产设备**和**材料**：

（i） 是**合同数据**中所列装运付费的物品；
（ii） 按照**合同**已运到**工程**所在国，在往**现场**的途中；

（iii） 已写入清洁装运提单或其他装运证明，此类提单或证明一并提交给**工程师**：

- 运费和保费的支付证据；
- **工程师**合理要求的任何其他文件；（以及）
- **承包商**的书面承诺书，**承包商**将（在提交下一份**报表**之前）向**雇主**提交由**雇主**同意的实体按**雇主**同意的格式出具、与根据**本款**规定应付金额和货币一致的银行保函（但此类同意不应免除**承包商**在本段规定的下列任何义务）。此保函可具有与**第 14.2.1 项**［*预付款保函*］中提到的格式相类似的格式，并应做到在**生产设备**和**材料**已在**现场**妥善储存并做好防止损失、损害或变质的保护以前，一直有效； |

or

(c) the relevant Plant and Materials:

  (i) are those listed in the Contract Data for payment when delivered to the Site, and
  (ii) have been delivered to and are properly stored on the Site, are protected against loss, damage or deterioration, and appear to be in accordance with the Contract.

The amount so agreed or determined shall take account of the evidence and documents required under this Sub-Clause and of the contract value of the Plant and Materials. If sub-paragraph (b) above applies, the Engineer shall have no obligation to certify any payment under this Sub-Clause until the Employer has received the bank guarantee in accordance with sub-paragraph (b)(iii) above. The sum to be certified by the Engineer in an IPC shall be the equivalent of eighty percent (80%) of this agreed or determined amount. The currencies for this certified sum shall be the same as those in which payment will become due when the contract value is included under sub-paragraph (i) of Sub-Clause 14.3 [*Application for Interim Payment*]. At that time, the Payment Certificate shall include the applicable amount to be deducted which shall be equivalent to, and in the same currencies and proportions as, this additional amount for the relevant Plant and Materials.

14.6 Issue of IPC

No amount will be certified or paid to the Contractor until:

(a) the Employer has received the Performance Security in the form, and issued by an entity, in accordance with Sub-Clause 4.2.1 [*Contractor's obligations*]; and
(b) the Contractor has appointed the Contractor's Representative in accordance with Sub-Clause 4.3 [*Contractor's Representative*].

14.6.1 The IPC

The Engineer shall, within 28 days after receiving a Statement and supporting documents, issue an IPC to the Employer, with a copy to the Contractor:

(a) stating the amount which the Engineer fairly considers to be due; and
(b) including any additions and/or deductions which have become due under Sub-Clause 3.7 [*Agreement or Determination*] or under the Contract or otherwise,

with detailed supporting particulars (which shall identify any difference between a certified amount and the corresponding amount in the Statement and give the reasons for such difference).

14.6.2 Withholding (amounts in) an IPC

Before the issue of the Taking-Over Certificate for the Works, the Engineer may withhold an IPC in an amount which would (after retention and other deductions) be less than the minimum amount of the IPC (if any) stated in the Contract Data. In this event, the Engineer shall promptly give a Notice to the Contractor accordingly.

An IPC shall not be withheld for any other reason, although:

(a) if any thing supplied or work done by the Contractor is not in accordance with the Contract, the estimated cost of rectification or replacement may be withheld until rectification or replacement has been completed;

或

(c) 有关**生产设备**和**材料**：

（i） 是**合同数据**中所列运到**现场**时付款的物品；（以及）

（ii） 已运到**现场**和妥善储存，并已做好防止损失、损害或变质的保护，似乎已符合**合同**要求。

要商定或确定的金额，应考虑本款要求的证据和各项文件及**生产设备和材料**的合同价值。如果上述（b）段适用，在**雇主**收到上述（b）(iii) 段规定的银行保函之前，**工程师**没有义务证明本项所述的任何付款。**工程师**在**期中付款证书**中确认的金额应等于该商定或确定金额的百分之八十（80%）。该确认金额的货币，应与**第14.3款**［*期中付款的申请*］(i) 段包括的合同价值应付的货币相同。此时，**付款证书**应计入适当的减少额，该减少额应与相关**生产设备**和**材料**的增加额相等，并采用同样的货币和比例。

14.6 期中付款证书的签发

在以下情况之前，不得向**承包商**确认或支付任何款项：

(a) **雇主**已收到由实体出具的、符合**第4.2.1项**［*承包商的义务*］规定格式的**履约担保**；（以及）

(b) **承包商**已根据**第4.3款**［*承包商代表*］的规定指定了**承包商代表**。

14.6.1 期中付款证书

工程师应在收到有关**报表**和证明文件后28天内，向**雇主**签发**期中付款证书**，并抄送**承包商**：

(a) 说明**工程师**公正地确定应付的金额；（以及）

(b) 包括根据**第3.7款**［*商定或确定*］或根据**合同**或其他规定到期应付的任何增加和／或扣减。

并附详细的证明资料（该资料应证明经确认金额与**报表**中相应金额之间的任何差额，并说明差额的原因）。

14.6.2 期中付款证书暂扣（金额）

在签发**工程接收证书**前，**工程师**可扣留（扣除保留金和其他应扣款项后）少于**合同数据**中规定的**期中付款证书**的最低额（如果有）。在此情况下，**工程师**应立即**通知承包商**。

虽然存在以下情况，对**期中付款证书**不应因任何其他原因予以扣发：

(a) 如果**承包商**供应的任何物品或完成的工作不符合**合同**要求，在完成修正和更换前，可以扣发该修正和更换所需的估计费用；

(b) if the Contractor was or is failing to perform any work, service or obligation in accordance with the Contract, the value of this work or obligation may be withheld until the work or obligation has been performed. In this event, the Engineer shall promptly give a Notice to the Contractor describing the failure and with detailed supporting particulars of the value withheld; and/or

(c) if the Engineer finds any significant error or discrepancy in the Statement or supporting documents, the amount of the IPC may take account of the extent to which this error or discrepancy has prevented or prejudiced proper investigation of the amounts in the Statement until such error or discrepancy is corrected in a subsequent Statement.

For each amount so withheld, in the supporting particulars for the IPC the Engineer shall detail his/her calculation of the amount and state the reasons for it being withheld.

14.6.3 Correction or modification

The Engineer may in any Payment Certificate make any correction or modification that should properly be made to any previous Payment Certificate. A Payment Certificate shall not be deemed to indicate the Engineer's acceptance, approval, consent or Notice of No-objection to any Contractor's Document or to (any part of) the Works.

If the Contractor considers that an IPC does not include any amounts to which the Contractor is entitled, these amounts shall be identified in the next Statement (the "identified amounts" in this paragraph). The Engineer shall then make any correction or modification that should properly be made in the next Payment Certificate. Thereafter, to the extent that:

(a) the Contractor is not satisfied that this next Payment Certificate includes the identified amounts; and
(b) the identified amounts do not concern a matter for which the Engineer is already carrying out his/her duties under Sub-Clause 3.7 [*Agreement or Determination*]

the Contractor may, by giving a Notice, refer this matter to the Engineer and Sub-Clause 3.7 [*Agreement or Determination*] shall apply (and, for the purpose of Sub-Clause 3.7.3 [*Time limits*], the date the Engineer receives this Notice shall be the date of commencement of the time limit for agreement under Sub-Clause 3.7.3).

14.7 Payment

The Employer shall pay to the Contractor:

(a) the amount certified in each Advance Payment Certificate within the period stated in the Contract Data (if not stated, 21 days) after the Employer receives the Advance Payment Certificate;
(b) the amount certified in each IPC issued under:

   (i) Sub-Clause 14.6 [*Issue of IPC*], within the period stated in the Contract Data (if not stated, 56 days) after the Engineer receives the Statement and supporting documents; or
   (ii) Sub-Clause 14.13 [*Issue of FPC*], within the period stated in the Contract Data (if not stated, 28 days) after the Employer receives the IPC; and

(c) the amount certified in the FPC within the period stated in the Contract Data (if not stated, 56 days) after the Employer receives the FPC.

Payment of the amount due in each currency shall be made into the bank account, nominated by the Contractor, in the payment country (for this currency) specified in the Contract.

（b） 如果**承包商**未能按照**合同**要求履行任何工作、服务或义务，在该项工作或义务完成前，可以扣留该工作或义务的价值。在这种情况下，**工程师**应立即向**承包商**发出**通知**，说明未履行工作等的情况，并附上扣留价值的详细证明资料；(和／或)

（c） 如果**工程师**在**报表**或证明文件中发现任何重大错误或差异，**期中付款证书**的金额可考虑该错误或差异对**报表**中金额进行适当调查的妨碍或损害的程度，直到该错误或差异在随后的**报表**中得到改正为止。

对于每一笔如此扣留的金额，**工程师**应在**期中付款证书**的证明材料中详细说明其对该笔金额的计算，并说明扣留的原因。

14.6.3 改正或修改

工程师可在任一次付款证书中，对以前任何**付款证书**做出任何适当的改正或修改。**付款证书**不应被视为表明**工程师**的接受、批准、同意或对任何**承包商文件**或**工程**（任何部分）的**不反对通知**。

如果**承包商**认为**期中付款证书**没有包括**承包商**有权获得的任何金额，则应在下一份**报表**中确认这些金额（本项中"确认的金额"）。**工程师**应在下一份**付款证书**中做出任何适当的改正或修改。此后，只要：

（a） **承包商**对下一份**付款证书**中包括确认的金额不满意；（以及）

（b） 所确认的金额不涉及**工程师**已根据**第 3.7 款**［*商定或确定*］履行其职责的事项。

承包商可发**通知**将此事项提交给**工程师**，**第 3.7 款**［*商定或确定*］应适用（以及，就 3.7.3 项［*时限*］而言，工程师收到本**通知**的日期，应为**第 3.7.3 项**规定的商定时限的开始日期）。

14.7 付款

雇主应向**承包商**支付：

（a） 在**雇主**收到**预付款证书**后，在**合同数据**中规定的期限内（如未规定，应为 21 天）各**预付款证书**确认的金额；

（b） 各**期中付款证书**确认的金额根据如下：

（i） 在工程师收到按照**第 14.6 款**［*期中付款证书的签发*］和**合同数据**中规定提交**报表**和证明文件的期限内（如未规定，应为 56 天）；(或)

（ii） 在**雇主**收到**第 14.13 款**［*最终付款证书的签发*］和**合同数据**中规定的**最终付款证书**的期限内（如未规定，应为 28 天）；(以及)

（c） 在**雇主**收到**合同数据**中规定的**最终付款证书**确认的金额的期限内（如未规定，应为 56 天）。

每种货币的应付款额，应汇入**合同**（为此货币）规定的付款国境内的**承包商**指定的银行账户。

14.8
Delayed Payment

If the Contractor does not receive payment in accordance with Sub-Clause 14.7 [*Payment*], the Contractor shall be entitled to receive financing charges compounded monthly on the amount unpaid during the period of delay. This period shall be deemed to commence on the expiry of the time for payment specified in Sub-Clause 14.7 [*Payment*], irrespective (in the case of sub-paragraph (b) of Sub-Clause 14.7) of the date on which any IPC is issued.

Unless otherwise stated in the Contract Data, these financing charges shall be calculated at the annual rate of three percent (3%) above:

(a) the average bank short-term lending rate to prime borrowers prevailing for the currency of payment at the place of payment, or
(b) where no such rate exists at that place, the same rate in the country of the currency of payment, or
(c) in the absence of such a rate at either place, the appropriate rate fixed by the law of the country of the currency of payment.

The Contractor shall by request be entitled to payment of these financing charges by the Employer, without:

(i) the need for the Contractor to submit a Statement or any formal Notice (including any requirement to comply with Sub-Clause 20.2 [*Claims For Payment and/or EOT*]) or certification; and
(ii) prejudice to any other right or remedy.

14.9
Release of Retention Money

After the issue of the Taking-Over Certificate for:

(a) the Works, the Contractor shall include the first half of the Retention Money in a Statement; or
(b) a Section, the Contractor shall include the relevant percentage of the first half of the Retention Money in a Statement.

After the latest of the expiry dates of the Defects Notification Periods, the Contractor shall include the second half of the Retention Money in a Statement promptly after such latest date. If a Taking-Over Certificate was (or was deemed to have been) issued for a Section, the Contractor shall include the relevant percentage of the second half of the Retention Money in a Statement promptly after the expiry date of the DNP for the Section.

In the next IPC after the Engineer receives any such Statement, the Engineer shall certify the release of the corresponding amount of Retention Money. However, when certifying any release of Retention Money under Sub-Clause 14.6 [*Issue of IPC*], if any work remains to be executed under Clause 11 [*Defects after Taking Over*], the Engineer shall be entitled to withhold certification of the estimated cost of this work until it has been executed.

The relevant percentage for each Section shall be the percentage value of the Section as stated in the Contract Data. If the percentage value of a Section is not stated in the Contract Data, no percentage of either half of the Retention Money shall be released under this Sub-Clause in respect of such Section.

14.10
Statement at Completion

Within 84 days after the Date of Completion of the Works, the Contractor shall submit to the Engineer a Statement at completion with supporting documents, in accordance with Sub-Clause 14.3 [*Application for Interim Payment*], showing:

14.8
延误的付款

如果**承包商**没有在按照**第 14.7 款**[*付款*]规定的时间收到付款，**承包商**应有权就未付款额按月计算复利，收取延期的融资费用。该延误期应视为从**第 14.7 款**[*付款*]规定的支付期限届满时算起，而不考虑[如**第 14.7 款**（b）段的情况]签发任何**期中付款证书**的日期。

除非**合同数据**中另有规定，上述融资费用应以 3% 的年利率进行计算：

（a） 对主要借款人的平均银行短期贷款利率，以付款地点的付款货币为准；（或）

（b） 如果在该地没有这样的汇率，与付款货币所在国的汇率相同；（或）

（c） 如果在任何地方都没有这样的汇率，由付款货币所在国法律规定的适当汇率。

承包商应有权要求雇主支付这些融资费用，而无须：

（i） **承包商**提供一份**报表**或任何正式**通知**（包括遵守**第 20.2 款**[*付款和/或竣工时间延长的索赔*]规定的任何要求）或证明；（以及）

（ii） 损害任何其他权利或补偿。

14.9
保留金的发放

签发工程接收证书后：

（a） 对**工程**，**承包商**应在**报表**中包含**保留金**的前半部分；（或）

（b） 对**分项工程**，**承包商**应在**报表**中包含**保留金**前半部分的相关百分比。

在各**缺陷通知期限**的最末一个期满日期后，**承包商**应在最末到期日后立即将**保留金**的后半部分包含在**报表**中。如对某**分项工程**签发了（或被视为已签发）**接收证书**，**承包商**应在该分项工程的**缺陷通知期限**期满日期后，立即在**报表**中包含**保留金**后半部分的相关百分比。

在**工程师**收到任何此类**报表**后的下一次**期中付款证书**中，**工程师**应确认相应金额的**保留金**已发放。但是，在根据**第 14.6 款**[*期中付款证书的签发*]的规定，确认任何**保留金**的发放时，如果根据**第 11 条**[*接收后的缺陷*]的规定，还有任何工作需要做的，**工程师**应有权在该项工作完成前，暂不签发该工作估算费用的证书。

各**分项工程**的相关百分比应为**合同数据**中规定的该**分项工程**的百分比价值。如果**合同数据**中没有规定**分项工程**的百分比价值，则不得根据本款规定，就该**分项工程**发放任何一半**保留金**的百分比。

14.10
竣工报表

在**工程**竣工后 84 天内，**承包商**应按照**第 14.3 款**[*期中付款的申请*]的规定，向**工程师**提交竣工**报表**并附证明文件，列出：

(a) the value of all work done in accordance with the Contract up to the Date of Completion of the Works
(b) any further sums which the Contractor considers to be due at the Date of Completion of the Works; and
(c) an estimate of any other amounts which the Contractor considers have or will become due after the Date of Completion of the Works under the Contract or otherwise. These estimated amounts shall be shown separately (to those of sub-paragraphs (a) and (b) above) and shall include estimated amounts for:

   (i) Claims for which the Contractor has submitted a Notice under Sub-Clause 20.2 [*Claims For Payment and/or EOT*];
   (ii) any matter referred to the DAAB under Sub-Clause 21.4 [*Obtaining DAAB's Decision*]; and
   (iii) any matter for which a NOD has been given under Sub-Clause 21.4 [*Obtaining DAAB's Decision*].

The Engineer shall then issue an IPC in accordance with Sub-Clause 14.6 [*Issue of IPC*].

14.11 Final Statement

Submission by the Contractor of any Statement under the following provisions of this Sub-Clause shall not be delayed by reason of any referral under Sub-Clause 21.4 [*Obtaining DAAB's Decision*] or any arbitration under Sub-Clause 21.6 [*Arbitration*].

14.11.1 Draft Final Statement

Within 56 days after the issue of the Performance Certificate, the Contractor shall submit to the Engineer, a draft final Statement.

This Statement shall:

(a) be in the same form as Statements previously submitted under Sub-Clause 14.3 [*Application for Interim Payment*];
(b) be submitted in one paper-original, one electronic copy and additional paper copies (if any) as stated in the Contract Data; and
(c) show in detail, with supporting documents:

   (i) the value of all work done in accordance with the Contract;
   (ii) any further sums which the Contractor considers to be due at the date of the issue of the Performance Certificate, under the Contract or otherwise; and
   (iii) an estimate of any other amounts which the Contractor considers have or will become due after the issue of the Performance Certificate, under the Contract or otherwise, including estimated amounts, by reference to the matters described in sub-paragraphs (c) (i) to (iii) of Sub-Clause 14.10 [*Statement at Completion*]. These estimated amounts shall be shown separately (to those of sub-paragraphs (i) and (ii) above).

Except for any amount under sub-paragraph (iii) above, if the Engineer disagrees with or cannot verify any part of the draft final Statement, the Engineer shall promptly give a Notice to the Contractor. The Contractor shall then submit such further information as the Engineer may reasonably require within the time stated in this Notice, and shall make such changes in the draft as may be agreed between them.

14.11.2 Agreed Final Statement

If there are no amounts under sub-paragraph (iii) of Sub-Clause 14.11.1 [*Draft Final Statement*], the Contractor shall then prepare and submit to the Engineer the final Statement as agreed (the "Final Statement" in these Conditions).

（a） 截止**工程竣工**的**日期**，根据合同要求完成的所有工作的价值；

（b） **承包商**认为在**工程竣工**之日应付的任何其他款项；（以及）

（c） **承包商**认为的、在根据合同或其他规定的**工程竣工日期**后已到期或将要到期任何其他款项的估算款额。估算款额应单独列出（与上述（a）和（b）段相同），并应包括以下估算款额：

　（i） **承包商**已根据**第 20.2 款**［*付款和/或竣工时间延长的索赔*］的规定提交通知的**索赔**；
　（ii） 根据**第 21.4 款**［*取得争端避免/裁决委员会的决定*］的规定，提交给**争端避免/裁决委员会**的任何事项；（以及）
　（iii） 根据**第 21.4 款**［*取得争端避免/裁决委员会的决定*］的规定，提出**不满意通知**的任何事项。

然后，**工程师**应按照**第 14.6 款**［*期中付款证书的签发*］的规定，签发**期中付款证书**。

14.11 最终报表

承包商根据本款下列规定提交的任何**报表**，不得因**第 21.4 款**［*取得争端避免/裁决委员会的决定*］规定的任何提交，或**第 21.6 款**［*仲裁*］规定的任何仲裁而延误。

14.11.1 最终报表草案

在签发**履约证书**后 56 天内，**承包商**应向**工程师**提交一份最终**报表**草案。

该**报表**应：

（a） 与之前根据**第 14.3 款**［*期中付款的申请*］提交的**报表**格式相同；

（b） 按**合同数据**中的规定，提交一份纸质原件、一份电子副本和附加纸质副本（如果有）；（以及）

（c） 附证明文件，详细说明：

　（i） 根据合同完成的所有工作的价值；
　（ii） **承包商**认为根据合同或其他规定在签发**履约证书**日期到期的任何其他款额；（以及）
　（iii） **承包商**认为根据合同或其他规定在签发**履约证书**后已到期或将要到期的任何其他款项的估算款额，包括根据**第 14.10 款**［*竣工报表*］（c）（i）至（iii）段所述事项估算款额。估算款额应单独列出［与上述（i）和（ii）段相同］。

除上述（iii）段的任何款额外，如果**工程师**不同意或无法核实最终**报表**草案中的任何部分，**工程师**应立即向**承包商**发出通知。然后，**承包商**应在本通知规定的时间内提交工程师可能合理要求的补充资料，并应按双方可能商定的意见，对该草案进行修改。

14.11.2 商定的最终报表

如果**第 14.11.1 项**［*最终报表草案*］（iii）段没有规定款额，**承包商**应按商定的意见编制并向**工程师**提交最终报表（本条件中称为"**最终报表**"）。

However if:

(a) there are amounts under sub-paragraph (iii) of Sub-Clause 14.11.1 [*Draft Final Statement*]; and/or
(b) following discussions between the Engineer and the Contractor, it becomes evident that they cannot agree any amount(s) in the draft final Statement,

the Contractor shall then prepare and submit to the Engineer a Statement, identifying separately: the agreed amounts, the estimated amounts and the disagreed amount(s) (the "Partially Agreed Final Statement" in these Conditions).

14.12 Discharge

When submitting the Final Statement or the Partially Agreed Final Statement (as the case may be), the Contractor shall submit a discharge which confirms that the total of such Statement represents full and final settlement of all moneys due to the Contractor under or in connection with the Contract. This discharge may state that the total of the Statement is subject to any payment that may become due in respect of any Dispute for which a DAAB proceeding or arbitration is in progress under Sub-Clause 21.6 [*Arbitration*] and/or that it becomes effective after the Contractor has received:

(a) full payment of the amount certified in the FPC; and
(b) the Performance Security.

If the Contractor fails to submit this discharge, the discharge shall be deemed to have been submitted and to have become effective when the conditions of sub-paragraphs (a) and (b) have been fulfilled.

A discharge under this Sub-Clause shall not affect either Party's liability or entitlement in respect of any Dispute for which a DAAB proceeding or arbitration is in progress under Clause 21 [*Disputes and Arbitration*].

14.13 Issue of FPC

Within 28 days after receiving the Final Statement or the Partially Agreed Final Statement (as the case may be), and the discharge under Sub-Clause 14.12 [*Discharge*], the Engineer shall issue to the Employer (with a copy to the Contractor), the Final Payment Certificate which shall state:

(a) the amount which the Engineer fairly considers is finally due, including any additions and/or deductions which have become due under Sub-Clause 3.7 [*Agreement or Determination*] or under the Contract or otherwise; and
(b) after giving credit to the Employer for all amounts previously paid by the Employer and for all sums to which the Employer is entitled, and after giving credit to the Contractor for all amounts (if any) previously paid by the Contractor and/or received by the Employer under the Performance Security, the balance (if any) due from the Employer to the Contractor or from the Contractor to the Employer, as the case may be.

If the Contractor has not submitted a draft final Statement within the time specified under Sub-Clause 14.11.1 [*Draft Final Statement*], the Engineer shall request the Contractor to do so. Thereafter, if the Contractor fails to submit a draft final Statement within a period of 28 days, the Engineer shall issue the FPC for such an amount as the Engineer fairly considers to be due.

If:

但是如果：

（a） 第 14.11.1 项 [*最终报表草案*]（iii）段规定有款额；（和/或）

（b） 经讨论之后，**工程师**和**承包商**很明显不能就最终**报表**草案中的任何款额达成商定。

则**承包商**应编制并向**工程师**提交一份**报表**，分别说明：商定的款额、估算的款额和不同意的款额（本条件中的"部分商定的最终报表"）。

14.12 结清证明

承包商在提交**最终报表**或部分商定的**最终报表**（视情况而定）时，应提交一份结清证明，确认**最终报表**上的总额代表了合同规定的或与合同有关的事项，应付给**承包商**的所有款项的全部和最终的结算总额。该结清证明可注明该总额是根据与**争端避免/裁决委员会**程序有关的任何**争端**，或根据**第 21.6 款** [*仲裁*] 有关的正在进行仲裁的任何到期应付款项，和/或在**承包商**收到下列文件后生效：⊖

（a） 全额支付**最终付款证书**确认的款额；（以及）
（b） **履约担保**。

如果**承包商**未能提交结清证明，则该结清证明应被视为已提交，并在满足（a）和（b）段条件时生效。

本款规定的结清证明，不应影响任一方根据**第 21 条** [*争端和仲裁*] 正在进行**争端避免/裁决委员会**程序或仲裁的任何**争端**方面的责任或权利。

14.13 最终付款证书的签发

工程师在收到**第 14.12 款** [*结清证明*] 规定的**最终报表**或部分商定的**最终报表**和结清证明后 28 天内，应向**雇主**签发**最终付款证书**（抄送**承包商**），说明：

（a） **工程师**应公平地考虑最终应付款额，包括根据**第 3.7 款** [*商定或确定*] 或根据合同或其他规定应付的任何增加和/或扣减款额；（以及）

（b） 确认**雇主**先前已付的所有款额以及**雇主**有权得到的所有款额后，确认**承包商**先前已付的所有款额和/或**雇主**根据**履约担保**得到的所有款额后，**雇主**尚需付给**承包商**，或**承包商**尚需付给**雇主**的余额（如果有），视情况而定。

如果**承包商**未按**第 14.11.1 项** [*最终报表草案*] 规定的时间内提交**最终报表草案**，**工程师**应要求**承包商**提交。此后，如**承包商**在 28 天期限内未能提交**最终报表草案**，**工程师**应按其公正确定的应付款额签发**最终付款证书**。

如果：

⊖ 此处中文按照勘误表改正后的英文翻译。——译者注

(i) the Contractor has submitted a Partially Agreed Final Statement under Sub-Clause 14.11.2 [*Agreed Final Statement*]; or

(ii) no Partially Agreed Final Statement has been submitted by the Contractor but, to the extent that a draft final Statement submitted by the Contractor is deemed by the Engineer to be a Partially Agreed Final Statement

the Engineer shall proceed in accordance with Sub-Clause 14.6 [Issue of IPC] to issue an IPC.

14.14 Cessation of Employer's Liability

The Employer shall not be liable to the Contractor for any matter or thing under or in connection with the Contract or execution of the Works, except to the extent that the Contractor shall have included an amount expressly for it in:

(a) the Final Statement or Partially Agreed Final Statement; and
(b) (except for matters or things arising after the issue of the Taking-Over Certificate for the Works) the Statement under Sub-Clause 14.10 [*Statement at Completion*].

Unless the Contractor makes or has made a Claim under Sub-Clause 20.2 [*Claims For Payment and/or EOT*] in respect of an amount or amounts under the FPC within 56 days of receiving a copy of the FPC the Contractor shall be deemed to have accepted the amounts so certified. The Employer shall then have no further liability to the Contractor, other than to pay the amount due under the FPC and return the Performance Security to the Contractor.

However, this Sub-Clause shall not limit the Employer's liability under the Employer's indemnification obligations, or the Employer's liability in any case of fraud, gross negligence, deliberate default or reckless misconduct by the Employer.

14.15 Currencies of Payment

The Contract Price shall be paid in the currency or currencies named in the Contract Data. If more than one currency is so named, payments shall be made as follows:

(a) if the Accepted Contract Amount was expressed in Local Currency only or in Foreign Currency only:

(i) the proportions or amounts of the Local and Foreign Currencies, and the fixed rates of exchange to be used for calculating the payments, shall be as stated in the Contract Data, except as otherwise agreed by both Parties;

(ii) payments and deductions under Sub-Clause 13.4 [*Provisional Sums*] and Sub-Clause 13.6 [*Adjustments for Changes in Laws*] shall be made in the applicable currencies and proportions; and

(iii) other payments and deductions under sub-paragraphs (i) to (iv) of Sub-Clause 14.3 [*Application for Interim Payment*] shall be made in the currencies and proportions specified in sub-paragraph (a)(i) above;

(b) whenever an adjustment is agreed or determined under Sub-Clause 13.2 [*Value Engineering*] or Sub-Clause 13.3 [*Variation Procedure*], the amount payable in each of the applicable currencies shall be specified. For this purpose, reference shall be made to the actual or expected currency proportions of the Cost of the varied work, and to the proportions of various currencies specified in sub-paragraph (a)(i) above;

（i）承包商根据第 14.11.2 项［*商定的最终报表*］的规定提交了部分商定的最终报表；（或）

（ii）承包商未提交部分商定的最终报表，但如果承包商提交的最终报表草案被工程师视为部分商定的最终报表。

工程师应按照第 14.6 款［*期中付款证书的签发*］的规定，签发期中付款证书。

14.14 雇主责任的中止

除了承包商在下列文件中，因合同或工程实施造成的或与之有关的任何问题或事项，明确提出款项要求以外，雇主应不再为之对承包商承担责任：

（a）最终报表或部分商定的最终报表；（以及）
（b）第 14.10 款［*竣工报表*］规定的报表（签发工程接收证书后发生的问题或事项除外）。

除非承包商在收到最终付款证书副本后 56 天内，根据第 20.2 款［*付款和/或竣工时间延长的索赔*］的规定，提出一笔或多笔款项的索赔，否则承包商应被视为已接受经确认的款项。除非支付最终付款证书规定的到期款额，并将履约担保金退还给承包商，雇主不应对承包商承担进一步的责任。

但是，本款不应减少雇主因其保障义务，或因其任何欺骗、重大过失、故意违约或轻率不当行为等情况引起的责任。

14.15 支付的货币

合同价格应按合同数据中规定的货币或几种货币支付。如果规定了一种以上货币，应按以下办法支付：

（a）如果中标合同金额只是用当地货币或外币表示的：

（i）当地货币和外币的比例或款额，以及计算付款采用的固定汇率，除双方另有商定外，应按合同数据中的规定；

（ii）根据第 13.4 款［*暂列金额*］和第 13.6 款［*因法律改变的调整*］规定的付款和扣减，应按适用的货币和比例；（以及）

（iii）根据第 14.3 款［*期中付款的申请*］（i）至（iv）段做出的其他支付和扣减，应按上述（a）（i）段中规定的货币和比例。

（b）当根据第 13.2 款［*价值工程*］或第 13.3 款［*变更程序*］的规定进行商定或确定调整时，应规定以每种适用货币支付的款额。为此，应参考各项工作费用的实际或预期货币比例，以及上述（a）（i）段中规定的各种货币的比例；

(c) payment of Delay Damages shall be made in the currencies and proportions specified in the Contract Data;
(d) other payments to the Employer by the Contractor shall be made in the currency in which the sum was expended by the Employer, or in such currency as may be agreed by both Parties;
(e) if any amount payable by the Contractor to the Employer in a particular currency exceeds the sum payable by the Employer to the Contractor in that currency, the Employer may recover the balance of this amount from the sums otherwise payable to the Contractor in other currencies; and
(f) if no rates of exchange are stated in the Contract Data, they shall be those prevailing on the Base Date and published by the central bank of the Country.

15 Termination by Employer

15.1 Notice to Correct

If the Contractor fails to carry out any obligation under the Contract the Engineer may, by giving a Notice to the Contractor, require the Contractor to make good the failure and to remedy it within a specified time ("Notice to Correct" in these Conditions).

The Notice to Correct shall:

(a) describe the Contractor's failure;
(b) state the Sub-Clause and/or provisions of the Contract under which the Contractor has the obligation; and
(c) specify the time within which the Contractor shall remedy the failure, which shall be reasonable, taking due regard of the nature of the failure and the work and/or other action required to remedy it.

After receiving a Notice to Correct the Contractor shall immediately respond by giving a Notice to the Engineer describing the measures the Contractor will take to remedy the failure, and stating the date on which such measures will be commenced in order to comply with the time specified in the Notice to Correct.

The time specified in the Notice to Correct shall not imply any extension of the Time for Completion.

15.2 Termination for Contractor's Default

Termination of the Contract under this Clause shall not prejudice any other rights of the Employer under the Contract or otherwise.

15.2.1 Notice

The Employer shall be entitled to give a Notice (which shall state that it is given under this Sub-Clause 15.2.1) to the Contractor of the Employer's intention to terminate the Contract or, in the case of sub-paragraph (f), (g) or (h) below a Notice of termination, if the Contractor:

(a) fails to comply with:

(i) a Notice to Correct;
(ii) a binding agreement, or final and binding determination, under Sub-Clause 3.7 [*Agreement or Determination*]; or
(iii) a decision of the DAAB under 21.4 [*Obtaining DAAB's Decision*] (whether binding or final and binding)

(c) **误期损害赔偿费**的支付应按**合同数据**中规定的货币和比例；

(d) **承包商**付给**雇主**的其他款项应以**雇主**花费该款实际用的货币，或双方可能商定的货币；

(e) 如果**承包商**以某种特定货币应付给**雇主**的任何款额，超过**雇主**以该货币应付给**承包商**的款额，**雇主**可从以其他货币应付给**承包商**的款额中收回该项差额；（以及）

(f) 如果**合同数据**中未规定汇率，应采用**基准日期**当天和**工程**所在国中央银行公布的汇率。

15 由雇主终止

15.1 通知改正

如果**承包商**未能根据**合同**履行任何义务，**工程师**可通过向**承包商**发出**通知**，要求其在规定时间内，纠正并补救上述未履约（本**条件**中的"**通知改正**"）。

通知改正应：

(a) 说明**承包商**的未履约；
(b) 述明**承包商**具有履约义务的**条款**和/或合同条款；（以及）
(c) 在适当考虑未履约的性质及纠正未履约所需的工作和/或其他行动的情况下，明确**承包商**应纠正未履约的合理时间。

收到**通知改正**后，**承包商**应立即做出回应，向**工程师**发出**通知**，描述其将采取的纠正未履约的措施，并说明开始采取这些措施的日期，以便与**通知改正**中规定的时间一致。

通知改正中规定的时间并不表示**竣工时间**的任何延长。

15.2 因承包商违约的终止

根据本**条款**终止合同不应损害合同规定的或**雇主**享有的任何其他权利。

15.2.1 通知

雇主应有权向**承包商**发出**通知**（应说明是按照 **15.2.1 项**的规定发出的），告知**承包商**其打算终止**合同**的意向；（或），如果**承包商**有下述（f）（g）或（h）段的情况，**雇主**应有权终止**合同**：

(a) 未能遵守：

 (i) **通知改正**；
 (ii) 第 **3.7 款** [*商定或确定*] 规定的具有约束力的商定或具有约束力的最终确定；（或）
 (iii) **争端避免/裁决委员会**根据 **21.4 款** [*取得争端避免/裁决委员会的决定*] 规定做出的决定（无论是具有约束力的还是最终并具有约束力的）。

and such failure constitutes a material breach of the Contractor's obligations under the Contract;

(b) abandons the Works or otherwise plainly demonstrates an intention not to continue performance of the Contractor's obligations under the Contract;

(c) without reasonable excuse fails to proceed with the Works in accordance with Clause 8 [*Commencement, Delays and Suspension*] or, if there is a maximum amount of Delay Damages stated in the Contract Data, his failure to comply with Sub-Clause 8.2 [*Time for Completion*] is such that the Employer would be entitled to Delay Damages that exceed this maximum amount;

(d) without reasonable excuse fails to comply with a Notice of rejection given by the Engineer under Sub-Clause 7.5 [*Defects and Rejection*] or an Engineer's instruction under Sub-Clause 7.6 [*Remedial Work*], within 28 days after receiving it;

(e) fails to comply with Sub-Clause 4.2 [*Performance Security*];

(f) subcontracts the whole, or any part of, the Works in breach of Sub-Clause 5.1 [*Subcontractors*], or assigns the Contract without the required agreement under Sub-Clause 1.7 [*Assignment*];

(g) becomes bankrupt or insolvent; goes into liquidation, administration, reorganisation, winding-up or dissolution; becomes subject to the appointment of a liquidator, receiver, administrator, manager or trustee; enters into a composition or arrangement with the Contractor's creditors; or any act is done or any event occurs which is analogous to or has a similar effect to any of these acts or events under applicable Laws;

or if the Contractor is a JV:

(i) any of these matters apply to a member of the JV, and

(ii) the other member(s) do not promptly confirm to the Employer that, in accordance with Sub-Clause 1.14(a) [*Joint and Several Liability*], such member's obligations under the Contract shall be fulfilled in accordance with the Contract; or

(h) is found, based on reasonable evidence, to have engaged in corrupt, fraudulent, collusive or coercive practice at any time in relation to the Works or to the Contract.

15.2.2 Termination

Unless the Contractor remedies the matter described in a Notice given under Sub-Clause 15.2.1 [*Notice*] within 14 days of receiving the Notice, the Employer may by giving a second Notice to the Contractor immediately terminate the Contract. The date of termination shall be the date the Contractor receives this second Notice.

However, in the case of sub-paragraph (f), (g) or (h) of Sub-Clause 15.2.1 [*Notice*], the Employer may by giving a Notice under Sub-Clause 15.2.1 immediately terminate the Contract and the date of termination shall be the date the Contractor receives this Notice.

15.2.3 After termination

After termination of the Contract under Sub-Clause 15.2.2 [*Termination*], the Contractor shall:

(a) comply immediately with any reasonable instructions included in a Notice given by the Employer under this Sub-Clause:

(i) for the assignment of any subcontract; and

(ii) for the protection of life or property or for the safety of the Works;

这种未履约构成对合同规定的**承包商**义务的严重违反；

(b) 放弃**工程**，或以其他方式明确表现出不愿继续履行**合同**规定的**承包商**义务的意向；

(c) 在无合理解释的情况下，未能按照第 8 条 [*开工、延误和暂停*] 的规定实施**工程**，或，如果**合同数据**中规定了**误期损害赔偿费**的最大金额，其未能遵守第 8.2 款 [*竣工时间*] 的规定，则可使**雇主**有权获得高于该最大金额的**误期损害赔偿费**；

(d) 在无合理解释的情况下，收到通知后的 28 天内未遵守**工程师**根据第 7.5 款 [*缺陷和拒收*] 发出的拒收通知，或**工程师**根据第 7.6 款 [*修补工作*] 发出的指示；

(e) 未能遵守第 4.2 款 [*履约担保*] 的规定；
(f) 违反第 5.1 款 [*分包商*] 的规定，将全部或部分工程分包，或未获得第 1.7 款 [*权益转让*] 规定同意的情况下转让**合同**；

(g) 破产或无力偿债；进入清算、管理、重组、清理或解散等程序；委托了清算人、接管人、管理人、经理或受托人；与**承包商**的债权人达成和解或安排；(或) 根据适用**法律**，做出了与这些行为或事件类似，或具有类似作用的任何行为或发生的任何事件。

或，如果**承包商**是**联营体**：

(i) 这些事项中的任何一项均适用于**联营体**的成员；(以及)
(ii) 其他成员未按第 1.14 款 [*共同的和各自的责任*] (a) 段的规定立即向**雇主**确认，**合同**规定的此类成员义务应按**合同**规定履行；(或)

(h) 根据合理证据，发现参与了与**工程**或**合同**有关的腐败、欺诈、串通或胁迫行为。

15.2.2
终止

除非**承包商**在收到**通知** 14 天内，对按第 15.2.1 项 [*通知*] 规定发出的**通知**中所述的事项进行补救，否则**雇主**可以向**承包商**再次发出**通知**，立即终止合同。终止日期为**承包商**收到该再次**通知**的日期。

但是，在第 15.2.1 项 [*通知*] 的 (f)(g) 或 (h) 段所述情况下，**雇主**可根据第 15.2.1 项的规定发出**通知**，立即终止合同，终止日期为**承包商**收到该**通知**的日期。

15.2.3
终止后

根据第 15.2.2 项 [*终止*] 规定终止合同后，**承包商**应：

(a) 立即遵守**雇主**根据本款在**通知**中包含的任何合理指示：

(i) 任何分包合同的转让；(以及)
(ii) 保护生命或财产或**工程**安全。

(b) deliver to the Engineer:

    (i) any Goods required by the Employer,
    (ii) all Contractor's Documents, and
    (iii) all other design documents made by or for the Contractor to the extent, if any, that the Contractor is responsible for the design of part of the Permanent Works under Sub-Clause 4.1 [*Contractor's General Obligations*]; and

(c) leave the Site and, if the Contractor does not do so, the Employer shall have the right to expel the Contractor from the Site.

15.2.4 Completion of the Works

After termination under this Sub-Clause, the Employer may complete the Works and/or arrange for any other entities to do so. The Employer and/or these entities may then use any Goods and Contractor's Documents (and other design documents, if any) made by or on behalf of the Contractor to complete the Works.

After such completion of the Works, the Employer shall give another Notice to the Contractor that the Contractor's Equipment and Temporary Works will be released to the Contractor at or near the Site. The Contractor shall then promptly arrange their removal, at the risk and cost of the Contractor. However, if by this time the Contractor has failed to make a payment due to the Employer, these items may be sold (to the extent permitted by applicable Laws) by the Employer in order to recover this payment. Any balance of the proceeds shall then be paid to the Contractor.

15.3 Valuation after Termination for Contractor's Default

After termination of the Contract under Sub-Clause 15.2 [*Termination for Contractor's Default*], the Engineer shall proceed under Sub-Clause 3.7 [*Agreement or Determination*] to agree or determine the value of the Permanent Works, Goods and Contractor's Documents, and any other sums due to the Contractor for work executed in accordance with the Contract (and, for the purpose of Sub-Clause 3.7.3 [*Time limits*], the date of termination shall be the date of commencement of the time limit for agreement under Sub-Clause 3.7.3).

This valuation shall include any additions and/or deductions, and the balance due (if any), by reference to the matters described in sub-paragraphs (a) and (b) of Sub-Clause 14.13 [*Issue of FPC*].

This valuation shall not include the value of any Contractor's Documents, Materials, Plant and Permanent Works to the extent that they do not comply with the Contract.

15.4 Payment after Termination for Contractor's Default

The Employer may withhold payment to the Contractor of the amounts agreed or determined under Sub-Clause 15.3 [*Valuation after Termination for Contractor's Default*] until all the costs, losses and damages (if any) described in the following provisions of this Sub-Clause have been established.

After termination of the Contract under Sub-Clause 15.2 [*Termination for Contractor's Default*], the Employer shall be entitled subject to Sub-Clause 20.2 [*Claims For Payment and/or EOT*] to payment by the Contractor of:

(a) the additional costs of execution of the Works, and all other costs reasonably incurred by the Employer (including costs incurred in clearing, cleaning and reinstating the Site as described under Sub-Clause 11.11 [*Clearance of Site*]), after allowing for any sum due to the Contractor under Sub-Clause 15.3 [*Valuation after Termination for Contractor's Default*];

(b) 向**工程师**交付：

(i) **雇主**要求的任何**货物**；
(ii) 所有**承包商文件**；（以及）
(iii) 如果是**承包商**根据第 4.1 款 [*承包商的一般义务*] 的规定设计的部分**永久工程**，由**承包商**编制或为**承包商**编制的所有其他设计文件；（以及）

(c) 离开**现场**，如果**承包商**没有离开**现场**，则**雇主**有权将**承包商**驱离**现场**。

15.2.4
工程竣工

根据本款终止后，**雇主**可以完成**工程**和／或安排任何其他实体完成**工程**。**雇主**和／或这些实体之后可以使用任何**货物**，和**承包商**或代表**承包商**编制的**承包商文件**（以及其他设计文件，如果有）以完成**工程**。

工程竣工后，**雇主**应再次向**承包商**发出**通知**，说明**承包商设备**和**临时工程**将在**现场**或**现场**附近交还给**承包商**。**承包商**应立即安排将其运离，风险和费用由**承包商**承担。但如果此时**承包商**还有应付**雇主**的款项没有付清，**雇主**可以出售这些物品（在适用**法律**允许的范围内），以收回该款项。收益的任何余款应随后支付给**承包商**。

15.3
因承包商违约终止后的估价

在根据第 15.2 款 [*因承包商违约的终止*] 的规定终止合同后，**工程师**应按照第 3.7 款 [*商定或确定*] 的规定继续商定或确定**永久工程**、**货物**和**承包商文件**的价值，以及**承包商**按照合同实施的工作应得的任何其他款项（以及，就第 3.7.3 项 [*时限*] 而言，终止日期应为根据第 3.7.3 项商定的时限的开始日期）。

该估价应包括任何增加和／或扣减，以及应付余额（如果有），并参照第 14.13 款 [*最终付款证书的签发*] 第（a）和（b）段所述的事项。

如果**承包商文件**、材料、生产设备及**永久工程**不符合合同规定，则该估价不包括其价值。

15.4
因承包商违约终止后的付款

在本款以下规定中所述的所有成本、损失和损害（如果有）确定前，**雇主**可暂扣按照第 15.3 款 [*因承包商违约终止后的估价*] 规定的商定或确定支付**承包商**的金额。

按照第 15.2 款 [*因承包商违约的终止*] 的规定终止合同后，**雇主**应有权根据第 20.2 款 [*付款和／或竣工时间延长的索赔*] 的规定要求**承包商**支付以下款项：

(a) 在根据第 15.3 款 [*因承包商违约终止的估价*] 的规定允许付给**承包商**的任何款额后，实施**工程**的额外费用，以及**雇主**合理发生的所有其他费用（包括第 11.11 款 [*现场清理*] 规定中所述的清理、清洗和恢复**现场**发生的费用）；

(b) any losses and damages suffered by the Employer in completing the Works; and
(c) Delay Damages, if the Works or a Section have not been taken over under Sub-Clause 10.1 [*Taking Over the Works and Sections*] and if the date of termination under Sub-Clause 15.2 [*Termination for Contractor's Default*] occurs after the date corresponding to the Time for Completion of the Works or Section (as the case may be). Such Delay Damages shall be paid for every day that has elapsed between these two dates.

15.5 Termination for Employer's Convenience

The Employer shall be entitled to terminate the Contract at any time for the Employer's convenience, by giving a Notice of such termination to the Contractor (which Notice shall state that it is given under this Sub-Clause 15.5).

After giving a Notice to terminate under this Sub-Clause, the Employer shall immediately:

(a) have no right to further use any of the Contractor's Documents, which shall be returned to the Contractor, except those for which the Contractor has received payment or for which payment is due under a Payment Certificate;
(b) if Sub-Clause 4.6 [*Co-operation*] applies, have no right to allow the continued use (if any) of any Contractor's Equipment, Temporary Works, access arrangements and/or other of the Contractor's facilities or services; and
(c) make arrangements to return the Performance Security to the Contractor.

Termination under this Sub-Clause shall take effect 28 days after the later of the dates on which the Contractor receives this Notice or the Employer returns the Performance Security. Unless and until the Contractor has received payment of the amount due under Sub-Clause 15.6 [*Valuation after Termination for Employer's Convenience*], the Employer shall not execute (any part of) the Works or arrange for (any part of) the Works to be executed by any other entities.

After this termination, the Contractor shall proceed in accordance with Sub-Clause 16.3 [*Contractor's Obligations After Termination*].

15.6 Valuation after Termination for Employer's Convenience

After termination under Sub-Clause 15.5 [*Termination for Employer's Convenience*] the Contractor shall, as soon as practicable, submit detailed supporting particulars (as reasonably required by the Engineer) of:

(a) the value of work done, which shall include:

   (i) the matters described in sub-paragraphs (a) to (e) of Sub-Clause 18.5 [*Optional Termination*], and
   (ii) any additions and/or deductions, and the balance due (if any), by reference to the matters described in sub-paragraphs (a) and (b) of Sub-Clause 14.13 [*Issue of FPC*]; and

(b) the amount of any loss of profit or other losses and damages suffered by the Contractor as a result of this termination.

The Engineer shall then proceed under Sub-Clause 3.7 [*Agreement or Determination*] to agree or determine the matters described in sub-paragraphs (a) and (b) above (and, for the purpose of Sub-Clause 3.7.3 [*Time limits*], the date the Engineer receives the Contractor's particulars under this Sub-Clause shall be the date of commencement of the time limit for agreement under Sub-Clause 3.7.3).

(b) 在**工程**竣工过程中，**雇主**遭受的任何损失和损害；（以及）

(c) 还未按照第 10.1 款［*工程和分项工程的接收*］的规定接收**工程**或分项**工程**，以及第 15.2 款［*因承包商违约的终止*］规定的终止日期在**工程**或分项**工程**（视情况而定）的竣工时间相对应的日期后发生，而造成的误期损害赔偿费。该误期损害赔偿费应在上述两个日期之间按日支付。

| 15.5 为雇主便利的终止 | |

雇主应有权在对其便利的任何时候，通过向**承包商**发出终止**通知**，终止**合同**（该**通知**应说明是根据第 15.5 款的规定发出的）。

根据本款发出终止**通知**后，**雇主**应立即：

(a) 无权进一步使用应交还给**承包商**的任何**承包商文件**，但**承包商**已收到付款或**付款证书**规定应付款的文件除外；

(b) 如果第 4.6 款［*合作*］适用，则无权继续使用（如果有）任何**承包商设备**、临时**工程**、进出安排和/或**承包商**的其他设施或服务；（以及）

(c) 安排将**履约担保**退还给**承包商**。

本**款**规定的终止应在**承包商**收到该**通知**或**雇主**退还**履约担保**两者中较晚的日期后第 28 天生效。除非且直到**承包商**收到第 15.6 款［*为雇主便利终止后的估价*］规定的应付款额，否则**雇主**不得实施**工程**（任何部分）或安排其他实体实施**工程**（任何部分）。

终止后，**承包商**应按照第 16.3 款［*终止后承包商的义务*］的规定继续进行。

| 15.6 为雇主便利终止后的估价 | |

按照第 15.5 款［*为雇主便利的终止*］的规定终止后，**承包商**应在切实可行的情况下，尽快（按照**工程师**的合理要求）提交以下详细的证明资料：

(a) 已完成工作的价值，其中应包括：

(i) 第 18.5 款［*自主选择终止*］(a) 至 (e) 段中所述的事项；（以及）
(ii) 任何增加和/或扣减额，以及应付余额（如果有），并参照第 14.13 款［*最终付款证书的签发*］(a) 和 (b) 段中所述的事项；（以及）

(b) **承包商**因终止**合同**而遭受的任何利润损失或其他损失和损害的金额。

然后，**工程师**应按照第 3.7 款［*商定或确定*］的规定，商定或确定上述 (a) 和 (b) 段中所述的事项（以及，就第 3.7.3 项的［*时限*］而言，**工程师**根据本款收到**承包商**的详细资料的日期应为按照第 3.7.3 款商定的时限的开始日期）。

The Engineer shall issue a Payment Certificate for the amount so agreed or determined, without the need for the Contractor to submit a Statement.

15.7
Payment after Termination for Employer's Convenience

The Employer shall pay the Contractor the amount certified in the Payment Certificate under Sub-Clause 15.6 [*Valuation after Termination for Employer's Convenience*] within 112 days after the Engineer receives the Contractor's submission under that Sub-Clause.

16 Suspension and Termination by Contractor

16.1
Suspension by Contractor

If:

(a) the Engineer fails to certify in accordance with Sub-Clause 14.6 [*Issue of IPC*];
(b) the Employer fails to provide reasonable evidence in accordance with Sub-Clause 2.4 [*Employer's Financial Arrangements*];
(c) the Employer fails to comply with Sub-Clause 14.7 [*Payment*]; or
(d) the Employer fails to comply with:

　(i) a binding agreement, or final and binding determination under Sub-Clause 3.7 [*Agreement or Determination*]; or
　(ii) a decision of the DAAB under 21.4 [*Obtaining DAAB's Decision*] (whether binding or final and binding)

and such failure constitutes a material breach of the Employer's obligations under the Contract,

the Contractor may, not less than 21 days after giving a Notice to the Employer (which Notice shall state that it is given under this Sub-Clause 16.1), suspend work (or reduce the rate of work) unless and until the Employer has remedied such a default.

This action shall not prejudice the Contractor's entitlements to financing charges under Sub-Clause 14.8 [*Delayed Payment*] and to termination under Sub-Clause 16.2 [*Termination by Contractor*].

If the Employer subsequently remedies the default as described in the above Notice before the Contractor gives a Notice of termination under Sub-Clause 16.2 [*Termination by Contractor*], the Contractor shall resume normal working as soon as is reasonably practicable.

If the Contractor suffers delay and/or incurs Cost as a result of suspending work (or reducing the rate of work) in accordance with this Sub-Clause, the Contractor shall be entitled subject to Sub-Clause 20.2 [*Claims For Payment and/or EOT*] to EOT and/or payment of such Cost Plus Profit.

16.2
Termination by Contractor

Termination of the Contract under this Clause shall not prejudice any other rights of the Contractor, under the Contract or otherwise.

16.2.1
Notice

The Contractor shall be entitled to give a Notice (which shall state that it is given under this Sub-Clause 16.2.1) to the Employer of the Contractor's intention to terminate the Contract or, in the case of sub-paragraph (g)(ii), (h), (i) or (j) below a Notice of termination, if:

工程师应按商定或确定的金额签发**付款证书**，而**承包商**无须提交**报表**。

| 15.7 为雇主便利终止后的付款 | **雇主**应在**工程师**收到**承包商**按照第15.6款［*为雇主便利终止后的估价*］规定提交文件后112天内，向**承包商**支付**付款证书**中确认的金额。 |

16　由承包商暂停和终止

16.1 由承包商暂停

如果：

（a）**工程师**未按照第14.6款［*期中付款证书的签发*］的规定确认发证；
（b）**雇主**未能按照第2.4款［*雇主的资金安排*］的规定提供合理证据；
（c）**雇主**未遵守第14.7款［*付款*］的规定；（或）
（d）**雇主**未遵守：

　　（i）具有约束力的商定，或第3.7款［*商定或确定*］规定的最终并具有约束力的确定；（或）
　　（ii）争端避免/裁决委员会根据第21.4款［*取得争端避免/裁决委员会的决定*］的规定做出的决定（无论是具有约束力的还是最终并具有约束力的）。

以及此类未履约构成了对**合同**规定的**雇主**应承担义务的重大违约，

承包商可以在向**雇主**发出**通知**后不少于21天（**通知**应说明是根据第16.1款的规定发出的），暂停工作（或放慢工作进度），除非并直到**雇主**对此项违约进行了补救。

该行动不应影响**承包商**根据第14.8款［*延误的付款*］的规定获得融资费用，以及按照第16.2款［*由承包商终止*］的规定提出终止的权利。

如果**雇主**在**承包商**根据第16.2款［*由承包商终止*］的规定发出终止**通知**前，按照上述**通知**中的说明纠正了违约行为，则**承包商**应在合理可行的范围内尽快恢复正常工作。

如果因按照本款暂停工作（或放慢工作进度），使**承包商**遭受延误和/或招致增加**费用**，**承包商**应有权根据第20.2款［*付款和/或竣工时间延长的索赔*］的规定，获得**竣工时间**的延长和/或此类**成本加利润**的支付。

16.2 由承包商终止

根据本**条**终止**合同**不应损害**承包商**根据**合同**或有关规定的任何其他权利。

16.2.1 通知

承包商应有权向**雇主**发出**通知**（应说明**通知**是根据第16.2.1款的规定发出的），告知其终止**合同**的意向，或者在以下（g）（ii）、（h）（i）或（j）段的情况下，**通知**终止，如果：

(a) the Contractor does not receive the reasonable evidence within 42 days after giving a Notice under Sub-Clause 16.1 [*Suspension by Contractor*] in respect of a failure to comply with Sub-Clause 2.4 [*Employer's Financial Arrangements*];
(b) the Engineer fails, within 56 days after receiving a Statement and supporting documents, to issue the relevant Payment Certificate;
(c) the Contractor does not receive the amount due under any Payment Certificate within 42 days after the expiry of the time stated in Sub-Clause 14.7 [*Payment*];
(d) the Employer fails to comply with:

   (i) a binding agreement, or final and binding determination under Sub-Clause 3.7 [*Agreement or Determination*]; or
   (ii) a decision of the DAAB under 21.4 [*Obtaining DAAB's Decision*] (whether binding or final and binding)

and such failure constitutes a material breach of the Employer's obligations under the Contract;

(e) the Employer substantially fails to perform, and such failure constitutes a material breach of, the Employer's obligations under the Contract;
(f) the Contractor does not receive a Notice of the Commencement Date under Sub-Clause 8.1 [*Commencement of Works*] within 84 days after receiving the Letter of Acceptance;
(g) the Employer:

   (i) fails to comply with Sub-Clause 1.6 [*Contract Agreement*], or
   (ii) assigns the Contract without the required agreement under Sub-Clause 1.7 [*Assignment*];

(h) a prolonged suspension affects the whole of the Works as described in sub-paragraph (b) of Sub-Clause 8.12 [*Prolonged Suspension*];
(i) the Employer becomes bankrupt or insolvent; goes into liquidation, administration, reorganisation, winding-up or dissolution; becomes subject to the appointment of a liquidator, receiver, administrator, manager or trustee; enters into a composition or arrangement with the Employer's creditors; or any act is done or any event occurs which is analogous to or has a similar effect to any of these acts or events under applicable Laws; or
(j) the Employer is found, based on reasonable evidence, to have engaged in corrupt, fraudulent, collusive or coercive practice at any time in relation to the Works or to the Contract.

16.2.2 Termination

Unless the Employer remedies the matter described in a Notice given under Sub-Clause 16.2.1 [*Notice*] within 14 days of receiving the Notice, the Contractor may by giving a second Notice to the Employer immediately terminate the Contract. The date of termination shall then be the date the Employer receives this second Notice.

However, in the case of sub-paragraph (g)(ii), (h), (i) or (j) of Sub-Clause 16.2.1 [*Notice*], by giving a Notice under Sub-Clause 16.2.1 the Contractor may terminate the Contract immediately and the date of termination shall be the date the Employer receives this Notice.

If the Contractor suffers delay and/or incurs Cost during the above period of 14 days, the Contractor shall be entitled subject to Sub-Clause 20.2 [*Claims For Payment and/or EOT*] to EOT and/or payment of such Cost Plus Profit.

（a）承包商在根据第 16.1 款［由承包商暂停］的规定，就未能遵守第 2.4 款［雇主的资金安排］规定的事项发出通知后 42 天内，仍未收到合理的证据；

（b）工程师未能在收到报表和证明文件后 56 天内签发有关付款证书；

（c）在第 14.7 款［付款］规定的付款时间到期后 42 天内，承包商仍未收到付款证书规定的应付款额；

（d）雇主未能遵守：

　　（i）具有约束力的商定，或第 3.7 款［商定或确定］规定的最终并具有约束力的确定；（或）
　　（ii）争端避免／裁决委员会根据第 21.4 款［取得争端避免／裁决委员会的决定］的规定做出的决定（无论是具有约束力的还是最终且具有约束力的）。

以及此类未履约构成了对合同规定的雇主义务的严重违约；

（e）雇主未履约，以及此类未履约构成了对合同规定的雇主义务的严重违约；

（f）承包商在收到中标函后 84 天内，未收到第 8.1 款［工程的开工］规定的开工日期通知；

（g）雇主：

　　（i）未遵守第 1.6 款［合同协议书］的规定；（或）
　　（ii）在未经第 1.7 款［权益转让］要求的商定的情况下，转让合同。

（h）第 8.12 款［拖长的暂停］（b）段所述的拖长的暂停影响了整个工程；

（i）雇主破产或无力偿债；进入清算、管理、重组、清理或解散等程序；委托了清算人、接管人、管理人、经理或受托人；与雇主的债权人达成和解或安排；或根据适用法律，做出了与这些行为或事件类似，或具有类似作用的任何行为或发生的任何事件；（或）

（j）根据合理的证据，发现雇主从事了与工程或合同有关的腐败、欺诈、串通或胁迫行为。

16.2.2 终止

除非雇主在收到通知 14 天内，对按第 16.2.1 项［通知］规定发出的通知中所述的事项予以改正，否则承包商可以通过向雇主再次发出通知，立即终止合同。终止日期应为雇主收到该通知的日期。

但是，在第 16.2.1 项［通知］（g）（ii）、（h）、（i）或（j）段所述情况下，承包商可根据第 16.2.1 项的规定发出通知，立即终止合同，终止日期应为雇主收到该通知的日期。

如果承包商在上述 14 天内遭受延误和／或招致增加费用，则承包商应有权根据第 20.2 款［付款和／或竣工时间延长的索赔］的规定，获得竣工时间的延长和／或此类成本加利润的支付。

16.3
Contractor's Obligations After Termination

After termination of the Contract under Sub-Clause 15.5 [*Termination for Employer's Convenience*], Sub-Clause 16.2 [*Termination by Contractor*] or Sub-Clause 18.5 [*Optional Termination*], the Contractor shall promptly:

(a) cease all further work, except for such work as may have been instructed by the Engineer for the protection of life or property or for the safety of the Works. If the Contractor incurs Cost as a result of carrying out such instructed work the Contractor shall be entitled subject to Sub-Clause 20.2 [*Claims For Payment and/or EOT*] to be paid such Cost Plus Profit;

(b) deliver to the Engineer all Contractor's Documents, Plant, Materials and other work for which the Contractor has received payment; and

(c) remove all other Goods from the Site, except as necessary for safety, and leave the Site.

16.4
Payment after Termination by Contractor

After termination under Sub-Clause 16.2 [*Termination by Contractor*], the Employer shall promptly:

(a) pay the Contractor in accordance with Sub-Clause 18.5 [*Optional Termination*]; and

(b) subject to the Contractor's compliance with Sub-Clause 20.2 [*Claims For Payment and/or EOT*], pay the Contractor the amount of any loss of profit or other losses and damages suffered by the Contractor as a result of this termination.

17 Care of the Works and Indemnities

17.1
Responsibility for Care of the Works

Unless the Contract is terminated in accordance with these Conditions or otherwise, subject to Sub-Clause 17.2 [*Liability for Care of the Works*] the Contractor shall take full responsibility for the care of the Works, Goods and Contractor's Documents from the Commencement Date until the Date of Completion of the Works, when responsibility for the care of the Works shall pass to the Employer. If a Taking-Over Certificate is issued (or is deemed to be issued) for any Section or Part, responsibility for the care of the Section or Part shall then pass to the Employer.

If the Contract is terminated in accordance with these Conditions or otherwise, the Contractor shall cease to be responsible for the care of the Works from the date of termination.

After responsibility has accordingly passed to the Employer, the Contractor shall take responsibility for the care of any work which is outstanding on the Date of Completion, until this outstanding work has been completed.

If any loss or damage occurs to the Works, Goods or Contractor's Documents, during the period when the Contractor is responsible for their care, from any cause whatsoever except as stated in Sub-Clause 17.2 [*Liability for Care of the Works*], the Contractor shall rectify the loss or damage at the Contractor's risk and cost, so that the Works, Goods or Contractor's Documents (as the case may be) comply with the Contract.

17.2
Liability for Care of the Works

The Contractor shall be liable for any loss or damage caused by the Contractor to the Works, Goods or Contractor's Documents after the issue of a Taking-Over Certificate. The Contractor shall also be liable for any loss

16.3 终止后承包商的义务

根据第 15.5 款 [*为雇主便利的终止*]、第 16.2 款 [*由承包商终止*] 或第 18.5 款 [*自主选择终止*] 的规定终止合同后，**承包商**应立即：

(a) 停止所有进一步的工作，但**工程师**为保护生命或财产或**工程**安全可能指示的工作除外。如果**承包商**因执行该指示的工作而产生**费用**，则**承包商**应有权根据第 20.2 款 [*付款和/或竣工时间延长的索赔*] 的规定，获得此类**成本加利润**的支付；

(b) 将**承包商**已得到付款的所有**承包商文件**、**生产设备**、**材料**和其他**工作**交付给**工程师**；（以及）

(c) 从**现场**运走除安全需要以外的所有其他**货物**，并撤离**现场**。

16.4 由承包商终止后的付款

根据第 16.2 款 [*由承包商终止*] 的规定终止后，**雇主**应立即：

(a) 按照第 18.5 款 [*自主选择终止*] 的规定，向**承包商**付款；（以及）

(b) 在**承包商**遵守第 20.2 款 [*付款和/或竣工时间延长的索赔*] 规定的前提下，付给**承包商**因此项终止而遭受的任何利润损失，或其他损失或损害的款额。

17 工程照管和保障

17.1 工程照管的职责

除非根据本**条件**或其他条款终止**合同**，否则在遵守第 17.2 款 [*工程照管的责任*] 的前提下，**承包商**应自**工程开工日期**起至签发**工程接收证书**止，承担对**工程**、**货物**和**承包商文件**的全部照管责任，直至**工程**的照管责任转移给**雇主**。如果对任何**分项工程**或**部分工程**签发了（或被视为已签发）**接收证书**，则应将照管该**分项工程**或**部分工程**的责任转移给**雇主**。⊖

如果根据本**条件**或其他条款终止了**合同**，则**承包商**应自终止之日起不再负责**工程**的照管。

同样，**承包商**将责任转移给**雇主**后，应负责照管**竣工日期**前尚未完成的扫尾工作，直到完成尚未完成的扫尾工作。

承包商负责照管期间，如果**工程**、**货物**或**承包商文件**发生任何损失或损害，除因第 17.2 款 [*工程照管的责任*] 所述的情况外，**承包商**应弥补损失或损害，风险和费用由其自行承担，使**工程**、**货物**或**承包商文件**（视情况而定）符合**合同**规定。

17.2 工程照管的责任

签发**接收证书**后，**承包商**应对其**工程**、**货物**或**承包商文件**造成的任何损失或损害承担责任。对于签发**接收证书**后发生的，以及在签发**接收**

⊖ 此处中文按照勘误表改正后的英文翻译。——译者注

or damage, which occurs after the issue of a Taking-Over Certificate and which arose from an event which occurred before the issue of this Taking-Over Certificate, for which the Contractor was liable.

The Contractor shall have no liability whatsoever, whether by way of indemnity or otherwise, for loss or damage to the Works, Goods or Contractor's Documents caused by any of the following events (except to the extent that such Works, Goods or Contractor's Documents have been rejected by the Engineer under Sub-Clause 7.5 [*Defects and Rejection*] before the occurrence of any of the following events):

(a) interference, whether temporary or permanent, with any right of way, light, air, water or other easement (other than that resulting from the Contractor's method of construction) which is the unavoidable result of the execution of the Works in accordance with the Contract;

(b) use or occupation by the Employer of any part of the Permanent Works, except as may be specified in the Contract;

(c) fault, error, defect or omission in any element of the design of the Works by the Employer or which may be contained in the Specification and Drawings (and which an experienced contractor exercising due care would not have discovered when examining the Site and the Specification and Drawings before submitting the Tender), other than design carried out by the Contractor in accordance with the Contractor's obligations under the Contract;

(d) any operation of the forces of nature (other than those allocated to the Contractor in the Contract Data) which is Unforeseeable or against which an experienced contractor could not reasonably have been expected to have taken adequate preventative precautions;

(e) any of the events or circumstances listed under sub-paragraphs (a) to (f) of Sub-Clause 18.1 [*Exceptional Events*]; and/or

(f) any act or default of the Employer's Personnel or the Employer's other contractors.

Subject to Sub-Clause 18.4 [*Consequences of an Exceptional Event*], if any of the events described in sub-paragraphs (a) to (f) above occurs and results in damage to the Works, Goods or Contractor's Documents the Contractor shall promptly give a Notice to the Engineer. Thereafter, the Contractor shall rectify any such loss and/or damage that may arise to the extent instructed by the Engineer. Such instruction shall be deemed to have been given under Sub-Clause 13.3.1 [*Variation by Instruction*].

If the loss or damage to the Works or Goods or Contractor's Documents results from a combination of:

(i) any of the events described in sub-paragraphs (a) to (f) above, and
(ii) a cause for which the Contractor is liable,

and the Contractor suffers a delay and/or incurs Cost from rectifying the loss and/or damage, the Contractor shall subject to Sub-Clause 20.2 [*Claims for Payment and/or EOT*] be entitled to a proportion of EOT and/or Cost Plus Profit to the extent that any of the above events have contributed to such delays and/or Cost.

17.3 Intellectual and Industrial Property Rights

In this Sub-Clause, "infringement" means an infringement (or alleged infringement) of any patent, registered design, copyright, trademark, trade name, trade secret or other intellectual or industrial property right relating to the Works; and "claim" means a third party claim (or proceedings pursuing a third party claim) alleging an infringement.

证书之前发生的、**承包商**应承担责任的事件而引起的任何损失或损害，**承包商**也应承担责任。

对于因以下任何事件造成的**工程**、**货物**或**承包商文件**的损失或损害，**承包商**概不承担赔偿或其他有关的责任（除非此类**工程**、**货物**或**承包商文件**已经在以下任何事件发生之前，被**工程师**根据**第7.5款**[*缺陷和拒收*]的规定已经拒收）：

(a) 对任何通行权、光线、空气、水或其他地役权（**承包商**施工方法所致除外）的临时或永久干预，并且是根据**合同**实施**工程**所无法避免的结果；

(b) **雇主**使用或占用**永久工程**的任何部分，但**合同**另有规定的情况除外；

(c) **雇主**负责**工程**设计的要素或可能包含在**规范要求**和**图纸**中的过失、错误、缺陷或遗漏（以及在提交**投标书**之前检查**现场**和**规范要求**和**图纸**时，有经验的承包商在实施应有的照管时不会发现），但不是**承包商**按照**合同**规定的义务所进行的设计；

(d) 不可预见的或有经验的承包商无法合理预期，并采取充分预防措施应对的任何自然力的作用（而不是**合同数据**分配给**承包商**的）；

(e) **第18.1款**[*例外事件*](a)至(f)段所列的任何事件或情况；（和/或）

(f) **雇主人员**或**雇主**的其他承包商的任何作为或不作为。

按照**第18.4款**[*例外事件的后果*]的规定，如果发生以上(a)至(f)段所述的任何事件，并导致**工程**、**货物**或**承包商文件**的损害，**承包商**应立即**通知工程师**。然后，**承包商**应按照**工程师**的指示，纠正可能造成的任何此类损失和/或损害。该指示应视为已根据**第13.3.1项**[*指示变更*]的规定发出。

如果**工程**、**货物**或**承包商文件**的损失或损害是由于以下综合原因造成的：

(i) 上述(a)至(f)段所述的任何事件；（以及）
(ii) **承包商**应承担责任的原因，

以及**承包商**因纠正损失和/或损害而遭受延误和/或招致增加**费用**，**承包商**有权根据**第20.2款**[*付款和/或竣工时间延长的索赔*]的规定，获得**竣工时间**的延长和/或**成本加利润**的相应比例，前提是上述任何事件已导致此类延误和/或**成本**。

17.3 知识产权和工业产权

在本款中，"侵权"系指侵犯（或被指称侵犯）与**工程**有关的任何专利权、已登记的设计、版权、商标、商号商品名称、商业机密或其他知识产权或工业产权；"索赔"系指指控侵权的第三方索赔（或为第三方索赔进行的诉讼）。

Whenever a Party receives a claim but fails to give notice to the other Party of the claim within 28 days of receiving it, the first Party shall be deemed to have waived any right to indemnity under this Sub-Clause.

The Employer shall indemnify and hold the Contractor harmless against and from any claim (including legal fees and expenses) alleging an infringement which is or was:

(a) an unavoidable result of the Contractor's compliance with the Specification and Drawings and/or any Variation; or
(b) a result of any Works being used by the Employer:

　(i) for a purpose other than that indicated by, or reasonably to be inferred from, the Contract, or
　(ii) in conjunction with any thing not supplied by the Contractor, unless such use was disclosed to the Contractor before the Base Date or is stated in the Contract.

The Contractor shall indemnify and hold the Employer harmless against and from any other claim (including legal fees and expenses) alleging an infringement which arises out of or in relation to:

(i) the Contractor's execution of the Works; or
(ii) the use of Contractor's Equipment.

If a Party is entitled to be indemnified under this Sub-Clause, the indemnifying Party may (at the indemnifying Party's cost) assume overall responsibility for negotiating the settlement of the claim, and/or any litigation or arbitration which may arise from it. The other Party shall, at the request and cost of the indemnifying Party, assist in contesting the claim. This other Party (and the Contractor's Personnel or the Employer's Personnel, as the case may be) shall not make any admission which might be prejudicial to the indemnifying Party, unless the indemnifying Party failed to promptly assume overall responsibility for the conduct of any negotiations, litigation or arbitration after being requested to do so by the other Party.

17.4 Indemnities by Contractor

The Contractor shall indemnify and hold harmless the Employer, the Employer's Personnel, and their respective agents, against and from all third party claims, damages, losses and expenses (including legal fees and expenses) in respect of:

(a) bodily injury, sickness, disease or death of any person whatsoever arising out of or in the course of or by reason of the Contractor's execution of the Works, unless attributable to any negligence, wilful act or breach of the Contract by the Employer, the Employer's Personnel, or any of their respective agents; and
(b) damage to or loss of any property, real or personal (other than the Works), to the extent that such damage or loss:

　(i) arises out of or in the course of or by reason of the Contractor's execution of the Works, and
　(ii) is attributable to any negligence, wilful act or breach of the Contract by the Contractor, the Contractor's Personnel, their respective agents, or anyone directly or indirectly employed by any of them.

To the extent, if any, that the Contractor is responsible for the design of part of the Permanent Works under Sub-Clause 4.1 [*Contractor's General Obligations*], and/or any other design under the Contract, the Contractor shall also indemnify and hold harmless the Employer against all acts,

当一方收到索赔但未能在收到后的 28 天内，向另一方发出关于该索赔的**通知**时，该方应被认为已放弃根据**本款**规定的任何赔偿的权利。[一]

雇主应保障并保持**承包商**免受因以下情况提出的指称侵权的任何索赔（包括法律费用和支出）引起的损害，该索赔是或曾经是：

(a) 因**承包商**遵从**规范要求**和**图纸**和／或任何**变更**的要求而造成的不可避免的结果；（或）

(b) 因**雇主**为以下原因使用任何**工程**的结果：

（i） 为了**合同**中指明的，或根据**合同**可合理推断的事项以外的目的；（或）

（ii） 与非**承包商**提供的任何物品联合使用，除非此项使用已在**基准日期**前向**承包商**透露，或在**合同**中有规定。

承包商应保障并保持**雇主**免受由以下事项产生或与之有关的、指控侵权的任何其他索赔（包括法律费用和支出）引起的损害：

（i） **承包商**实施工程；（或）
（ii） 使用**承包商**的**设备**。

如果一方有权根据**本款**规定获得赔偿，赔偿方可（由其承担费用）承担协商解决索赔和／或由此产生的任何诉讼或仲裁的全部责任。另一方应在赔偿方请求并承担费用的情况下，协助对索赔进行抗辩。另一方（及**承包商人员**或**雇主人员**，视情况而定）不应做出可能损害赔偿方的任何承认，除非在另一方提出要求后，赔偿方未能立即对任何谈判、诉讼或仲裁事宜承担全部责任。

17.4 由承包商保障

承包商应保障和保持使**雇主**、**雇主人员**及其各自的代理人免受以下所有第三方索赔、损害赔偿费、损失和开支（包括法律费用和开支）带来的损害：

(a) 任何人员的人身伤害、患病、疾病或死亡，不论是由于**承包商**的施工引起，或在其过程中，或因其原因产生的，**雇主**、**雇主人员**或他们各自的任何代理人的任何疏忽、故意行为，或违反**合同**造成的情况除外；（以及）

(b) 由下列情况造成的对任何财产、不动产或人员（**工程**除外）的损害或损失：

（i） 由于**承包商**实施**工程**引起，或在其过程中，或因其原因产生的；（以及）

（ii） 由**承包商**、**承包商人员**、他们各自的代理人，或由他们中任何人员直接或间接雇用的任何人员的疏忽、故意行为，或违反**合同**造成的。

如果**承包商**按照第 4.1 款 [*承包商的一般义务*] 的规定，负责设计部分**永久工程**，和／或**合同**规定的任何其他设计，则**承包商**还应保障和

[一] 此处中文按照勘误表改正后的英文翻译。——译者注

errors or omissions by the Contractor in carrying out the Contractor's design obligations that result in the Works (or Section or Part or major item of Plant, if any), when completed, not being fit for the purpose(s) for which they are intended under Sub-Clause 4.1 [*Contractor's General Obligations*].

17.5
Indemnities by Employer

The Employer shall indemnify and hold harmless the Contractor, the Contractor's Personnel, and their respective agents, against and from all third party claims, damages, losses and expenses (including legal fees and expenses) in respect of:

(a) bodily injury, sickness, disease or death, or loss of or damage to any property other than the Works, which is attributable to any negligence, wilful act or breach of the Contract by the Employer, the Employer's Personnel, or any of their respective agents; and

(b) damage to or loss of any property, real or personal (other than the Works), to the extent that such damage or loss arises out of any event described under sub-paragraphs (a) to (f) of Sub-Clause 17.2 [*Liability for Care of the Works*].

17.6
Shared Indemnities

The Contractor's liability to indemnify the Employer, under Sub-Clause 17.4 [*Indemnities by Contractor*] and/or under Sub-Clause 17.3 [*Intellectual and Industrial Property Rights*], shall be reduced proportionately to the extent that any event described under sub-paragraphs (a) to (f) of Sub-Clause 17.2 [*Liability for Care of the Works*] may have contributed to the said damage, loss or injury.

Similarly, the Employer's liability to indemnify the Contractor, under Sub-Clause 17.5 [*Indemnities by Employer*], shall be reduced proportionately to the extent that any event for which the Contractor is responsible under Sub-Clause 17.1 [*Responsibility for Care of the Works*] and/or under Sub-Clause 17.3 [*Intellectual and Industrial Property Rights*] may have contributed to the said damage, loss or injury.

18 Exceptional Events

18.1
Exceptional Events

"**Exceptional Event**" means an event or circumstance which:

(i) is beyond a Party's control;
(ii) the Party could not reasonably have provided against before entering into the Contract;
(iii) having arisen, such Party could not reasonably have avoided or overcome; and
(iv) is not substantially attributable to the other Party.

An Exceptional Event may comprise but is not limited to any of the following events or circumstances provided that conditions (i) to (iv) above are satisfied:

(a) war, hostilities (whether war be declared or not), invasion, act of foreign enemies;
(b) rebellion, terrorism, revolution, insurrection, military or usurped power, or civil war;
(c) riot, commotion or disorder by persons other than the Contractor's Personnel and other employees of the Contractor and Subcontractors;

保持雇主免受承包商在履行其工程（或分项工程或部分工程或生产设备的主要部件，如果有）设计义务时的所有行为、错误或遗漏造成的伤害，竣工时，这些工程不符合第 4.1 款［**承包商的一般义务**］规定的预期目的。

17.5
由雇主保障

雇主应保障和保持使**承包商**、**承包商人员**以及他们各自的代理人免受以下所有第三方索赔、损害赔偿费、损失和开支（包括法律费用和开支）带来的损害：

（a） 由于**雇主**、**雇主人员**，或他们各自的任何代理人的任何疏忽、故意行为或违反**合同**造成的除**工程**以外的人身伤害、患病、疾病或死亡；（以及）

（b） 第 17.2 款［*工程照管的责任*］的（a）至（f）段所述情况造成的对任何财产、不动产或人员（工程除外）的损害或损失。

17.6
保障分担

按照第 17.4 款［*由承包商保障*］和/或第 17.3 款［*知识产权和工业产权*］的规定，如果第 17.2 款［*工程照管的责任*］（a）至（f）段所述的任何事件可能是造成上述损害、损失或伤害的原因，则**承包商**对**雇主**的赔偿责任应按比例减少。

同样，按照第 17.5 款［*由雇主保障*］的规定，**承包商**应根据第 17.1 款［*工程照管的职责*］和/或第 17.3 款［*知识产权和工业产权*］的规定负责的任何事件可能是造成上述损害、损失或伤害的原因，则**雇主**对**承包商**的赔偿责任应按比例减少。

18 例外事件

18.1
例外事件

"**例外事件**"系指以下事件或情况：

（i） 一方无法控制的；
（ii） 该方在签订**合同**前，不能对之进行合理预防的；

（iii） 发生后，该方不能合理避免或克服的；（以及）

（iv） 不能主要归责于另一方的。

如果满足上述（i）至（iv）段的条件，则**例外事件**可包括但不限于以下任何事件或情况：

（a） 战争、敌对行动（无论宣战与否）、入侵、外敌行为；

（b） 叛乱、恐怖主义、革命、暴动、军事政变或篡夺政权，或内战；

（c） **承包商人员**和**承包商**及其**分包商**的其他雇员以外的人员的骚动、喧闹或混乱；

(d) strike or lockout not solely involving the Contractor's Personnel and other employees of the Contractor and Subcontractors;

(e) encountering munitions of war, explosive materials, ionising radiation or contamination by radio-activity, except as may be attributable to the Contractor's use of such munitions, explosives, radiation or radio-activity; or

(f) natural catastrophes such as earthquake, tsunami, volcanic activity, hurricane or typhoon.

18.2 Notice of an Exceptional Event

If a Party is or will be prevented from performing any obligations under the Contract due to an Exceptional Event (the "affected Party" in this Clause), then the affected Party shall give a Notice to the other Party of such an Exceptional Event, and shall specify the obligations, the performance of which is or will be prevented (the "prevented obligations" in this Clause).

This Notice shall be given within 14 days after the affected Party became aware, or should have become aware, of the Exceptional Event, and the affected Party shall then be excused performance of the prevented obligations from the date such performance is prevented by the Exceptional Event. If this Notice is received by the other Party after this period of 14 days, the affected Party shall be excused performance of the prevented obligations only from the date on which this Notice is received by the other Party.

Thereafter, the affected Party shall be excused performance of the prevented obligations for so long as such Exceptional Event prevents the affected Party from performing them. Other than performance of the prevented obligations, the affected Party shall not be excused performance of all other obligations under the Contract.

However, the obligations of either Party to make payments due to the other Party under the Contract shall not be excused by an Exceptional Event.

18.3 Duty to Minimise Delay

Each Party shall at all times use all reasonable endeavours to minimise any delay in the performance of the Contract as a result of an Exceptional Event.

If the Exceptional Event has a continuing effect, the affected Party shall give further Notices describing the effect every 28 days after giving the first Notice under Sub-Clause 18.2 [*Notice of an Exceptional Event*].

The affected Party shall immediately give a Notice to the other Party when the affected Party ceases to be affected by the Exceptional Event. If the affected Party fails to do so, the other Party may give a Notice to the affected Party stating that the other Party considers that the affected Party's performance is no longer prevented by the Exceptional Event, with reasons.

18.4 Consequences of an Exceptional Event

If the Contractor is the affected Party and suffers delay and/or incurs Cost by reason of the Exceptional Event of which he/she gave a Notice under Sub-Clause 18.2 [*Notice of an Exceptional Event*], the Contractor shall be entitled subject to Sub-Clause 20.2 [*Claims For Payment and/or EOT*] to:

(a) EOT; and/or

(b) if the Exceptional Event is of the kind described in sub-paragraphs (a) to (e) of Sub-Clause 18.1 [*Exceptional Events*] and, in the case of sub-paragraphs (b) to (e) of that Sub-Clause, occurs in the Country, payment of such Cost.

(d) 不仅仅涉及**承包商人员**和**承包商**及其**分包商**的其他雇员的罢工或停工;

(e) 遭遇战争军火、爆炸物资、电离辐射或放射性污染,但可能因**承包商**使用此类军火、炸药、辐射或放射性引起的除外;(或)

(f) 自然灾害,如地震、海啸、火山活动、飓风或台风。

18.2 例外事件的通知

如果一方(本**条**中"受影响方")因**例外事件**已或将无法履行根据合同规定的任何义务,则受影响方应向另一方发出该**例外事件**的**通知**,并应明确说明已或将受到阻止履行的各项义务(本**条**中"被阻止的义务")。

该**通知**应在受影响方意识到或本应意识到**例外事件**后 14 天内发出。然后,受影响方应自**例外事件**阻止履行义务之日起,免除其履行被阻止的义务。如果另一方在 14 天的期限后收到该**通知**,受影响方应仅在从另一方收到**通知**之日起,履行被阻止的义务。

此后,只要该**例外事件**阻止受影响方履行义务,受影响方就应被免除履行被阻止的义务。除履行被阻止的义务外,受影响方不得免除履行根据本合同规定的所有其他义务。

但是,任一方根据**合同**向另一方支付款项的义务不得因任何**例外事件**而免除。

18.3 将延误减至最小的义务

各方都应在任何时间尽所有合理的努力,使**例外事件**对履行合同造成的任何延误减至最小。

如果**例外事件**具有持续影响,则受影响方应在根据第 18.2 款 [*例外事件的通知*] 发出第一次**通知**后,每 28 天发出进一步**通知**,说明受影响的情况。

受影响方不再受**例外事件**影响时,应立即向另一方发出**通知**。如果其未能发出**通知**,则另一方可以向受影响方发出**通知**,阐述理由,说明另一方认为该**例外事件**不再阻止受影响方履行义务。

18.4 例外事件的后果

如果**承包商**是受影响方,且由于其根据第 18.2 款 [*例外事件的通知*] 的规定发出**通知**的**例外事件**,而遭受延误和/或招致增加**费用**,**承包商**应有权根据第 20.2 款 [*付款和/或竣工时间延长的索赔*] 的规定要求:

(a) 竣工时间的延长;和/或
(b) 如果该**例外事件**属于第 18.1 款 [*例外事件*](a) 至 (e) 段所述的类型,且该款 (b) 至 (e) 段所述**例外事件**发生在工程所在国,获得此类费用的支付。

18.5
Optional Termination

If the execution of substantially all the Works in progress is prevented for a continuous period of 84 days by reason of an Exceptional Event of which Notice has been given under Sub-Clause 18.2 [*Notice of an Exceptional Event*], or for multiple periods which total more than 140 days due to the same Exceptional Event, then either Party may give to the other Party a Notice of termination of the Contract.

In this event, the date of termination shall be the date 7 days after the Notice is received by the other Party, and the Contractor shall proceed in accordance with Sub-Clause 16.3 [*Contractor's Obligations After Termination*].

After the date of termination the Contractor shall, as soon as practicable, submit detailed supporting particulars (as reasonably required by the Engineer) of the value of the work done, which shall include:

(a) the amounts payable for any work carried out for which a price is stated in the Contract;

(b) the Cost of Plant and Materials ordered for the Works which have been delivered to the Contractor, or of which the Contractor is liable to accept delivery. This Plant and Materials shall become the property of (and be at the risk of) the Employer when paid for by the Employer, and the Contractor shall place the same at the Employer's disposal;

(c) any other Cost or liability which in the circumstances was reasonably incurred by the Contractor in the expectation of completing the Works;

(d) the Cost of removal of Temporary Works and Contractor's Equipment from the Site and the return of these items to the Contractor's place of business in the Contractor's country (or to any other destination(s) at no greater cost); and

(e) the Cost of repatriation of the Contractor's staff and labour employed wholly in connection with the Works at the date of termination.

The Engineer shall then proceed under Sub-Clause 3.7 [*Agreement or Determination*] to agree or determine the value of work done (and, for the purpose of Sub-Clause 3.7.3 [*Time limits*], the date the Engineer receives the Contractor's particulars under this Sub-Clause shall be the date of commencement of the time limit for agreement under Sub-Clause 3.7.3).

The Engineer shall issue a Payment Certificate, under Sub-Clause 14.6 [*Issue of IPC*], for the amount so agreed or determined, without the need for the Contractor to submit a Statement.

18.6
Release from Performance under the Law

In addition to any other provision of this Clause, if any event arises outside the control of the Parties (including, but not limited to, an Exceptional Event) which:

(a) makes it impossible or unlawful for either Party or both Parties to fulfil their contractual obligations; or

(b) under the law governing the Contract, entitles the Parties to be released from further performance of the Contract,

and if the Parties are unable to agree on an amendment to the Contract that would permit the continued performance of the Contract, then after either Party gives a Notice to the other Party of such event:

(i) the Parties shall be discharged from further performance, and without prejudice to the rights of either Party in respect of any previous breach of the Contract; and

18.5 自主选择终止	如果由于已根据**第 18.2 款**［*例外事件的通知*］的规定发出**通知**的**例外事件**，导致进行中的全部**工程**的实施连续 84 天受阻，或由于同一**例外事件**导致多个时段累计受阻超过 140 天，任一方可向另一方发出**通知**，终止合同。

在此情况下，终止日期应为另一方收到**通知** 7 天后的日期，**承包商**应按照**第 16.3 款**［*终止后承包商的义务*］的规定继续进行。

终止日期后，**承包商**应在切实可行的范围内，尽快提交（**工程师**合理要求的）有关已完成工作价值的详细证明资料，其中包括：

(a) 已完成的、**合同**中有价格规定的任何工作的应付款额；

(b) 为**工程**订购的、已交付给**承包商**或**承包商**有责任接受交付的**生产设备和材料**的**费用**。当**雇主**支付上述费用后，此项**生产设备和材料**应成为**雇主**的财产（风险也由其承担），**承包商**应将其交由**雇主**处理；

(c) 在**承包商**原预期要完成**工程**的情况下，合理产生的任何其他**费用**或债务；

(d) 将**临时工程**和**承包商设备**撤离**现场**，并运回**承包商**本国工作地点的**费用**（或运往任何其他目的地，但其费用不得更高）；（以及）

(e) 将在终止日期时完全为**工程**雇用的**承包商**的员工送返回国的**费用**。

然后，**工程师**应按照**第 3.7 款**［*商定或确定*］的规定，商定或确定已完成工作的价值（就**第 3.7.3 项**［*时限*］而言，**工程师**根据本项规定收到**承包商**的详细资料的日期，应为根据**第 3.7.3 项**商定的时限的开工日期）。

工程师应根据**第 14.6 款**［*期中付款证书的签发*］的规定，就商定或确定的金额签发**付款证书**，而无须**承包商**提交**报表**。 |
| 18.6 依法解除履约 | 除本**条**的任何其他规定外，如果发生双方不能控制的任何事件（包括但不限于**例外事件**）：

(a) 使任一方或双方完成其合同义务成为不可能的或非法的；（或）

(b) 根据**合同**管辖法律规定，双方有权解除进一步履行**合同**的义务。

如果双方不能就允许继续履行**合同**的**合同**修改案达成一致，则在任一方将此类事件**通知**另一方后：

(i) 在不损害任一方任何先前违反**合同**的权利的情况下，双方应解除进一步履约的义务；（以及） |

(ii) the amount payable by the Employer to the Contractor shall be the same as would have been payable under Sub-Clause 18.5 [*Optional Termination*], and such amount shall be certified by the Engineer, as if the Contract had been terminated under that Sub-Clause.

Insurance 19

19.1 General Requirements

Without limiting either Party's obligations or responsibilities under the Contract, the Contractor shall effect and maintain all insurances for which the Contractor is responsible with insurers and in terms, both of which shall be subject to consent by the Employer. These terms shall be consistent with terms (if any) agreed by both Parties before the date of the Letter of Acceptance.

The insurances required to be provided under this Clause are the minimum required by the Employer, and the Contractor may, at the Contractor's own cost, add such other insurances that the Contractor may deem prudent.

Whenever required by the Employer, the Contractor shall produce the insurance policies which the Contractor is required to effect under the Contract. As each premium is paid, the Contractor shall promptly submit either a copy of each receipt of payment to the Employer (with a copy to the Engineer), or confirmation from the insurers that the premium has been paid.

If the Contractor fails to effect and keep in force any of the insurances required under Sub-Clause 19.2 [*Insurances to be provided by the Contractor*] then, and in any such case, the Employer may effect and keep in force such insurances and pay any premium as may be necessary and recover the same from the Contractor from time to time by deducting the amount(s) so paid from any moneys due to the Contractor or otherwise recover the same as a debt from the Contractor. The provisions of Clause 20 [*Employer's and Contractor's Claims*] shall not apply to this Sub-Clause.

If either the Contractor or the Employer fails to comply with any condition of the insurances effected under the Contract, the Party so failing to comply shall indemnify the other Party against all direct losses and claims (including legal fees and expenses) arising from such failure.

The Contractor shall also be responsible for the following:

(a) notifying the insurers of any changes in the nature, extent or programme for the execution of the Works; and
(b) the adequacy and validity of the insurances in accordance with the Contract at all times during the performance of the Contract.

The permitted deductible limits allowed in any policy shall not exceed the amounts stated in the Contract Data (if not stated, the amounts agreed with the Employer).

Where there is a shared liability the loss shall be borne by each Party in proportion to each Party's liability, provided the non-recovery from insurers has not been caused by a breach of this Clause by the Contractor or the Employer. In the event that non-recovery from insurers has been caused by such a breach, the defaulting Party shall bear the loss suffered.

（ii）雇主应付给**承包商**的款额，应与**第 18.5 款**［*自主选择终止*］规定的应付款额相同，并且该款额应由**工程师**确认，如同**合同**已根据本款终止。

19 保险

19.1 一般要求

在不限制任一方的**合同**规定义务或责任的情况下，**承包商**应在保险人处按条款办理并保持其负责的所有保险项目，保险人及条款应征得**雇主**的同意。这些条款应与双方在**中标函**日期前商定的条款（如果有）一致。

此**款**规定的应提供的保险是**雇主**的最低要求，**承包商**可自费增加其认为必要的保险。

无论任何时候，只要**雇主**提出要求，**承包商**应投保**合同**要求的其应投保的保险。每一笔保费支付后，**承包商**应立即将每一份付款收据的复印件交给**雇主**（一份给**工程师**），或提供保险人已收到保费的确认函。

如果**承包商**未能按照**第 19.2 款**［*由承包商提供的保险*］规定的要求办理保险，并使之保持有效，则在任何情况下，**雇主**均可办理此类保险并使之保持有效，并支付任何有必要支付的保险费，并可以随时在给**承包商**的任何到期应付款中扣除上述金额，或将上述金额作为**承包商**的债务。**第 20 条**［*雇主和承包商的索赔*］的规定应不适用本款。

如果**承包商**或**雇主**中的任一方未能遵守**合同**规定的保险的任何条件，未遵守**方**应赔偿另一方因这种不当而造成的所有直接损失和索赔（包括法律费用和开支）。

承包商还应对以下事项负责：

（a）通知保险人**工程**实施的性质、范围或程序的任何改变；

（b）履行**合同**期间的任何时候，根据合同规定保证保险的充分性和有效性。

任何保单中允许的可扣减额不应超过**合同数据**中规定的金额（如未规定，则按与**雇主**商定的金额）。

如果损失未能从保险人处获得补偿，且未获得补偿不是因**承包商**或**雇主**违反此**条**规定造成的，对于共同责任的损失，应由各方按比例分担。如果未能从保险人处获得补偿是由于此类违约行为造成的，则违约方应承担所遭受的损失。

19.2 Insurance to be provided by the Contractor

The Contractor shall provide the following insurances:

19.2.1 The Works

The Contractor shall insure and keep insured in the joint names of the Contractor and the Employer from the Commencement Date until the date of the issue of the Taking-Over Certificate for the Works:

(a) the Works and Contractor's Documents, together with Materials and Plant for incorporation in the Works, for their full replacement value. The insurance cover shall extend to include loss and damage of any part of the Works as a consequence of failure of elements defectively designed or constructed with defective material or workmanship; and

(b) an additional amount of fifteen percent (15%) of such replacement value (or such other amount as may be specified in the Contract Data) to cover any additional costs incidental to the rectification of loss or damage, including professional fees and the cost of demolition and removal of debris.

The insurance cover shall cover the Employer and the Contractor against all loss or damage from whatever cause arising until the issue of the Taking-Over Certificate for the Works. Thereafter, the insurance shall continue until the date of the issue of the Performance Certificate in respect of any incomplete work for loss or damage arising from any cause occurring before the date of the issue of the Taking-Over Certificate for the Works, and for any loss or damage occasioned by the Contractor in the course of any operation carried out by the Contractor for the purpose of complying with the Contractor's obligations under Clause 11 [*Defects after Taking Over*] and Clause 12 [*Tests after Completion*].

However, the insurance cover provided by the Contractor for the Works may exclude any of the following:

(i) the cost of making good any part of the Works which is defective (including defective material and workmanship) or otherwise does not comply with the Contract, provided that it does not exclude the cost of making good any loss or damage to any other part of the Works attributable to such defect or non-compliance;

(ii) indirect or consequential loss or damage including any reductions in the Contract Price for delay;

(iii) wear and tear, shortages and pilferages; and

(iv) unless otherwise stated in the Contract Data, the risks arising from Exceptional Events.

19.2.2 Goods

The Contractor shall insure, in the joint names of the Contractor and the Employer, the Goods and other things brought to Site by the Contractor to the extent specified and/or amount stated in the Contract Data (if not specified or stated, for their full replacement value including delivery to Site).

The Contractor shall maintain this insurance from the time the Goods are delivered to the Site until they are no longer required for the Works.

19.2.3 Liability for breach of professional duty

To the extent, if any, that the Contractor is responsible for the design of part of the Permanent Works under Sub-Clause 4.1 [*Contractor's General Obligations*], and/or any other design under the Contract, and consistent with the indemnities specified in Clause 17 [*Care of the Works and Indemnities*]:

19.2 由承包商提供的保险		承包商应提供以下保险：

19.2.1
工程

承包商应从开工日期至工程接收证书的签发日期内，以雇主和承包商的共同名义投保并保持保险：

（a） 工程和承包商文件，连同工程所含的材料和生产设备，保险金额为其全部重置价值。保险范围应扩大到由于设计缺陷或使用有缺陷的材料或工艺导致的部件故障而造成的工程任何部分的损失或损害；（以及）

（b） 此重置价值增加15%的附加金额（或合同数据中规定的此类其他金额），以涵盖修复损失或损害的额外费用，包括专业费用以及拆除、移除废弃物的费用。

保险范围应涵盖雇主和承包商，以防止在签发工程接收证书之前因任何原因造成的所有损失或损害。此后，对于因签发工程接收证书之前发生的任何原因造成的任何损失或损害，任何未完成的工作，以及承包商为履行第11条［接收后的缺陷］规定的承包商义务在承包商作业过程中造成的任何损失或损坏，保险应持续到签发履约证书之日。㊀

但是，承包商为工程提供的保险范围可不包括以下任何一项：

（i） 修复有缺陷的（包括有缺陷的材料和工艺）或不符合合同规定的工程任何部分的费用，但前提是不排除修复由于上述缺陷或不合规导致工程其他任何部分的损失或损害的费用；

（ii） 间接或结果性损失或损害，包括因延误而扣减的合同价格；

（iii） 磨损、短缺和盗窃；（以及）
（iv） 例外事件引起的风险，除非合同数据中另有说明。

19.2.2
货物

承包商应以承包商和雇主的共同名义，对合同数据中规定的和/或注明金额的（如果未规定或未注明，包括交付至现场的整个重置价值）、承包商交运到现场的货物和其他物品投保。

承包商应自货物交付至现场起一直保持该保险，直至工程不再需要。

19.2.3
违反职业职责的责任

如果承包商根据第4.1款［承包商的一般义务］的规定，负责部分永久工程的设计，和/或合同规定的其他设计，并与第17条［工程照管和保障］规定的赔偿相一致，则：

㊀ 此处中文按照勘误表改正后的英文翻译。——译者注

(a) the Contractor shall effect and maintain professional indemnity insurance against liability arising out of any act, error or omission by the Contractor in carrying out the Contractor's design obligations in an amount not less than that stated in the Contract Data (if not stated, the amount agreed with the Employer); and

(b) if stated in the Contract Data, such professional indemnity insurance shall also indemnify the Contractor against liability arising out of any act, error or omission by the Contractor in carrying out the Contractor's design obligations under the Contract that results in the Works (or Section or Part or major item of Plant, if any), when completed, not being fit for the purpose(s) for which they are intended under Sub-Clause 4.1 [*Contractor's General Obligations*].

The Contractor shall maintain this insurance for the period specified in the Contract Data.

19.2.4 Injury to persons and damage to property

The Contractor shall insure, in the joint names of the Contractor and the Employer, against liabilities for death or injury to any person, or loss of or damage to any property (other than the Works) arising out of the performance of the Contract and occurring before the issue of the Performance Certificate, other than loss or damage caused by an Exceptional Event.

The insurance policy shall include a cross liability clause such that the insurance shall apply to the Contractor and the Employer as separate insureds.

Such insurance shall be effected before the Contractor begins any work on the Site and shall remain in force until the issue of the Performance Certificate and shall be for not less than the amount stated in the Contract Data (if not stated, the amount agreed with the Employer).

19.2.5 Injury to employees

The Contractor shall effect and maintain insurance against liability for claims, damages, losses and expenses (including legal fees and expenses) arising out of the execution of the Works in respect of injury, sickness, disease or death of any person employed by the Contractor or any of the Contractor's other personnel.

The Employer and the Engineer shall also be indemnified under the policy of insurance, except that this insurance may exclude losses and claims to the extent that they arise from any act or neglect of the Employer or of the Employer's Personnel.

The insurance shall be maintained in full force and effect during the whole time that the Contractor's Personnel are assisting in the execution of the Works. For any person employed by a Subcontractor, the insurance may be effected by the Subcontractor, but the Contractor shall be responsible for the Subcontractor's compliance with this Sub-Clause.

19.2.6 Other insurances required by Laws and by local practice

The Contractor shall provide all other insurances required by the Laws of the countries where (any part of) the Works are being carried out, at the Contractor's own cost.

Other insurances required by local practice (if any) shall be detailed in the Contract Data and the Contractor shall provide such insurances in compliance with the details given, at the Contractor's own cost.

(a) 承包商应对其在履行承包商的设计义务时，因任何行为、错误或遗漏而产生的责任投保并保持职业责任保险，其金额不少于合同数据中规定的金额（如未说明，按与雇主商定的金额投保）；（以及）

(b) 如果合同数据中规定了职业责任保险，还应赔偿承包商因其履行工程（或生产设备的分项工程、部分工程或主要事项）合同规定的承包商的设计义务中的任何行为、错误或遗漏而产生的责任，竣工时，未能达到第4.1款［承包商的一般义务］规定的预期目的。

承包商应在合同数据规定的期限内，保持此保险。

19.2.4 人员伤害和财产损害

承包商应以雇主和承包商的共同名义，对因履行合同而产生的，以及在签发履约证书之前发生的，任何人员的死亡或伤害，或任何财产（工程除外）的损失或损害等责任投保，由例外事件导致的损失或损害除外。

保单应包括交叉责任条款，以便保险适用于作为单独被保险人的承包商和雇主。

此类保险应在承包商开始在现场进行任何工作之前投保，并直至签发履约证书为止保持有效，且保险金额不得少于合同数据中规定的金额（如未规定，则应按与雇主商定的金额计算）。

19.2.5 雇员的人身伤害

承包商有责任对其雇用的任何人员或任何其他承包商人员在实施工程过程中发生的伤害、患病、疾病或死亡而引起的索赔、损害、损失和开支（包括法律费用和开支）的责任办理并维持保险。

除该保险可不包括由雇主或雇主人员的任何行为或疏忽引起的损失和索赔的情况以外，雇主和工程师也应由该项保单得到保障。

此保险应在承包商人员参加工程实施的整个期间保持全面实施和有效。对于分包商雇用的任何人员，此类保险可以由分包商投保，但承包商应对其符合本款规定的保险负责。

19.2.6 法律和当地惯例要求的其他保险

承包商应提供工程（或工程的任何部分）所在国家的法律要求的所有其他保险，费用由承包商自行承担。

当地惯例要求的其他保险（如果有）应在合同数据中详细说明，承包商应根据给出的详细信息提供此类保险，费用由承包商承担。

20 Employer's and Contractor's Claims

20.1 Claims

A Claim may arise:

(a) if the Employer considers that the Employer is entitled to any additional payment from the Contractor (or reduction in the Contract Price) and/or to an extension of the DNP;

(b) if the Contractor considers that the Contractor is entitled to any additional payment from the Employer and/or to EOT; or

(c) if either Party considers that he/she is entitled to another entitlement or relief against the other Party. Such other entitlement or relief may be of any kind whatsoever (including in connection with any certificate, determination, instruction, Notice, opinion or valuation of the Engineer) except to the extent that it involves any entitlement referred to in sub-paragraphs (a) and/or (b) above.

In the case of a Claim under sub-paragraph (a) or (b) above, Sub-Clause 20.2 [*Claims For Payment and/or EOT*] shall apply.

In the case of a Claim under sub-paragraph (c) above, where the other Party or the Engineer has disagreed with the requested entitlement or relief (or is deemed to have disagreed if he/she does not respond within a reasonable time), a Dispute shall not be deemed to have arisen but the claiming Party may, by giving a Notice refer the Claim to the Engineer and Sub-Clause 3.7 [*Agreement or Determination*] shall apply. This Notice shall be given as soon as practicable after the claiming Party becomes aware of the disagreement (or deemed disagreement) and include details of the claiming Party's case and the other Party's or the Engineer's disagreement (or deemed disagreement).

20.2 Claims For Payment and/or EOT

If either Party considers that he/she is entitled to any additional payment by the other Party (or, in the case of the Employer, a reduction in the Contract Price) and/or to EOT (in the case of the Contractor) or an extension of the DNP (in the case of the Employer) under any Clause of these Conditions or otherwise in connection with the Contract, the following Claim procedure shall apply:

20.2.1 Notice of Claim

The claiming Party shall give a Notice to the Engineer, describing the event or circumstance giving rise to the cost, loss, delay or extension of DNP for which the Claim is made as soon as practicable, and no later than 28 days after the claiming Party became aware, or should have become aware, of the event or circumstance (the "Notice of Claim" in these Conditions).

If the claiming Party fails to give a Notice of Claim within this period of 28 days, the claiming Party shall not be entitled to any additional payment, the Contract Price shall not be reduced (in the case of the Employer as the claiming Party), the Time for Completion (in the case of the Contractor as the claiming Party) or the DNP (in the case of the Employer as the claiming Party) shall not be extended, and the other Party shall be discharged from any liability in connection with the event or circumstance giving rise to the Claim.

20.2.2 Engineer's initial response

If the Engineer considers that the claiming Party has failed to give the Notice of Claim within the period of 28 days under Sub-Clause 20.2.1 [*Notice of Claim*] the Engineer shall, within 14 days after receiving the Notice of Claim, give a Notice to the claiming Party accordingly (with reasons).

20 雇主和承包商的索赔

20.1 索赔

以下情况下可能会产生**索赔**：

(a) 如果**雇主**认为其有权从**承包商**处获得任何额外付款（或降低**合同价格**）和/或延长**缺陷通知期限**；

(b) 如果**承包商**认为其有权从**雇主**处获得任何额外付款，和/或**竣工时间**的延长；（或）

(c) 任一方认为自己有权获得另一方的另一种权利或责任的免除。此类其他应享权利或责任的免除可以是任何形式的（包括与**工程师**的任何证书、确定、指示、**通知**、意见或估价有关的），但以上（a）段和/或（b）段所涉及的任何应享权利除外。

对于根据上述（a）或（b）段提出的索赔，**第 20.2 款**［*付款和/或竣工时间延长的索赔*］应适用。

对于根据上述（c）段提出的**索赔**，如果另一方或**工程师**不同意所要求的应享权利或免除的责任（或，如果其在合理时间内未做出答复，则被视为不同意），不应认为已发生**争端**，但索赔方可以通过发出**通知**将**索赔**移交给**工程师**，且**第 3.7 款**［*商定或确定*］应适用。索赔方意识到不同意（或被视为不同意）后，应在切实可行的范围内尽快发出**通知**，并应包含提出索赔方的案件以及另一方或**工程师**的不同意（或被视为不同意）的详细信息。

20.2 付款和/或竣工时间延长的索赔

如果任一方认为其有权获得另一方的任何额外付款（或，如果是**雇主**降低**合同价格**）和/或**竣工时间**的延长（如果是**承包商**），或在本条件的任何条款下或与合同有关的其他情况下，**缺陷通知期限**的延长（如果是**雇主**），以下**索赔**程序应适用：

20.2.1 索赔通知

索赔方意识到或应已意识到事件或情况（本**条件**中的**索赔通知**）后的 28 天内，应尽快向**工程师**发出**通知**，描述该事件或情况造成的费用增加、损失、延误或**缺陷通知期限**的延长等。

如果索赔方在此 28 天内未发出**索赔通知**，则索赔方无权获得任何额外付款，**合同价格**也不得降低（如果**雇主**是索赔方），**竣工时间**（如果**承包商**为索赔方）或**缺陷通知期限**（如果**雇主**是索赔方）不得延长，并且应免除另一方与造成**索赔**的事件或情况有关的任何责任。

20.2.2 工程师的初步响应

如果**工程师**认为索赔方未能在**第 20.2.1 项**［*索赔通知*］规定的 28 天内发出**索赔通知**，则**工程师**应在收到**索赔通知**后 14 天内通知索赔方（并说明理由）。

If the Engineer does not give such a Notice within this period of 14 days, the Notice of Claim shall be deemed to be a valid Notice. If the other Party disagrees with such deemed valid Notice of Claim the other Party shall give a Notice to the Engineer which shall include details of the disagreement. Thereafter, the agreement or determination of the Claim under Sub-Clause 20.2.5 [*Agreement or determination of the Claim*] shall include a review by the Engineer of such disagreement.

If the claiming Party receives a Notice from the Engineer under this Sub-Clause and disagrees with the Engineer or considers there are circumstances which justify late submission of the Notice of Claim, the claiming Party shall include in its fully detailed Claim under Sub-Clause 20.2.4 [*Fully detailed claim*] details of such disagreement or why such late submission is justified (as the case may be).

20.2.3 Contemporary records

In this Sub-Clause 20.2, "contemporary records" means records that are prepared or generated at the same time, or immediately after, the event or circumstance giving rise to the Claim.

The claiming Party shall keep such contemporary records as may be necessary to substantiate the Claim.

Without admitting the Employer's liability, the Engineer may monitor the Contractor's contemporary records and/or instruct the Contractor to keep additional contemporary records. The Contractor shall permit the Engineer to inspect all these records during normal working hours (or at other times agreed by the Contractor), and shall if instructed submit copies to the Engineer. Such monitoring, inspection or instruction (if any) by the Engineer shall not imply acceptance of the accuracy or completeness of the Contractor's contemporary records.

20.2.4 Fully detailed Claim

In this Sub-Clause 20.2, "fully detailed Claim" means a submission which includes:

(a) a detailed description of the event or circumstance giving rise to the Claim;
(b) a statement of the contractual and/or other legal basis of the Claim;
(c) all contemporary records on which the claiming Party relies; and
(d) detailed supporting particulars of the amount of additional payment claimed (or amount of reduction of the Contract Price in the case of the Employer as the claiming Party), and/or EOT claimed (in the case of the Contractor) or extension of the DNP claimed (in the case of the Employer).

Within either:

(i) 84 days after the claiming Party became aware, or should have become aware, of the event or circumstance giving rise to the Claim, or
(ii) such other period (if any) as may be proposed by the claiming Party and agreed by the Engineer

the claiming Party shall submit to the Engineer a fully detailed Claim.

If within this time limit the claiming Party fails to submit the statement under sub-paragraph (b) above, the Notice of Claim shall be deemed to have lapsed, it shall no longer be considered as a valid Notice, and the Engineer shall, within 14 days after this time limit has expired, give a Notice to the claiming Party accordingly.

如果**工程师**在 14 天内未能发出此类**通知**，**索赔通知**应被视为有效**通知**。如果另一方不同意此类视为有效的**索赔通知**，则应向**工程师**发出**通知**，其中应包含不同意的详细说明。此后，根据第 20.2.5 项［*索赔的商定或确定*］规定的**索赔**的商定或确定，应包括**工程师**对此类不同意的审核。

如果索赔方根据本**款**规定收到了**工程师**发出的**通知**，并且不同意该**工程师**，或认为在某些情况下有理由延迟提交**索赔通知**，索赔方应在第 20.2.4 项［*充分详细的索赔*］规定的充分详细**索赔**中包括有关此类不同意的详细信息，或说明延迟提交的合理理由（视情况而定）。

20.2.3
同期记录

在第 **20.2 款**中，"同期记录"系指在引起**索赔**的事件或情况的同时或之后立即准备或生成的记录。

索赔方应保留可能用于支持**索赔**所必需的同期记录。

在不承认**雇主**责任的情况下，**工程师**可以监督**承包商**的同期记录和／或指示**承包商**保留其他同期记录。**承包商**应准许**工程师**在正常工作时间内（或在承包商同意的其他时间）检验所有这些记录，并应按指示将副本提交给**工程师**。**工程师**进行的此类监督、检验或指示（如果有），并不表示接受**承包商**同期记录的准确性或完整性。

20.2.4
充分详细的索赔

在第 **20.2 款**中，"充分详细的**索赔**"系指提交的文件，包括：

（a）对引起**索赔**的事件或情况的详细描述；
（b）有关**索赔**的合同和／或其他法律依据的声明；
（c）索赔方所依赖的所有同期记录；（以及）
（d）索赔的额外付款金额的详细证明资料（或如果**雇主**是索赔方，则为**合同价格**的减少额），和／或（如果是**承包商**）要求的**竣工时间**的延长或（如果是**雇主**）要求的**缺陷通知期限**的延长。

在以下任何情况下：

（i）索赔方意识到或将意识到引起**索赔**的事件或情况的 84 天后；（或）

（ii）索赔方提议并经**工程师**商定的此类其他期限（如果有）。

索赔方均应向**工程师**提交充分详细的**索赔**。

如果在此时限内索赔方未能按照上述（b）段提交声明，**索赔通知**应视为已失效，不再视为有效**通知**，**工程师**应在该时限到期后 14 天内，向索赔方发出相应**通知**。

If the Engineer does not give such a Notice within this period of 14 days, the Notice of Claim shall be deemed to be a valid Notice. If the other Party disagrees with such deemed valid Notice of Claim the other Party shall give a Notice to the Engineer which shall include details of the disagreement. Thereafter, the agreement or determination of the Claim under Sub-Clause 20.2.5 [*Agreement or determination of the Claim*] shall include a review by the Engineer of such disagreement.

If the claiming Party receives a Notice from the Engineer under this Sub-Clause 20.2.4 and if the claiming Party disagrees with such Notice or considers there are circumstances which justify late submission of the statement under sub-paragraph (b) above, the fully detailed claim shall include details of the claiming Party's disagreement or why such late submission is justified (as the case may be).

If the event or circumstance giving rise to the Claim has a continuing effect, Sub-Clause 20.2.6 [*Claims of continuing effect*] shall apply.

20.2.5 Agreement or determination of the Claim

After receiving a fully detailed Claim either under Sub-Clause 20.2.4 [*Fully detailed Claim*], or an interim or final fully detailed Claim (as the case may be) under Sub-Clause 20.2.6 [*Claims of continuing effect*], the Engineer shall proceed under Sub-Clause 3.7 [*Agreement or Determination*] to agree or determine:

(a) the additional payment (if any) to which the claiming Party is entitled or the reduction of the Contract Price (in the case of the Employer as the claiming Party); and/or

(b) the extension (if any) of the Time for Completion (before or after its expiry) under Sub-Clause 8.5 [*Extension of Time for Completion*] (in the case of the Contractor as the claiming Party), or the extension (if any) of the DNP (before its expiry) under Sub-Clause 11.3 [*Extension of Defects Notification Period*] (in the case of the Employer as the claiming Party),

to which the claiming Party is entitled under the Contract.

If the Engineer has given a Notice under Sub-Clause 20.2.2 [*Engineer's initial response*] and/or under Sub-Clause 20.2.4 [*Fully detailed Claim*], the Claim shall nevertheless be agreed or determined in accordance with this Sub-Clause 20.2.5. The agreement or determination of the Claim shall include whether or not the Notice of Claim shall be treated as a valid Notice taking account of the details (if any) included in the fully detailed claim of the claiming Party's disagreement with such Notice(s) or why late submission is justified (as the case may be). The circumstances which may be taken into account (but shall not be binding) may include:

- whether or to what extent the other Party would be prejudiced by acceptance of the late submission;
- in the case of the time limit under Sub-Clause 20.2.1 [*Notice of Claim*], any evidence of the other Party's prior knowledge of the event or circumstance giving rise to the Claim, which the claiming Party may include in its supporting particulars; and/or
- in the case of the time limit under Sub-Clause 20.2.4 [*Fully detailed Claim*], any evidence of the other Party's prior knowledge of the contractual and/or other legal basis of the Claim, which the claiming Party may include in its supporting particulars.

If, having received the fully detailed Claim under Sub-Clause 20.2.4 [*Fully detailed Claim*], or in the case of a Claim under Sub-Clause 20.2.6 [*Claims of continuing effect*] an interim or final fully detailed Claim (as the case may be), the Engineer requires necessary additional particulars:

如果工程师在 14 天内未能发出此类**通知**，则**索赔通知**应视为有效**通知**。如果另一方不同意该视为有效的**索赔通知**，其应向工程师发出**通知**，其中应包含不同意的详细说明。此后，根据第 20.2.5 项 [*索赔的商定或确定*] 的规定达成对**索赔**的商定或确定，应包括**工程师**对此类不同意的审核。

如果索赔方收到工程师根据第 20.2.4 项发出的**通知**，并且如果索赔方不同意该**通知**或认为在某些情况下有理由根据上述（b）段延迟提交该声明，则充分详细的索赔应包括索赔方不同意的详细信息，或延误提交的详细情况（视情况而定）。

如果引起**索赔**的事件或情况具有持续影响，则第 20.2.6 项 [*具有持续影响的索赔*] 应适用。

20.2.5
索赔的商定或确定

在收到第 20.2.4 项 [*充分详细的索赔*] 或第 20.2.6 项 [*具有持续影响的索赔*] 规定的临时或最终的充分详细的**索赔**（视情况而定）后，**工程师**应按照第 3.7 款 [*商定或确定*] 的规定进行商定或确定：

(a) 索赔方有权获得的额外付款（如果有）或降低**合同价格**（如果**雇主**是索赔方）；（和 / 或）

(b) 根据第 8.5 款 [*竣工时间的延长*] 的规定（如果**承包商**是索赔方）延长（如果有）**竣工时间**（在到期之前或之后），或按照第 11.3 款 [*缺陷通知期限的延长*] 的规定（如果**雇主**是**索赔方**）延长（如果有）**缺陷通知期限**（到期之前）。

根据**合同**规定索赔方有权获得以上权利。

如果**工程师**已根据第 20.2.2 项 [*工程师的初步响应*] 和 / 或根据第 20.2.4 项 [*充分详细的索赔*] 的规定发出**通知**，仍应根据第 20.2.5 项商定或确定**索赔**。**索赔**的商定或确定应包括考虑到，**索赔方**不同意该**通知**充分详细的**索赔**中所包含的详细信息（如果有），是否将**索赔通知**视为有效**通知**，或延迟提交的理由（视情况而定）。可考虑的情况（但不具有约束力）可能包括：

- 接受延迟提交是否或在多大程度上会损害另一方；

- 如果是第 20.2.1 项 [*索赔通知*] 规定的时限，另一方事先对引起**索赔**的事件或情况了解到的任何证据，索赔方可将其包括在证明资料中；（和 / 或）

- 如果是第 20.2.4 项 [*充分详细的索赔*] 规定的时限，应提供另一方对**索赔**的合同和 / 或其他法律依据事先了解到的任何证据，索赔方可将其包括在证明资料中。

如果已收到根据第 20.2.4 项 [*充分详细的索赔*] 规定的充分详细**索赔**，或出现第 20.2.6 项 [*具有持续影响的索赔*] 规定的**索赔**情况、临时或最终的充分详细的**索赔**（视情况而定），**工程师**应要求提供其他必要的补充说明：

(i) he/she shall promptly give a Notice to the claiming Party, describing the additional particulars and the reasons for requiring them;
(ii) he/she shall nevertheless give his/her response on the contractual or other legal basis of the Claim, by giving a Notice to the claiming Party, within the time limit for agreement under Sub-Clause 3.7.3 [*Time limits*];
(iii) as soon as practicable after receiving the Notice under sub-paragraph (i) above, the claiming Party shall submit the additional particulars; and
(iv) the Engineer shall then proceed under Sub-Clause 3.7 [*Agreement or Determination*] to agree or determine the matters under sub-paragraphs (a) and/or (b) above (and, for the purpose of Sub-Clause 3.7.3 [*Time limits*], the date the Engineer receives the additional particulars from the claiming Party shall be the date of commencement of the time limit for agreement under Sub-Clause 3.7.3).

20.2.6 Claims of continuing effect

If the event or circumstance giving rise to a Claim under this Sub-Clause 20.2 has a continuing effect:

(a) the fully detailed Claim submitted under Sub-Clause 20.2.4 [*Fully detailed Claim*] shall be considered as interim;
(b) in respect of this first interim fully detailed Claim, the Engineer shall give his/her response on the contractual or other legal basis of the Claim, by giving a Notice to the claiming Party, within the time limit for agreement under Sub-Clause 3.7.3 [*Time limits*];
(c) after submitting the first interim fully detailed Claim the claiming Party shall submit further interim fully detailed Claims at monthly intervals, giving the accumulated amount of additional payment claimed (or the reduction of the Contract Price, in the case of the Employer as the claiming Party), and/or extension of time claimed (in the case of the Contractor as the claiming Party) or extension of the DNP (in the case of the Employer as the claiming Party); and
(d) the claiming Party shall submit a final fully detailed Claim within 28 days after the end of the effects resulting from the event or circumstance, or within such other period as may be proposed by the claiming Party and agreed by the Engineer. This final fully detailed Claim shall give the total amount of additional payment claimed (or the reduction of the Contract Price, in the case of the Employer as the claiming Party), and/or extension of time claimed (in the case of the Contractor as the claiming Party) or extension of the DNP (in the case of the Employer as the claiming Party).

20.2.7 General requirements

After receiving the Notice of Claim, and until the Claim is agreed or determined under Sub-Clause 20.2.5 [*Agreement or determination of the Claim*], in each Payment Certificate the Engineer shall include such amounts for any Claim as have been reasonably substantiated as due to the claiming Party under the relevant provision of the Contract.

The Employer shall only be entitled to claim any payment from the Contractor and/or to extend the DNP, or set off against or make any deduction from any amount due to the Contractor, by complying with this Sub-Clause 20.2.

The requirements of this Sub-Clause 20.2 are in addition to those of any other Sub-Clause which may apply to the Claim. If the claiming Party fails to comply with this or any other Sub-Clause in relation to the Claim, any additional payment and/or any EOT (in the case of the Contractor as the claiming Party) or extension of the DNP (in the case of the Employer as the claiming Party), shall take account of the extent (if any) to which the failure has prevented or prejudiced proper investigation of the Claim by the Engineer.

（i）其应立即向索赔方发出**通知**，说明补充的证明资料及要求提交的原因；

（ii）尽管如此，其仍应在按照**第 3.7.3 项**［*时限*］商定的时限内，通过向索赔方发出**通知**，对**索赔**的合同或其他法律依据做出答复；

（iii）在收到上述（i）段规定的**通知**后，在切实可行的范围内，索赔方应尽快提交补充的证明资料；（以及）

（iv）**工程师**应按照**第 3.7 款**［*商定或确定*］的规定进行，同意或确定上述（a）段和／或（b）段规定的事项（以及，就**第 3.7.3 项**［*时限*］而言，**工程师**收到索赔方补充证明资料的日期应为按照**第 3.7.3 项**商定的时限的开始日期）。

20.2.6
具有持续影响的索赔

如果引起**第 20.2 款**规定的**索赔**的事件或情况具有持续影响，则：

（a）根据**第 20.2.4 项**［*充分详细的索赔*］规定提交的充分详细的**索赔**，应被视为临时性的；

（b）就第一个临时性充分详细的**索赔**，**工程师**应在按照**第 3.7.3 项**［*时限*］商定的时限内，向索赔方发出**通知**，就**索赔**的合同或其他法律依据做出答复；

（c）提出第一个临时性充分详细的**索赔**后，索赔方应每月提出进一步的临时性充分详细的**索赔**，给出索赔的累计额外付款数额（或，如果**雇主**是索赔方，**合同价格**的减少），和／或索赔时间的延长（如果**承包商**是索赔方）或**缺陷通知期限**的延长（如果**雇主**是索赔方）；（以及）

（d）索赔方应在事件或情况造成的影响终止后 28 天内，或在索赔方提议并经**工程师**商定的其他期限内，提交最终充分详细的**索赔**。该最终充分详细的**索赔**应提供索赔的额外付款总额（或，如果**雇主**是索赔方，降低**合同价格**），和／或索赔时间的延长（如果**承包商**是索赔方）或**缺陷通知期限**的延长（如果**雇主**是索赔方）。

20.2.7
一般要求

收到**索赔通知**后，根据**第 20.2.5 项**［*索赔的商定或确定*］的规定商定或确定**索赔**，**工程师**应在每份**付款证书**中明确已合理证实的、根据本**合同**相关条款应付给索赔方的任何索赔金额。

雇主只有权要求**承包商**支付任何款项，并根据**第 20.2 款**的规定，和／或延长**缺陷通知期限**，或抵消或从应付给**承包商**的任何金额中扣除。

第 20.2 款的要求是对可能适用于**索赔**的其他**条款**的补充。如果索赔方未能遵守本**款**或与**索赔**相关的任何其他**条款**，任何额外的付款和／或任何**竣工时间**的延长（如果**承包商**是索赔方）或**缺陷通知期限**的延长（如果**雇主**是索赔方），应考虑未履行义务在多大程度上（如果有）阻止或妨碍了**工程师**对**索赔**的适当调查。

21 Disputes and Arbitration

21.1 Constitution of the DAAB

Disputes shall be decided by a DAAB in accordance with Sub-Clause 21.4 [*Obtaining DAAB's Decision*]. The Parties shall jointly appoint the member(s) of the DAAB within the time stated in the Contract Data (if not stated, 28 days) after the date the Contractor receives the Letter of Acceptance.

The DAAB shall comprise, as stated in the Contract Data, either one suitably qualified member (the "sole member") or three suitably qualified members (the "members"). If the number is not so stated, and the Parties do not agree otherwise, the DAAB shall comprise three members.

The sole member or three members (as the case may be) shall be selected from those named in the list in the Contract Data, other than anyone who is unable or unwilling to accept appointment to the DAAB.

If the DAAB is to comprise three members, each Party shall select one member for the agreement of the other Party. The Parties shall consult both these members and shall agree the third member, who shall be appointed to act as chairperson.

The DAAB shall be deemed to be constituted on the date that the Parties and the sole member or the three members (as the case may be) of the DAAB have all signed a DAAB Agreement.

The terms of the remuneration of either the sole member or each of the three members, including the remuneration of any expert whom the DAAB consults, shall be mutually agreed by the Parties when agreeing the terms of the DAAB Agreement. Each Party shall be responsible for paying one-half of this remuneration.

If at any time the Parties so agree, they may appoint a suitably qualified person or persons to replace any one or more members of the DAAB. Unless the Parties agree otherwise, a replacement DAAB member shall be appointed if a member declines to act or is unable to act as a result of death, illness, disability, resignation or termination of appointment. The replacement member shall be appointed in the same manner as the replaced member was required to have been selected or agreed, as described in this Sub-Clause.

The appointment of any member may be terminated by mutual agreement of both Parties, but not by the Employer or the Contractor acting alone.

Unless otherwise agreed by both Parties, the term of the DAAB (including the appointment of each member) shall expire either:

(a) on the date the discharge shall have become, or deemed to have become, effective under Sub-Clause 14.12 [*Discharge*]; or
(b) 28 days after the DAAB has given its decision on all Disputes, referred to it under Sub-Clause 21.4 [*Obtaining DAAB's Decision*] before such discharge has become effective,

whichever is later.

However, if the Contract is terminated under any Sub-Clause of these Conditions or otherwise, the term of the DAAB (including the appointment of each member) shall expire 28 days after:

21 争端和仲裁

21.1 争端避免/裁决委员会的组成

争端应按照第21.4款［*取得争端避免/裁决委员会的决定*］的规定，由**争端避免/裁决委员会**裁决。双方应在**合同数据**中规定的时间内（如未规定，则为28天）在**承包商**收到**中标函**日期后，共同任命**争端避免/裁决委员会**的成员。

争端避免/裁决委员会应按**合同数据**中的规定，由具有适当资格的一名人员（"唯一成员"）或三名人员（"成员"）组成。如果对委员会人数没有规定，且双方未另行商定，**争端避免/裁决委员会**应由三人组成。

唯一成员或三名成员（视情况而定）应从**合同数据**中列出的名单中选出，但不能够或不愿意接受**争端避免/裁决委员会**任命的任何人除外。

如果**争端避免/裁决委员会**由三名成员组成，每方均应选择一名成员供另一方同意。双方应与这些成员协商，并商定第三名成员，该成员应任命为主席。

争端避免/裁决委员会应视为在双方和**争端避免/裁决委员会**的唯一成员或三名成员（视情况而定）签订**争端避免/裁决委员会协议书**之日成立。

唯一成员或三名成员中每一名成员的报酬条款，包括**争端避免/裁决委员会**咨询的任何专家的报酬，应由双方在商定**争端避免/裁决委员会协议书**条款时共同商定。各方应负责支付该报酬的一半。

如经双方商定，可在任何时候任命一名或几名有适当资格的人员，替代**争端避免/裁决委员会**的任何一名或几名成员。除非双方另有商定，在某一成员拒绝履行职责，或因其死亡、疾病、无行为能力、辞职或任命期满而不能履行职责时，应任命替代**争端避免/裁决委员会**成员。替代成员的任命方式与本**款**规定的要求选择或认可被替代成员的方式一致。

任何成员的任命，可经双方商定终止，但**雇主**或**承包商**都不能单独采取行动。

除非双方另有商定，**争端避免/裁决委员会**（包括每名成员的任命）的任期应在以下任一情况下到期：

(a) 根据第14.12款［*结算证明*］的规定，结算证明应已生效或视为生效之日；（或）
(b) **争端避免/裁决委员会**对所有**争端**做出决定后28天内，根据第21.4款［*取得争端避免/裁决委员会的决定*］的规定，此类结算证明已生效之前。

以较晚者为准。

但是，如果**合同**根据本**条件**的任何**条款**或其他条件终止，**争端避免/裁决委员会**（包括每位成员的任命）的任期应在以下任一情况28天后到期：

(i) the DAAB has given its decision on all Disputes, which were referred to it (under Sub-Clause 21.4 [*Obtaining DAAB's Decision*]) within 224 days after the date of termination; or
(ii) the date that the Parties reach a final agreement on all matters (including payment) in connection with the termination

whichever is earlier.

21.2
Failure to Appoint DAAB Member(s)

If any of the following conditions apply, namely:

(a) if the DAAB is to comprise a sole member, the Parties fail to agree the appointment of this member by the date stated in the first paragraph of Sub-Clause 21.1 [*Constitution of the DAAB*]; or
(b) if the DAAB is to comprise three persons, and if by the date stated in the first paragraph of Sub-Clause 21.1 [*Constitution of the DAAB*]:

   (i) either Party fails to select a member (for agreement by the other Party);
   (ii) either Party fails to agree a member selected by the other Party; and/or
   (iii) the Parties fail to agree the appointment of the third member (to act as chairperson) of the DAAB;

(c) the Parties fail to agree the appointment of a replacement within 42 days after the date on which the sole member or one of the three members declines to act or is unable to act as a result of death, illness, disability, resignation, or termination of appointment; or
(d) if, after the Parties have agreed the appointment of the member(s) or replacement, such appointment cannot be effected because one Party refuses or fails to sign a DAAB Agreement with any such member or replacement (as the case may be) within 14 days of the other Party's request to do so,

then the appointing entity or official named in the Contract Data shall, at the request of either or both Parties and after due consultation with both Parties, appoint the member(s) of the DAAB (who, in the case of sub-paragraph (d) above, shall be the agreed member(s) or replacement). This appointment shall be final and conclusive.

Thereafter, the Parties and the member(s) so appointed shall be deemed to have signed and be bound by a DAAB Agreement under which:

(i) the monthly services fee and daily fee shall be as stated in the terms of the appointment; and
(ii) the law governing the DAAB Agreement shall be the governing law of the Contract defined in Sub-Clause 1.4 [*Law and Language*].

Each Party shall be responsible for paying one-half of the remuneration of the appointing entity or official. If the Contractor pays the remuneration in full, the Contractor shall include one-half of the amount of such remuneration in a Statement and the Employer shall then pay the Contractor in accordance with the Contract. If the Employer pays the remuneration in full, the Engineer shall include one-half of the amount of such remuneration as a deduction under sub-paragraph (b) of Sub-Clause 14.6.1 [*The IPC*].

21.3
Avoidance of Disputes

If the Parties so agree, they may jointly request (in writing, with a copy to the Engineer) the DAAB to provide assistance and/or informally discuss and attempt to resolve any issue or disagreement that may have arisen

(i) 终止日期后224天内，**争端避免/裁决委员会**（根据**第21.4款**[*取得争端避免/裁决委员会的决定*]的规定）对提交的所有争端做出决定；（或）

(ii) 双方就与终止相关的所有事项（包括付款）达成最终商定的日期。

以较早者为准。

21.2 未能任命争端避免/裁决委员会成员

如下列任何情况适用，即：

(a) 如果**争端避免/裁决委员会**由一名唯一成员组成，则双方未能在**第21.1款**[*争端避免/裁决委员会的组成*]第一段所述日期前就该成员的任命达成一致；（或）

(b) 如果**争端避免/裁决委员会**由三人组成，在**第21.1款**[*争端避免/裁决委员会的组成*]第一段规定的日期之前：

(i) 任一方未能选择成员（经另一方同意）；

(ii) 任一方未能同意另一方选择的成员；（和/或）

(iii) 双方未能就**争端避免/裁决委员会**第三名成员（担任主席）的任命达成商定。

(c) 在唯一成员或三名成员中的一人拒绝履行职责，或因其死亡、疾病、无行为能力、辞职或任命期满而不能履行职责后42天内，双方未能就任命一位替代人员达成商定；（或）

(d) 如果在双方同意任命成员或替代人员后，由于一方拒绝或未能在另一方要求14天内与任何此类成员或替代人员（视情况而定）签署**争端避免/裁决委员会协议书**，则该任命不能生效。

然后，**合同数据**中指定的实体或官员，应根据一方或双方的要求，在与双方进行适当协商后，任命**争端避免/裁决委员会**的成员［根据上述（d）段的规定，该成员应为商定的成员或替代成员］。此任命应是最终的和决定性的。

此后，双方和如此任命的成员应被视为已签署**争端避免/裁决委员会协议书**并受其约束，根据该协议书：

(i) 服务的月费和日费应符合任命条款中的规定；（以及）

(ii) 管辖**争端避免/裁决委员会协议书**的法律应为**第1.4款**[*法律和语言*]规定的合同的适用法律。

每一方应负责支付任命实体或官员的一半报酬。如果**承包商**全额支付了报酬，则**承包商**应在**报表**中包括该报酬一半的数额，然后雇主应根据合同向**承包商**支付。如果雇主全额支付了报酬，**工程师**应按照**第14.6.1项**[*期中付款证书*]（b）段的规定，将报酬的一半数额作为扣减额。

21.3 争端避免

如双方同意，可共同请求（以书面形式，并抄送工程师）**争端避免/裁决委员会**提供协助，和/或非正式讨论并试图解决双方在履行合同期间可能出现的任何问题或分歧。如果**争端避免/裁决委员会**意识到

between them during the performance of the Contract. If the DAAB becomes aware of an issue or disagreement, it may invite the Parties to make such a joint request.

Such joint request may be made at any time, except during the period that the Engineer is carrying out his/her duties under Sub-Clause 3.7 [*Agreement or Determination*] on the matter at issue or in disagreement unless the Parties agree otherwise.

Such informal assistance may take place during any meeting, Site visit or otherwise. However, unless the Parties agree otherwise, both Parties shall be present at such discussions. The Parties are not bound to act on any advice given during such informal meetings, and the DAAB shall not be bound in any future Dispute resolution process or decision by any views or advice given during the informal assistance process, whether provided orally or in writing.

21.4 Obtaining DAAB's Decision

If a Dispute arises between the Parties then either Party may refer the Dispute to the DAAB for its decision (whether or not any informal discussions have been held under Sub-Clause 21.3 [*Avoidance of Disputes*]) and the following provisions shall apply.

21.4.1 Reference of a Dispute to the DAAB

The reference of a Dispute to the DAAB (the "reference" in this Sub-Clause 21.4) shall:

(a) if Sub-Clause 3.7 [*Agreement or Determination*] applied to the subject matter of the Dispute, be made within 42 days of giving or receiving (as the case may be) a NOD under Sub-Clause 3.7.5 [*Dissatisfaction with Engineer's determination*]. If the Dispute is not referred to the DAAB within this period of 42 days, such NOD shall be deemed to have lapsed and no longer be valid;
(b) state that it is given under this Sub-Clause;
(c) set out the referring Party's case relating to the Dispute;
(d) be in writing, with copies to the other Party and the Engineer; and
(e) for a DAAB of three persons, be deemed to have been received by the DAAB on the date it is received by the chairperson of the DAAB.

The reference of a Dispute to the DAAB under this Sub-Clause shall, unless prohibited by law, be deemed to interrupt the running of any applicable statute of limitation or prescription period.

21.4.2 The Parties' obligations after the reference

Both Parties shall promptly make available to the DAAB all information, access to the Site, and appropriate facilities, as the DAAB may require for the purposes of making a decision on the Dispute.

Unless the Contract has already been abandoned or terminated, the Parties shall continue to perform their obligations in accordance with the Contract.

21.4.3 The DAAB's decision

The DAAB shall complete and give its decision within:

(a) 84 days after receiving the reference; or
(b) such period as may be proposed by the DAAB and agreed by both Parties.

However, if at the end of this period, the due date(s) for payment of any DAAB member's invoice(s) has passed but such invoice(s) remains unpaid, the DAAB shall not be obliged to give its decision until such outstanding invoice(s) have been paid in full, in which case the DAAB shall give its decision as soon as practicable after payment has been received.

问题或分歧，可以邀请双方提出共同请求。

除双方另有商定，否则可以随时提出共同请求，**工程师**根据**第 3.7 款**［**商定或确定**］规定就有争议的或有分歧的事项履行其职责的期间除外。

此类非正式协助可在任何会议、**现场**考察或其他期间进行。但是，除非双方另有商定，否则双方应出席此类讨论。双方没有义务按照非正式会议期间提供的任何建议采取措施，**争端避免 / 裁决委员会**在今后的任何**争端**解决过程或决定中，不受非正式协助过程中提供的任何意见或建议的约束，无论是口头的还是书面的。

21.4
取得争端避免 / 裁决委员会的决定

如果双方间发生了**争端**，任一方可将该争端提交**争端避免 / 裁决委员会**决定（不论是否根据**第 21.3 款**［**争端避免**］进行了任何正式讨论），且以下条款应适用。

21.4.1
争端提交给争端避免 / 裁决委员会

将**争端**提交给**争端避免 / 裁决委员会**（第 21.4 款中的"提交"）应：

(a) 如果**第 3.7 款**［**商定或确定**］适用于**争端**标的，则应在根据**第 3.7.5 项**［**对工程师的确定不满意**］的规定发出或收到**不满意通知**（视情况而定）的 42 天内做出。如果未在 42 天内将**争端**提交给**争端避免 / 裁决委员会**，则该**不满意通知**应视为已期满且不再有效；

(b) 说明是根据本款发出的；
(c) 列出提交方与**争端**有关的案件；
(d) 采用书面形式，并向另一方和**工程师**提供副本；（以及）
(e) 对于三名成员的**争端避免 / 裁决委员会**，在**争端避免 / 裁决委员会**主席收到之日即视为**争端避免 / 裁决委员会**已收到。

除非法律禁止，否则根据本款将**争端**提交**争端避免 / 裁决委员会**应被视为中断任何适用的时效或时效期限。

21.4.2
提交后双方的义务

双方应立即向**争端避免 / 裁决委员会**提供其对**争端**做出决定而可能需要的所有信息、**现场**进入权和相应设施。

除**合同**已被放弃或终止外，双方应继续按照**合同**履行其义务。

21.4.3
争端避免 / 裁决委员会的决定

争端避免 / 裁决委员会应在以下时间内完成并做出决定：

(a) 收到提交后 84 天；（或）
(b) **争端避免 / 裁决委员会**提议并经双方商定的期限。

但是，如果在该期限结束时，任何**争端避免 / 裁决委员会**成员的发票的付款到期日已过，但仍未得到支付，则**争端避免 / 裁决委员会**没有义务在此类未付发票全额支付之前做出决定。在这种情况下，**争端避免 / 裁决委员会**应在收到付款后尽快做出决定。

The decision shall be given in writing to both Parties with a copy to the Engineer, shall be reasoned, and shall state that it is given under this Sub-Clause.

The decision shall be binding on both Parties, who shall promptly comply with it whether or not a Party gives a NOD with respect to such decision under this Sub-Clause. The Employer shall be responsible for the Engineer's compliance with the DAAB decision.

If the decision of the DAAB requires a payment of an amount by one Party to the other Party

(i) subject to sub-paragraph (ii) below, this amount shall be immediately due and payable without any certification or Notice; and
(ii) the DAAB may (as part of the decision), at the request of a Party but only if there are reasonable grounds for the DAAB to believe that the payee will be unable to repay such amount in the event that the decision is reversed under Sub-Clause 21.6 [*Arbitration*], require the payee to provide an appropriate security (at the DAAB's sole discretion) in respect of such amount.

The DAAB proceeding shall not be deemed to be an arbitration and the DAAB shall not act as arbitrator(s).

21.4.4 Dissatisfaction with DAAB's decision

If either Party is dissatisfied with the DAAB's decision:

(a) such Party may give a NOD to the other Party, with a copy to the DAAB and to the Engineer;
(b) this NOD shall state that it is a "Notice of Dissatisfaction with the DAAB's Decision" and shall set out the matter in Dispute and the reason(s) for dissatisfaction; and
(c) this NOD shall be given within 28 days after receiving the DAAB's decision.

If the DAAB fails to give its decision within the period stated in Sub-Clause 21.4.3 [*The DAAB's decision*], then either Party may, within 28 days after this period has expired, give a NOD to the other Party in accordance with sub-paragraphs (a) and (b) above.

Except as stated in the last paragraph of Sub-Clause 3.7.5 [*Dissatisfaction with Engineer's determination*], in Sub-Clause 21.7 [*Failure to Comply with DAAB's Decision*] and in Sub-Clause 21.8 [*No DAAB In Place*], neither Party shall be entitled to commence arbitration of a Dispute unless a NOD in respect of that Dispute has been given in accordance with this Sub-Clause 21.4.4.

If the DAAB has given its decision as to a matter in Dispute to both Parties, and no NOD under this Sub-Clause 21.4.4 has been given by either Party within 28 days after receiving the DAAB's decision, then the decision shall become final and binding on both Parties.

If the dissatisfied Party is dissatisfied with only part(s) of the DAAB's decision:

(i) this part(s) shall be clearly identified in the NOD;
(ii) this part(s), and any other parts of the decision that are affected by such part(s) or rely on such part(s) for completeness, shall be deemed to be severable from the remainder of the decision; and
(iii) the remainder of the decision shall become final and binding on both Parties as if the NOD had not been given.

该决定应以书面形式提供给双方，并抄送给**工程师**一份，并应说明该决定是根据本款做出的。

该决定对双方均具有约束力，无论一方根据本款做出的该决定是否发出**不满意通知**，双方均应立即予以遵守。雇主应对**工程师**遵守**争端避免/裁决委员会**决定负责。

如果**争端避免/裁决委员会**的决定要求一方向另一方支付一笔款项

（i）除下述第（ii）段另有规定外，该笔款项应立即到期应付，无须任何证明或**通知**；（以及）

（ii）在**争端避免/裁决委员会**有合理的理由相信，如果根据第21.6款[*仲裁*]的规定撤销决定，收款人将无法偿还该金额的情况下，应一方请求，**争端避免/裁决委员会**可以（作为决定的一部分）要求收款人就此类金额提供适当的担保（由**争端避免/裁决委员会**自行决定）。

争端避免/裁决委员会程序不应被视为仲裁，**争端避免/裁决委员会**不应担任仲裁员。

21.4.4 对争端避免/裁决委员会的决定不满意

如果任一方对**争端避免/裁决委员会**的决定不满意，则：

(a) 该方可以将**不满意通知**发给另一方，并抄送给**争端避免/裁决委员会**和**工程师**；

(b) **不满意通知**应说明其为"对**争端避免/裁决委员会**的决定的不满意通知"，并应在**争端**中阐明问题及不满意的原因；（以及）

(c) **不满意通知**应在收到**争端避免/裁决委员会**的决定后28天发出。

如果**争端避免/裁决委员会**未能在第21.4.3项[*争端避免/裁决委员会的决定*]规定的期限内做出决定，则任一方均可在该期限届满后的28天内，根据以上（a）和（b）段，向另一方发出**不满意通知**。

除第3.7.5项[*对工程师的决定不满意*]最后一段、第21.7款[*未能遵守争端避免/裁决委员会的决定*]和第21.8款[*未设立争端避免/裁决委员会*]规定的情况外，任一方均应无权着手**争端**的仲裁，按第21.4.4项规定就该**争端**已发出**不满意通知**的情况除外。

如果**争端避免/裁决委员会**已就**争端**事项提交了其决定，而任一方在收到**争端避免/裁决委员会**决定28天内，未根据第21.4.4项发出**不满意通知**，则该决定应成为最终的，对双方均具有约束力。

如果不满意的一方仅对**争端避免/裁决委员会**决定的某部分不满意：

（i）**不满意通知**应对该部分加以明确；

（ii）该部分及受该部分影响或依赖该部分完整性的决定的任何其他部分，应被视为与该决定的其余部分分开；（以及）

（iii）该决定的其余部分应成为最终决定，并对双方具有约束力，如同没有发出**不满意通知**。

21.5
Amicable Settlement

Where a NOD has been given under Sub-Clause 21.4 [*Obtaining DAAB's Decision*], both Parties shall attempt to settle the Dispute amicably before the commencement of arbitration. However, unless both Parties agree otherwise, arbitration may be commenced on or after the twenty-eighth (28th) day after the day on which this NOD was given, even if no attempt at amicable settlement has been made.

21.6
Arbitration

Unless settled amicably, and subject to Sub-Clause 3.7.5 [*Dissatisfaction with Engineer's determination*], Sub-Clause 21.4.4 [*Dissatisfaction with DAAB's decision*], Sub-Clause 21.7 [*Failure to Comply with DAAB's Decision*] and Sub-Clause 21.8 [*No DAAB In Place*], any Dispute in respect of which the DAAB's decision (if any) has not become final and binding shall be finally settled by international arbitration. Unless otherwise agreed by both Parties:

(a) the Dispute shall be finally settled under the Rules of Arbitration of the International Chamber of Commerce;
(b) the Dispute shall be settled by one or three arbitrators appointed in accordance with these Rules; and
(c) the arbitration shall be conducted in the ruling language defined in Sub-Clause 1.4 [*Law and Language*].

The arbitrator(s) shall have full power to open up, review and revise any certificate, determination (other than a final and binding determination), instruction, opinion or valuation of the Engineer, and any decision of the DAAB (other than a final and binding decision) relevant to the Dispute. Nothing shall disqualify the Engineer from being called as a witness and giving evidence before the arbitrator(s) on any matter whatsoever relevant to the Dispute.

In any award dealing with costs of the arbitration, the arbitrator(s) may take account of the extent (if any) to which a Party failed to cooperate with the other Party in constituting a DAAB under Sub-Clause 21.1 [*Constitution of the DAAB*] and/or Sub-Clause 21.2 [*Failure to Appoint DAAB Member(s)*].

Neither Party shall be limited in the proceedings before the arbitrator(s) to the evidence or arguments previously put before the DAAB to obtain its decision, or to the reasons for dissatisfaction given in the Party's NOD under Sub-Clause 21.4 [*Obtaining DAAB's Decision*]. Any decision of the DAAB shall be admissible in evidence in the arbitration.

Arbitration may be commenced before or after completion of the Works. The obligations of the Parties, the Engineer and the DAAB shall not be altered by reason of any arbitration being conducted during the progress of the Works.

If an award requires a payment of an amount by one Party to the other Party, this amount shall be immediately due and payable without any further certification or Notice.

21.7
Failure to Comply with DAAB's Decision

In the event that a Party fails to comply with any decision of the DAAB, whether binding or final and binding, then the other Party may, without prejudice to any other rights it may have, refer the failure itself directly to arbitration under Sub-Clause 21.6 [*Arbitration*] in which case Sub-Clause 21.4 [*Obtaining DAAB's Decision*] and Sub-Clause 21.5 [*Amicable Settlement*] shall not apply to this reference. The arbitral tribunal (constituted under Sub-Clause 21.6 [*Arbitration*]) shall have the power, by way of summary or other expedited procedure, to order, whether by an interim or provisional measure or an award (as may be appropriate under applicable law or otherwise), the enforcement of that decision.

21.5	
友好解决	如果已按照第 21.4 款［*取得争端避免／裁决委员会的决定*］发出了**不满意通知**，双方应在仲裁开始前，设法友好解决**争端**。但是，除非双方另有商定，即使没有设法友好解决，也可在发出**不满意通知**第 28 天或其后开始仲裁。

21.6	
仲裁	除非友好解决，并根据第 3.7.5 项［*对工程师的决定不满意*］、第 21.4.4 项［*对争端避免／裁决委员会的决定不满意*］、第 21.7 款［*未能遵守争端避免／裁决委员会的决定*］和第 21.8 款［*未设立争端避免／裁决委员会*］的规定，**争端避免／裁决委员会**的决定（如果有）尚未成为最终和具有约束力的任何**争端**，应通过国际仲裁最终解决。除非双方另有商定：

    （a） **争端**应根据**国际商会仲裁规则**最终解决；

    （b） **争端**应由按上述**规则**任命的一位或三位仲裁员负责解决；（以及）

    （c） 仲裁应根据第 1.4 款［*法律和语言*］规定的主导语言进行。

仲裁员应有充分的权利公开、审查和修改与该**争端**有关的**工程师**的任何证书、确定（最终的并具有约束力的确定除外）、指示、意见或估价，以及**争端避免／裁决委员会**的任何决定（最终的并具有约束力的决定除外）。**工程师**被传为证人，并向仲裁员就任何与**争端**有关的事项提供证据的资格不受任何影响。

在涉及仲裁费用的任何裁决中，仲裁员可考虑一方未能根据第 21.1 款［*争端避免／裁决委员会的组成*］和／或第 21.2 款［*未能任命争端避免／裁决委员会成员*］的规定与另一方合作组建**争端避免／裁决委员会**的程度（如果有）。

在仲裁员面前的程序中，任一方都不应局限于先前为取得**争端避免／裁决委员会**的决定而向**争端避免／裁决委员会**提供的证据或论据，或该方根据第 21.4 款［*取得争端避免／裁决委员会的决定*］的不满意通知中提出的不满意理由。**争端避免／裁决委员会**的任何决定均应在仲裁中被接受为证据。

仲裁在**工程**竣工前或竣工后均可开始。双方、**工程师**和**争端避免／裁决委员会**的义务，不得因在**工程**进行中正在执行的任何仲裁而改变。

如果裁决要求一方向另一方支付一笔金额，则该金额应立即支付，无须任何进一步的证明或**通知**。

21.7	
未能遵守争端避免／裁决委员会的决定	在一方未能遵守**争端避免／裁决委员会**的决定的情况下，无论是具有约束力的或最终且具有约束力的，另一方可以在不损害其可能拥有的其他权利的情况下，根据第 21.6 款［*仲裁*］的规定，将上述未遵守决定的事项直接提交仲裁，在此情况下，第 21.4 款［*取得争端避免／裁决委员会的决定*］和第 21.5 款［*友好解决*］的规定应不适用。仲裁庭（根据第 21.6 款［*仲裁*］的规定成立）应有权以简易程序或其他快速程序，通过临时措施或裁决（根据适用法律或其他法律）而命令执行该决定。

In the case of a binding but not final decision of the DAAB, such interim or provisional measure or award shall be subject to the express reservation that the rights of the Parties as to the merits of the Dispute are reserved until they are resolved by an award.

Any interim or provisional measure or award enforcing a decision of the DAAB which has not been complied with, whether such decision is binding or final and binding, may also include an order or award of damages or other relief.

21.8
No DAAB In Place

If a Dispute arises between the Parties in connection with, or arising out of, the Contract or the execution of the Works and there is no DAAB in place (or no DAAB is being constituted), whether by reason of the expiry of the DAAB's appointment or otherwise:

(a) Sub-Clause 21.4 [*Obtaining DAAB's Decision*] and Sub-Clause 21.5 [*Amicable Settlement*] shall not apply; and
(b) the Dispute may be referred by either Party directly to arbitration under Sub-Clause 21.6 [*Arbitration*] without prejudice to any other rights the Party may have.

在**争端避免／裁决委员会**做出具有约束力但不是最终决定的情况下，该临时措施或裁决应服从于明示权益保留，即保留双方对**争端**的是非曲直的权利，直到通过裁决解决为止。

执行未遵守**争端避免／裁决委员会**决定的任何临时的或临时措施或裁决，无论其是具有约束力的还是最终且具有约束力的，也可包括损害赔偿或其他救济的命令或裁决。

21.8 未设立争端避免／裁决委员会	如果双方之间发生与**合同**或**工程**实施有关或产生的**争端**，且未设立**争端避免／裁决委员会**（或**争端避免／裁决委员会**未组成），无论是由于**争端避免／裁决委员会**的任命到期还是其他原因，则： （a）第 21.4 款［*取得争端避免／裁决委员会的决定*］和第 21.5 款［*友好解决*］应不适用；（以及） （b）任一方可根据第 21.6 款［*仲裁*］的规定直接将争端提交仲裁，但不影响该方可能拥有的任何其他权利。

APPENDIX

General Conditions of Dispute Avoidance/Adjudication Agreement

1
Definitions

1.1 "**General Conditions of Dispute Avoidance/Adjudication Agreement**" or "**GCs**" means this document entitled "General Conditions of Dispute Avoidance/Adjudication Agreement", as published by FIDIC.

1.2 In the DAA Agreement (as defined below) and in the GCs, words and expressions which are not otherwise defined shall have the meanings assigned to them in the Contract (as defined in the DAA Agreement).

1.3 "Dispute Avoidance/Adjudication Agreement" or "DAA Agreement" means a tripartite agreement by and between:

(a) the Employer;
(b) the Contractor; and
(c) the DAAB Member who is defined in the DAA Agreement as being either:

(i) the sole member of the DAAB, or
(ii) one of the three members (or the chairman) of the DAAB.

1.4 "**DAAB's Activities**" means the activities carried out by the DAAB in accordance with the Contract and the GCs, including all Informal Assistance, meetings (including meetings and/or discussions between the DAAB members in the case of a three-member DAAB), Site visits, hearings and decisions.

1.5 "**DAAB Rules**" means the document entitled "**DAAB Procedural Rules**" published by FIDIC which are annexed to, and form part of, the GCs.

1.6 "**Informal Assistance**" means the informal assistance given by the DAAB to the Parties when requested jointly by the Parties under Sub-Clause 21.3 [*Avoidance of Disputes*] of the Conditions of Contract.

1.7 "**Term of the DAAB**" means the period starting on the Effective Date (as defined in Sub-Clause 2.1 below) and finishing on the date that the term of the DAAB expires in accordance with Sub-Clause 21.1 [*Constitution of the DAAB*] of the Conditions of Contract.

1.8 "**Notification**" means a notice in writing given under the GCs, which shall be:

(a) (i) a paper-original signed by the DAAB Member or the authorised representative of the Contractor or of the Employer (as the case may be); or
(ii) an electronic original generated from the system of electronic transmission agreed between the Parties and the DAAB, which electronic original is transmitted by the electronic address uniquely assigned to the DAAB Member or each such authorised representative (as the case may be);

(b) delivered by hand (against receipt), or sent by mail or courier (against receipt), or transmitted using the system of electronic transmission under sub-paragraph (a)(ii) above; and

附录

争端避免/裁决委员会协议书一般条件[一]

1
定义

1.1 "争端避免/裁决委员会协议书一般条件"或"GCs"系指由菲迪克（FIDIC）出版的标题为"争端避免/裁决委员会协议书一般条件"的文件。

1.2 在**争端避免/裁决委员会协议书**（定义见下文）和**一般条件**中，未另行定义的词和短语应具有合同（如**争端避免/裁决委员会协议书**中的定义）中赋予它们的含义。[一]

1.3 "**争端避免/裁决委员会协议书**"系指本合同定义的、由以下各方签订的三方协议书：[一]

（a） 雇主；
（b） 承包商；（以及）
（c） 争端避免/裁决委员会协议书中定义的下述两者中任一**争端避免/裁决委员会成员**：[一]

　　（i） **争端避免/裁决委员会**的唯一成员；（或）
　　（ii）**争端避免/裁决委员会**的三名成员之一（或主席）。[一]

1.4 "**争端避免/裁决委员会的活动**"系指**争端避免/裁决委员会**根据合同和**一般条件**进行的活动，包括所有非正式协助、会议（如为三名成员组成的**争端避免/裁决委员会**，则包括**争端避免/裁决委员会成员**之间的会议和/或讨论）、**现场考察**、意见听取会和决定。

1.5 "**争端避免/裁决委员会规则**"系指由菲迪克（FIDIC）发布的标题为"争端避免/裁决委员会程序规则"的文件，该文件是**一般条件**的附件，并成为**一般条件**的组成部分。

1.6 "**非正式协助**"系指**争端避免/裁决委员会**应各方根据合同条件第21.3款[*争端避免*]提出的请求，向各方提供的非正式协助。

1.7 "**争端避免/裁决委员会的任期**"系指根据合同条件第21.1款[*争端避免/裁决委员会的组成*]的规定，从生效日期（见下文**第2.1款**规定）到**争端避免/裁决委员会**任期结束的期间。

1.8 "**通知**"系指根据**一般条件**规定发出的书面**通知**，应为：

（a） （i） 由**争端避免/裁决委员会成员**或承包商或雇主的授权代表（视情况而定）签署的书面原件；（或）
　　 （ii）由各方与**争端避免/裁决委员会**商定的由电子传输系统生成的电子原件，该电子原件是通过特别分配给**争端避免/裁决委员会成员**或每个授权代表（视情况而定）的电子邮箱发送；

（b） 由专人递送（并取得收据），或邮寄，或由信使送达（并取得收据），或按上述（a）（ii）段通过电子传输系统传输；（以及）

[一] 此处中文按照勘误表改正后的英文翻译。——译者注

(c) delivered, sent or transmitted to the address for the recipient's communications as stated in the DAA Agreement. However, if the recipient gives a Notification of another address, all Notifications shall be delivered accordingly after the sender receives such Notification.

2 General Provisions

2.1 The DAA Agreement shall take effect:

(a) in the case of a sole-member DAAB, on the date when the Employer, the Contractor and the DAAB Member have each signed (or, under the Contract, are deemed to have signed) the DAA Agreement; or
(b) in the case of a three-member DAAB, on the date when the Employer, the Contractor, the DAAB Member and the Other Members have each signed (or under the Contract are deemed to have signed) a DAA Agreement.

(the "**Effective Date**" in the GCs).

2.2 Immediately after the Effective Date, either or both Parties shall give a Notification to the DAAB Member that the DAA Agreement has come into effect. If the DAAB Member does not receive such a notice within 182 days after entering into the DAA Agreement, it shall be void and ineffective.

2.3 The employment of the DAAB Member is a personal appointment. No assignment of, or subcontracting or delegation of the DAAB Member's rights and/or obligations under the DAA Agreement is permitted.

3 Warranties

3.1 The DAAB Member warrants and agrees that he/she is, and will remain at all times during the Term of the DAAB, impartial and independent of the Employer, the Contractor, the Employer's Personnel and the Contractor's Personnel (including in accordance with Sub-Clause 4.1 below).

3.2 If, after signing the DAA Agreement (or after he/she is deemed to have signed the DAA Agreement under the Contract), the DAAB Member becomes aware of any fact or circumstance which might:

(a) call into question his/her independence or impartiality; and/or
(b) be, or appear to be, inconsistent with his/her warranty and agreement under Sub-Clause 3.1 above

the DAAB Member warrants and agrees that he/she shall immediately disclose this in writing to the Parties and the Other Members (if any).

3.3 When appointing the DAAB Member, each Party relies on the DAAB Member's representations that he/she is:

(a) experienced and/or knowledgeable in the type of work which the Contractor is to carry out under the Contract;
(b) experienced in the interpretation of construction and/or engineering contract documentation; and
(c) fluent in the language for communications stated in the Contract Data (or the language as agreed between the Parties and the DAAB).

4 Independence and Impartiality

4.1 Further to Sub-Clauses 3.1 and 3.2 above, the DAAB Member shall:

(a) have no financial interest in the Contract, or in the project of which the Works are part, except for payment under the DAA Agreement;
(b) have no interest whatsoever (financial or otherwise) in the Employer, the Contractor, the Employer's Personnel or the Contractor's Personnel;

(c) 递交、邮寄或传输到**争端避免/裁决委员会协议书**规定的收件人的通信地址。但是，如果收件人发出了更改地址的**通知**，则在发件人收到该**通知**后，所有**通知**均应投送到相应的地址。⊖

2 一般规定

2.1 **争端避免/裁决委员会协议书**应在以下情况下生效：⊖

(a) 如果是唯一成员的**争端避免/裁决委员会**，则在**雇主、承包商**和**争端避免/裁决委员会成员**分别签署（或根据合同规定被视为签署）**争端避免/裁决委员会协议书**的日期；（或）⊖

(b) 如果是三名成员的**争端避免/裁决委员会**，则在**雇主、承包商、争端避免/裁决委员会成员**和**其他成员**分别签署（或根据合同规定被视为签署）**争端避免/裁决委员会协议书**的日期。⊖

（一般条件中的"生效日期"）。

2.2 生效日期后，任一方或双方应立即向**争端避免/裁决委员会成员**发出有关**争端避免/裁决委员会协议书**已生效的**通知**。如果**争端避免/裁决委员会成员**在签订**争端避免/裁决委员会协议书**后的 182 天内未收到此类**通知**，则该协议书应作废和无效。⊖

2.3 **争端避免/裁决委员会成员**的聘任属对个人的任命。**争端避免/裁决委员会协议书**规定的权利和/或义务不得转让，或分包或委派。⊖

3 保证

3.1 **争端避免/裁决委员会成员**保证并同意，其将在**争端避免/裁决委员会**任期内时刻对**雇主、承包商、雇主人员**和**承包商人员**（包括下述第 4.1 款规定的人员）保持公正和独立。

3.2 在签署**争端避免/裁决委员会协议书**后（或按合同规定被视为已签署**争端避免/裁决委员会协议书**后），如果**争端避免/裁决委员会成员**意识到下述任何事实或情况：⊖

(a) 可能会质疑其独立性或公正性；（和/或）
(b) 看来可能与其上述**第 3.1 款**的担保和同意不相符。

那么，**争端避免/裁决委员会成员**保证并同意，其应立即以书面形式向各方和**其他成员**（如果有）披露此信息。

3.3 任命**争端避免/裁决委员会成员**时，各方依据**成员**的下列表现：

(a) 具有**承包商**根据合同要进行工作类型的经验和/或知识；

(b) 具有解释施工和/或工程合同文件的经验；（以及）

(c) 能流利地使用**合同数据**规定的交流语言（或各方与**争端避免/裁决委员会**商定的语言）。

4 独立和公正

4.1 除上述**第 3.1 款**和 **3.2 款**外，**争端避免/裁决委员会成员**应：

(a) 除根据**争端避免/裁决委员会协议书**规定的付款外，在合同中或在**工程**为所属部分的项目中没有任何财务利益；⊖

(b) 与**雇主、承包商、雇主人员**或**承包商人员**没有任何利益关系（无论是财务方面的还是其他方面的）；

⊖ 此处中文按照勘误表改正后的英文翻译。——译者注

(c) in the ten years before signing the DAA Agreement, not have been employed as a consultant or otherwise by the Employer, the Contractor, the Employer's Personnel or the Contractor's Personnel;

(d) not previously have acted, and shall not act, in any judicial or arbitral capacity in relation to the Contract;

(e) have disclosed in writing to the Employer, the Contractor and the Other Members (if any), before signing the DAA Agreement (or before he/she is deemed to have signed the DAA Agreement under the Contract) and to his/her best knowledge and recollection, any:

　　(i) existing and/or past professional or personal relationships with any director, officer or employee of the Employer, the Contractor, the Employer's Personnel or the Contractor's Personnel (including as a dispute resolution practitioner on another project),

　　(ii) facts or circumstances which might call into question his/her independence or impartiality, and

　　(iii) previous involvement in the project of which the Contract forms part;

(f) not, while a DAAB Member and for the Term of the DAAB:

　　(i) be employed as a consultant or otherwise by, and/or

　　(ii) enter into discussions or make any agreement regarding future employment with

the Employer, the Contractor, the Employer's Personnel or the Contractor's Personnel, except as may be agreed by the Employer, the Contractor and the Other Members (if any); and/or

(g) not solicit, accept or receive (directly or indirectly) any gift, gratuity, commission or other thing of value from the Employer, the Contractor, the Employer's Personnel or the Contractor's Personnel, except for payment under the DAA Agreement.

5 General Obligations of the DAAB Member

5.1 The DAAB Member shall:

(a) comply with the GCs, the DAAB Rules and the Conditions of Contract that are relevant to the DAAB's Activities;

(b) not give advice to the Employer, the Contractor, the Employer's Personnel or the Contractor's Personnel concerning the conduct of the Contract, except as required to carry out the DAAB's Activities;

(c) ensure his/her availability during the Term of the DAAB (except in exceptional circumstances, in which case the DAAB Member shall give a Notification without delay to the Parties and the Other Members (if any) detailing the exceptional circumstances) for all meetings, Site visits, hearings and as is necessary to comply with sub-paragraph (a) above;

(d) become, and shall remain for the duration of the Term of the DAAB, knowledgeable about the Contract and informed about:

　　(i) the Parties' performance of the Contract;

　　(ii) the Site and its surroundings; and

　　(iii) the progress of the Works (and of any other parts of the project of which the Contract forms part)

including by visiting the Site, meeting with the Parties and by studying all documents received from either Party under Rule 4.3 of the DAAB Rules (which shall be maintained in a current working file, in hard-copy or electronic format at the DAAB Member's discretion); and

(e) be available to give Informal Assistance when requested jointly by the Parties.

(c) 签署**争端避免/裁决委员会协议书**之前十年内，未曾被**雇主**、**承包商**、**雇主人员**或**承包商人员**聘任为咨询顾问或其他职务；[1]

(d) 以前未曾且未来也不会以与**合同**有关的任何司法或仲裁身份行事；

(e) 签署**争端避免/裁决委员会协议书**之前（或根据**合同**其被视为已签署**争端避免/裁决委员会协议书**之前），已向**雇主**、**承包商**和**其他成员**（如果有）书面披露了，以及凭其知识和记忆了，任何：[1]

    (i) 与**雇主**、**承包商**、**雇主人员**或**承包商人员**的任何董事、高级职员或雇员的现有和/或过去的职业或个人关系（包括作为另一个项目的争端解决从业者）；

    (ii) 可能质疑其独立性或公正性的事实或情况；（以及）

    (iii) 曾参与**合同**所属部分的项目。

(f) 任**争端避免/裁决委员会成员**期间且在**争端避免/裁决委员会**的任期内，没有：

    (i) 被聘任为咨询顾问或以其他方式受聘；（和/或）
    (ii) 就未来受聘于下述情况[2]进行讨论或达成任何协议，

雇主、**承包商**、**雇主人员**或**承包商人员**，**雇主**、**承包商**和**其他成员**（如果有）同意的情况除外；和/或

(g) 除根据**争端避免/裁决委员会协议书**规定的报酬外，未从**雇主**、**承包商**、**雇主人员**或**承包商人员**处索取、接受（直接或间接）或接收任何礼物、酬金、佣金或其他有价物。[1]

5 争端避免/裁决委员会成员的一般义务

5.1 **争端避免/裁决委员会成员**应：

(a) 遵守与**争端避免/裁决委员会**活动相关的一般条件、**争端避免/裁决委员会规则**和**合同**条件；

(b) 除按要求执行**争端避免/裁决委员会**活动外，不向**雇主**、**承包商**、**雇主人员**或**承包商人员**提供有关执行**合同**的建议；

(c) 确保其在**争端避免/裁决委员会**任期内［特殊情况除外，在此情况下，**争端避免/裁决委员会成员**应立即**通知**各方和**其他成员**（如果有），详细说明特殊情况］能够参加上述（a）段所述的所有必要的会议、**现场**视察和意见听取会；

(d) **争端避免/裁决委员会**的任期内，保持对**合同**的了解和：

    (i) 各方履行**合同**的情况；
    (ii) **现场**及其周围环境；（以及）
    (iii) **工程**（以及**合同**为所属部分的项目的任何其他部分）的进度。

包括通过**现场**视察、会见各方以及研究根据**争端避免/裁决委员会规则**第4.3款从任一方收到的所有文件（应由**争端避免/裁决委员会成员**酌情以纸质或电子格式保存在当前工作文件中）；（以及）

(e) 应各方共同要求，提供非正式协助。

---

[1] 此处中文按照勘误表改正后的英文翻译。——译者注
[2] "下述情况"由译者添加。——译者注

6 General Obligations of the Parties

6.1 Each Party shall comply with the GCs, the DAAB Rules and the Conditions of Contract that are relevant to the DAAB's Activities. The Employer and the Contractor shall be responsible for compliance with this provision by the Employer's Personnel and the Contractor's Personnel, respectively.

6.2 Each Party shall cooperate with the other Party in constituting the DAAB, under Sub-Clause 21.1 [*Constitution of the DAAB*] and/or Sub-Clause 21.2 [*Failure to Appoint DAAB Member(s)*] of the Conditions of Contract, without delay.

6.3 In connection with the DAAB's Activities, each Party shall:

(a) cooperate in good faith with the DAAB; and
(b) fulfil its duties, and exercise any right or entitlement, under the Contract, the GCs and the DAAB Rules and/or otherwise

in the manner necessary to achieve the objectives under Rule 1 of the DAAB Rules.

6.4 The Parties, the Employer's Personnel and the Contractor's Personnel shall not request advice from or consultation with the DAAB Member regarding the Contract, except as required for the DAAB Member to carry out the DAAB's Activities.

6.5 At all times when interacting with the DAAB, each Party shall not compromise the DAAB's warranty of independence and impartiality under Sub-Clause 3.1 above.

6.6 In addition to providing documents under Rule 4.3 of the DAAB Rules, each Party shall ensure that the DAAB Member remains informed as is necessary to enable him/her to comply with sub-paragraph (d) of Sub-Clause 5.1 above.

7 Confidentiality

7.1 Subject to Sub-Clause 7.4 below, the DAAB Member shall treat the details of the Contract, all the DAAB's Activities and the documents provided under Rule 4.3 of the DAAB Rules as private and confidential, and shall not publish or disclose them without the prior written consent of the Parties and the Other Members (if any).

7.2 Subject to Sub-Clause 7.4 below, the Employer, the Contractor, the Employer's Personnel and the Contractor's Personnel shall treat the details of all the DAAB's Activities as private and confidential.

7.3 Each person's obligation of confidentiality under Sub-Clause 7.1 or Sub-Clause 7.2 above (as the case may be) shall not apply where the information:

(a) was already in that person's possession without an obligation of confidentiality before receipt under the DAA Agreement;
(b) becomes generally available to the public through no breach of the GCs; or
(c) is lawfully obtained by the person from a third party which is not bound by an obligation of confidentiality.

7.4 If a DAAB Member is replaced under the Contract, the Parties and/or the Other Members (if any) shall disclose details of the Contract, the documents provided under Rule 4.3 of the DAAB Rules and previous DAAB's Activities (including decisions, if any) to the replacement DAAB Member as is necessary in order to:

6		
双方的一般义务	6.1	各方均应遵守与**争端避免/裁决委员会**活动相关的一般条件、**争端避免/裁决委员会规则**及合同条件。雇主和承包商应分别对**雇主人员**和**承包商人员**遵守本规定负责。
	6.2	各方应根据合同条件第21.1款［**争端避免/裁决委员会的组成**］和/或第21.2款［**未能任命争端避免/裁决委员会成员**］的规定与另一方合作，迅速组建**争端避免/裁决委员会**。
	6.3	就**争端避免/裁决委员会**活动而言，各方应：

(a) 与**争端避免/裁决委员会**真诚合作；(以及)
(b) 按照**合同**、**一般条件**和**争端避免/裁决委员会规则**和/或其他规定，履行其职责并行使任何权利或授权

以实现**争端避免/裁决委员会规则**第1条中的目标所必需的方式。

	6.4	除**争端避免/裁决委员会成员**执行**争端避免/裁决委员会**活动所需的情况外，双方、**雇主人员**和**承包商人员**不得向**争端避免/裁决委员会成员**寻求有关合同的建议或意见。
	6.5	任何时候与**争端避免/裁决委员会**进行互动时，各方均不得损害**争端避免/裁决委员会**根据上述第3.1款规定的独立性和公正性的保证。
	6.6	除根据**争端避免/裁决委员会规则**第4.3款提供文件外，各方还应确保**争端避免/裁决委员会成员**保持必要的知情权，以使其能够遵守上述第5.1款（d）段的规定。
7		
保密	7.1	根据下述第7.4款的规定，**争端避免/裁决委员会成员**应将合同的详细信息、所有**争端避免/裁决委员会**活动，以及根据**争端避免/裁决委员会规则**第4.3款所提供的文件视为私人密件，未经各方及**其他成员**（如果有）事先书面同意，不得发布或披露。
	7.2	根据下述第7.4款的规定，雇主、承包商、**雇主人员**和**承包商人员**应将所有**争端避免/裁决委员会**活动的细节视为私人密件。
	7.3	如果信息具有以下几种情况，上述第7.1款或第7.2款（视情况而定）规定的每个人的保密义务应不适用：

(a) 根据**争端避免/裁决委员会协议书**规定接收前，在没有保密义务的前提下，该人已拥有的；[⊖]
(b) 在不违反**一般条件**的情况下，普遍向公众开放的；(或)
(c) 该人从不受保密义务约束的第三方合法获得的。

	7.4	如果根据合同规定替代了**争端避免/裁决委员会成员**，则双方和/或**其他成员**（如果有）应向替代的**争端避免/裁决委员会成员**透露合同必要的详细信息、根据**争端避免/裁决委员会规则**的第4.3款的规定提供的文件，以及此前的**争端避免/裁决委员会**活动（包括决定，如果有），以便：

[⊖] 此处中文按照勘误表改正后的英文翻译。——译者注

(a) enable the replacement DAAB Member to comply with sub-paragraph (d) of Sub-Clause 5.1 above; and

(b) ensure consistency in the manner in which the DAAB's Activities are conducted following such replacement.

8 The Parties' undertaking and indemnity

8.1 The Employer and the Contractor undertake to each other and to the DAAB Member that the DAAB Member shall not:

(a) be appointed as an arbitrator in any arbitration under the Contract;

(b) be called as a witness to give evidence concerning any Dispute in any arbitration under the Contract; or

(c) be liable for any claims for anything done or omitted in the discharge or purported discharge of the DAAB Member's functions, except in any case of fraud, gross negligence, deliberate default or reckless misconduct by him/her.

8.2 The Employer and the Contractor hereby jointly and severally indemnify and hold the DAAB Member harmless against and from all damages, losses and expenses (including legal fees and expenses) resulting from any claim from which he/she is relieved from liability under Sub-Clause 8.1 above.

9 Fees and Expenses

9.1 The DAAB Member shall be paid as follows, in the currency named in the DAA Agreement:

(a) a monthly fee, which shall be a fixed fee as payment in full for:

   (i) being available on 28-days' notice for all meetings, Site visits and hearings under the DAAB Rules (and, in the event of a request under Rule 3.6 of the DAAB Rules, being available for an urgent meeting or Site visit);

   (ii) becoming and remaining knowledgeable about the Contract, informed about the progress of the Works and maintaining a current working file of documents, in accordance with sub-paragraph (d) of Clause 5.1 above;

   (iii) all office and overhead expenses including secretarial services, photocopying and office supplies incurred in connection with his/her duties; and

   (iv) all services performed hereunder except those referred to in sub-paragraphs (b) and (c) of this Clause.

This fee shall be paid monthly with effect from the last day of the month in which the Effective Date occurs until the end of the month in which the Term of the DAAB expires, or the DAAB Member declines to act or is unable to act as a result of death, illness, disability, resignation or termination of his/her DAA Agreement.

If no monthly fee is stated in the DAA Agreement, the matters described in sub-paragraphs (i) to (iv) above shall be deemed to be covered by the daily fee under sub-paragraph (b) below;

(b) a daily fee, which shall be considered as payment in full for each day:

   (i) or part of a day, up to a maximum of two days' travel time in each direction, for the journey between the DAAB Member's home and the Site, or another location of a meeting with the Parties and/or the Other Members (if any);

   (ii) spent on attending meetings and making Site visits in accordance with Rule 3 of the DAAB Rules, and writing reports in accordance with Rule 3.10 of the DAAB Rules;

   (iii) spent on giving Informal Assistance;

(a) 使替代的**争端避免/裁决委员会**成员遵守上述**第 5.1 款**（d）段；（以及）

(b) 确保成员替代后**争端避免/裁决委员会**活动方式的一致性。

8
双方的承诺与赔偿

8.1 **雇主**和**承包商**相互承诺并对**争端避免/裁决委员会**成员保证，**争端避免/裁决委员会**成员不得：

(a) 在**合同**规定的任何仲裁中被任命为仲裁员；
(b) 被要求作为证人就**合同**规定的任何仲裁中的**争端**提供证据；（或）
(c) 对在履行或声称履行**争端避免/裁决委员会**成员职责过程中，所做的和未做的任何索赔负责，但其任何欺诈、严重过失、故意违约或不当行为除外。

8.2 **雇主**和**承包商**特此共同和各自赔偿并保护**争端避免/裁决委员会**成员，使其免受根据上述**第 8.1 款**免除其责任的任何索赔而造成的所有损害、损失和费用（包括律师费和开支）。

9
费用与支出

9.1 **争端避免/裁决委员会**成员应以**争端避免/裁决委员会协议书**中规定的货币，按以下方式获得报酬：⊖

(a) 月费，应作为固定费用全额支付：

(i) 可以参加**争端避免/裁决委员会规则**规定的、提前 28 天通知的所有会议、**现场**考察和意见听取会（以及，如果根据**争端避免/裁决委员会规则第 3.6 款**提出要求，则应参加紧急会议或**现场**考察）；
(ii) 根据上述**第 5.1 款**（d）段，熟悉**合同**内容、了解工程进展情况，并保持当前文件的工作档案；
(iii) 与其职责相关的所有办公及管理费用，包括秘书服务、复印和办公用品；（以及）
(iv) 除本**条**（b）和（c）段所提及的服务外，根据本协议提供的所有服务。

该费用应从**生效日期**当月的最后一天起每月支付，直至**争端避免/裁决委员会**任期届满，或其由于死亡、疾病、残疾、辞职或**争端避免/裁决委员会协议书**终止等原因，拒绝履行或无法履行其义务的当月月底为止。⊖

如果**争端避免/裁决委员会协议书**中未规定月费，则上述（i）至（iv）段中所述的事项应视为由下述（b）段中的日费所涵盖；⊖

(b) 日费，应视为每天全额支付：

(i) **争端避免/裁决委员会**成员的住所与**现场**之间的旅程，或与各方和/或其他成员（如果有）进行的会议的其他地点之间不足一天，最多两天的单程旅行的时间；

(ii) 按照**争端避免/裁决委员会规则第 3 条**规定参加会议和**现场**考察，并根据**争端避免/裁决委员会规则第 3.10 款**撰写报告的时间；

(iii) 提供非正式协助的时间；

⊖ 此处中文按照勘误表改正后的英文翻译。——译者注

(iv) spent on attending hearings (and, in case of a three-member DAAB, attending meeting(s) between the DAAB Members in accordance with sub-paragraph (a) of Rule 8.2 of the DAAB Rules, and communicating with the Other Members), and preparing decisions; and

(v) spent in preparation for a hearing, and studying written documentation and arguments from the Parties submitted in accordance with sub-paragraph (c) of Rule 7.1 of the DAAB Rules;

(c) all reasonable expenses, including necessary travel expenses (air fare in business class or equivalent, hotel and subsistence and other direct travel expenses, including visa charges) incurred in connection with the DAAB Member's duties, as well as the cost of telephone calls (and video conference calls, if any, and internet access), courier charges and faxes. The DAAB Member shall provide the Parties with a receipt for each item of expenses;

(d) any taxes properly levied in the Country on payments made to the DAAB Member (unless a national or permanent resident of the Country) under this Sub-Clause 9.1.

9.2 Subject to Sub-Clause 9.3 below, the amounts of the DAAB Member's monthly fee and daily fee, under Sub-Clause 9.1 above, shall be as specified in the DAA Agreement signed (or, under the Contract, deemed to have been signed) by the Parties and the DAAB Member.

9.3 If the Parties and the DAAB Member have agreed all other terms of the DAA Agreement but fail to jointly agree the amount of the monthly fee or the daily fee in the DAA Agreement (the "non-agreed fee" in this Sub-Clause):

(a) the DAA Agreement shall nevertheless be deemed to have been signed by the Parties and the DAAB Member, except that the fee proposed by the DAAB Member shall only temporarily apply;

(b) either Party or the DAAB Member may apply to the appointing entity or official named in the Contract Data to set the amount of the non-agreed fee;

(c) the appointing entity or official shall, as soon as practicable and in any case within 28 days after receiving any such application, set the amount of the non-agreed fee, which amount shall be reasonable taking due regard of the complexity of the Works, the experience and qualifications of the DAAB Member, and all other relevant circumstances;

(d) once the appointing entity or official has set the amount of the non-agreed fee, this amount shall be final and conclusive, shall replace the fee under sub-paragraph (a) above, and shall be deemed to have applied from the Effective Date; and

(e) thereafter, after giving credit to the Parties for all amounts previously paid in respect of the non-agreed fee, the balance (if any) due from the DAAB Member to the Parties or from the Parties to the DAAB Member, as the case may be, shall be paid.

9.4 The DAAB Member shall submit invoices for payment of the monthly fee and air fares quarterly in advance. Invoices for other expenses and for daily fees shall be submitted following the conclusion of a meeting, Site visit or hearing; and following the giving of a decision or an informal written note (under Rule 2.1 of the DAAB Rules). All invoices shall be accompanied by a brief description of the DAAB's Activities performed during the relevant period and shall be addressed to the Contractor.

9.5 The Contractor shall pay each of the DAAB Member's invoices in full within 28 days after receiving each invoice. Thereafter, the Contractor shall apply to the Employer (in the Statements under the Contract) for reimbursement of one-half of the amounts of these invoices. The Employer shall then pay the Contractor in accordance with the Contract.

(ⅳ) 出席意见听取会［如为三人争端避免/裁决委员会，则根据争端避免/裁决委员会规则第8.2款（a）段规定参加争端避免/裁决委员会成员之间的会议，并与其他成员进行沟通］，并起草决定的时间；（以及）

(ⅴ) 准备意见听取会、研究各方根据争端避免/裁决委员会规则第7.1款（c）段提交的书面文件和论据。

(c) 与**争端避免/裁决委员会成员**职责相关的所有合理支出，包括必要的差旅费用（商务舱或同等票价、酒店和生活补贴以及其他直接的差旅费用，包括签证费用），以及电话费用（和视频电话会议、如果有，及互联网连接）、快递费和传真等。**争端避免/裁决委员会成员**应向各方提供每项支出的收据；

(d) 根据第9.1款向工程所在国**争端避免/裁决委员会成员**（工程所在国的国民或永久居民除外）支付后，应在该国适当征收的任何税款。

9.2 根据下述第9.3款规定，第9.1款所述的**争端避免/裁决委员会成员**的月费和日费数额，应与各方与**争端避免/裁决委员会成员**签订的**争端避免/裁决委员会协议书**中规定的相符（或根据合同规定，被视为已签订）。⊖

9.3 如果各方和**争端避免/裁决委员会成员**已商定**争端避免/裁决委员会协议书**的所有其他条款，但未共同商定**争端避免/裁决委员会协议书**中的月费或日费金额（**本款中的"未商定的费用"**），则：⊖

(a) **争端避免/裁决委员会协议书**应被视为已由各方与**争端避免/裁决委员会成员**签署，但**争端避免/裁决委员会成员**建议的费用仅应临时适用；⊖

(b) 任一方或**争端避免/裁决委员会成员**均可向**合同数据**中指定的任命实体或官员申请设定未商定的费用金额；

(c) 任命实体或官员应在切实可行的范围内，并在收到该申请的28天内，尽快设定未商定的费用金额。该金额应在适当考虑**工程**的复杂性、**争端避免/裁决委员会成员**的经验和素质，以及所有其他相关情况的基础上合理设定；

(d) 一旦任命实体或官员设定了未商定的费用金额，此金额应为最终和决定性金额，应取代上述（a）段规定的费用，并应被视为已从**生效日期**开始适用；（以及）

(e) 认可各方全额支付前未商定的费用后，应视情况支付**争端避免/裁决委员会成员**应付双方，或双方应付**争端避免/裁决委员会成员**的余额（如果有）。

9.4 **争端避免/裁决委员会成员**应每季度提前提交发票，以支付月酬金和机票。会议、**现场**考察或意见听取会结束，且在做出决定或非正式书面说明之后（根据**争端避免/裁决委员会规则**第2.1款的规定），应提交其他费用和日常支出的发票。所有发票均应附有相关时间段内**争端避免/裁决委员会**活动的简要说明，并应寄给**承包商**。

9.5 **承包商**应在收到每张发票后的28天内全额支付**争端避免/裁决委员会成员**的费用。此后，**承包商**应向**雇主**（在合同的报表中）申请付还这些发票金额的一半。雇主应根据合同向**承包商**支付。

⊖ 此处中文按照勘误表改正后的英文翻译。——译者注

9.6 If the Contractor fails to pay to the DAAB Member the amount to which he/she is entitled under the DAA Agreement within the period of 28 days stated at Sub-Clause 9.5 above, the DAAB Member shall inform the Employer who shall promptly pay the amount due to the DAAB Member and any other amount which may be required to maintain the function of the DAAB. Thereafter the Employer shall, by written request, be entitled to payment from the Contractor of:

(a) all sums paid in excess of one-half of these amounts;
(b) the reasonable costs of recovering these amounts from the Contractor; and
(c) financing charges calculated at the rate specified in Sub-Clause 14.8 [*Delayed Payment*] of the Conditions of Contract.

The Employer shall be entitled to such payment from the Contractor without any requirement to comply with Sub-Clause 20.2 [*Claims For Payment and/or EOT*] of the Conditions of Contract, and without prejudice to any other right or remedy.

9.7 If the DAAB Member does not receive payment of the amount due within 56 days after submitting a valid invoice, the DAAB Member may:

(a) not less than 7 days after giving a Notification to the Parties and the Other Members (if any), suspend his/her services until the payment is received; and/or
(b) resign his/her appointment by giving a Notification under Sub-Clause 10.1 below.

10 Resignation and Termination

10.1 The DAAB Member may resign at any time for any reason, by giving a Notification of not less than 28 days (or other period as may be agreed by the Parties) to the Parties and to the Other Members (if any). During this period the Parties shall take the necessary steps to appoint, without delay, a replacement DAAB Member in accordance with Sub-Clause 21.1 [*Constitution of the DAAB*] of the Conditions of Contract (and, if applicable, Sub-Clause 21.2 [*Failure to Appoint DAAB Member(s)*] of the Conditions of Contract).

10.2 On expiry of the period stated in Sub-Clause 10.1 above, the resigning DAAB Member's DAA Agreement shall terminate with immediate effect. However (except if the DAAB Member is unable to act as a result of illness or disability) if, on the date of the DAAB Member's notice under Sub-Clause 10.1 above, the DAAB is dealing with any Dispute under Sub-Clause 21.4 [*Obtaining DAAB's Decision*] of the Conditions of Contract, the DAAB Member's resignation shall not take effect and his/her DAA Agreement shall not terminate until after the DAAB has given all the corresponding decisions in accordance with the Contract.

10.3 At any time the Parties may jointly terminate the DAA Agreement by giving a Notification of not less than 42 days to the DAAB Member.

10.4 If the DAAB Member fails, without justifiable excuse, to comply with Sub-Clause 5.1 above, the Parties may, without prejudice to their other rights or remedies, jointly terminate his/her DAA Agreement by giving a Notification (by recorded delivery) to the DAAB Member. This notice shall take effect when it is received by the DAAB Member.

10.5 If either Party fails, without justifiable excuse, to comply with Clause 6 above, the DAAB Member may, without prejudice to his/her other rights or remedies, terminate the DAA Agreement by giving a Notification to the Parties. This notice shall take effect when received by both Parties.

9.6 如果**承包商**未能在上述**第 9.5 款**所述的 28 天内向**争端避免/裁决委员会成员**支付其根据**争端避免/裁决委员会协议书**的规定应得的款项，**争端避免/裁决委员会成员**应通知**雇主**，**雇主**应立即支付应付给**争端避免/裁决委员会成员**的款额，以及维持**争端避免/裁决委员会**运行所需的任何其他款项。随后，**雇主**应通过书面形式，有权要求**承包商**支付以下费用：⊖

(a) 超过已付金额一半的所有款额；
(b) 从**承包商**处收追回这些金额的合理成本；（以及）
(c) 按合同条件**第 14.8 款**[*延误的付款*]规定的费率计算的财务费用。

雇主有权从**承包商**处获得这种款额，而无须遵守合同条件**第 20.2 款**[*付款和/或竣工时间延长的索赔*]的规定，并且不影响任何其他权利或补救。

9.7 如果**争端避免/裁决委员会成员**在提交有效发票后 56 天内未收到应付款，**争端避免/裁决委员会成员**可以：

(a) 在向双方和**其他成员**（如果有）发出**通知**后不少于 7 天内，暂停其服务，直到收到付款为止；（和/或）

(b) 根据下述**第 10.1 款**发出**通知**，辞去其任命。

10 辞职与终止

10.1 **争端避免/裁决委员会成员**可在任何时候，在不少于 28 天前（或双方可能商定的其他期限）向双方和**其他成员**（如果有）发出**通知**，以任何理由辞职。在此期间，双方应采取必要措施，按照合同条件**第 21.1 款**[*争端避免/裁决委员会的组成*]（以及，如适用，**第 21.2 款**[*未能任命争端避免/裁决委员会成员*]）的规定，立即任命替代的**争端避免/裁决委员会成员**。

10.2 上述**第 10.1 款**规定的期限届满后，辞职的**争端避免/裁决委员会成员**的**争端避免/裁决委员会协议书**应立即终止。但是（除非**争端避免/裁决委员会成员**因疾病或残疾而无法采取行动），如果**争端避免/裁决委员会成员**根据上述**第 10.1 款**发出通知之日，**争端避免/裁决委员会**正在根据合同条件**第 21.4 款**[*取得争端避免/裁决委员会的决定*]的规定处理任何争端，**争端避免/裁决委员会成员**的辞职将不生效，并且其**争端避免/裁决委员会协议书**将在**争端避免/裁决委员会**根据合同做出所有相应决定后才终止。⊖

10.3 在任何时候，双方可通过向**争端避免/裁决委员会成员**发出不少于 42 天的**通知**，共同终止**争端避免/裁决委员会协议书**。⊖

10.4 如果**争端避免/裁决委员会成员**在无合理理由的情况下，未能遵守上述**第 5.1 款**的规定，则双方可在不损害其他权利或补救措施的情况下，通过向其（通过挂号邮寄）发出**通知**共同终止其**争端避免/裁决委员会协议书**。该通知应自**争端避免/裁决委员会成员**收到后生效。⊖

10.5 如果任一方在没有合理理由的情况下，未能遵守上述**第 6 条**，则**争端避免/裁决委员会成员**可以在不损害其他权利或补救措施的情况下，通过向双方发出**通知**终止**争端避免/裁决委员会协议书**。双方收到后，该通知即生效。⊖

⊖ 此处中文按照勘误表改正后的英文翻译。——译者注

10.6 Any resignation or termination under this Clause shall be final and binding on the Parties and the DAAB Member. However, a notice given under Sub-Clause 10.3 or 10.4 above by either the Employer or the Contractor, but not by both, shall be of no effect.

10.7 Subject to sub-paragraph (b) of Sub-Clause 11.5 below, in the event of resignation or termination under this Clause the DAAB Member shall nevertheless be entitled to payment of any fees and/or expenses under his/her DAA Agreement that remain outstanding as of the date of termination of his/her DAA Agreement.

10.8 After resignation by the DAAB Member or termination of his/her DAA Agreement under this Clause, the DAAB Member shall:

(a) remain bound by his/her obligation of confidentiality under Sub-Clause 7.1 above; and

(b) return the original of any document in his/her possession to the Party who submitted such document in connection with the DAAB's Activities, at that Party's written request and cost.

10.9 Subject to any mandatory requirements under the governing law of the DAA Agreement, termination of the DAA Agreement under this Clause shall require no action of whatsoever kind by the Parties or the DAAB Member (as the case may be) other than as stated in this Clause.

11 Challenge

11.1 The Parties shall not object against the DAAB Member, except that either Party, or in the case of a three-member DAAB the Other Members jointly, shall be entitled to do so for an alleged lack of independence or impartiality or otherwise in which case Rule 10 Objection Procedure and Rule 11 Challenge Procedure of the DAAB Rules shall apply.

11.2 The decision issued under Rule 11 of the DAAB Rules (the "Decision on the Challenge" in the GCs) shall be final and conclusive.

11.3 At any time before the Decision on the Challenge is issued, the challenged DAAB Member may resign under Sub-Clause 10.1 above and, in such case, the challenging Party shall inform the International Chamber of Commerce (ICC). However, Sub-Clause 10.2 shall not apply to such resignation and the resigning DAAB Member's DAA Agreement shall terminate with immediate effect.

11.4 Unless the challenged DAAB Member has resigned, or his/her DAA Agreement has been terminated under Sub-Clause 10.3 above, the DAAB Member and the Other Members (if any) shall continue with the DAAB's Activities until the Decision on the Challenge is issued.

11.5 If the Decision on the Challenge is that the challenge is successful:

(a) the challenged DAAB Member's appointment, and his/her DAA Agreement, shall be deemed to have been terminated with immediate effect on the date of the notification of the Decision on the Challenge by ICC;

(b) the challenged DAAB Member shall not be entitled to any fees or expenses under his/her DAA Agreement from the date of the notification of the Decision on the Challenge by ICC;

(c) any decision under Sub-Clause 21.4.3 [*The DAAB's decision*] of the Conditions of Contract, given by the DAAB:

(i) after the challenge was referred under Rule 11 of the DAAB Rules; and

10.6 本条规定的任何辞职或终止都是最终决定，对双方和**争端避免／裁决委员会成员**均具有约束力。但是，雇主或承包商单方，而非双方共同地，根据上述**第 10.3 款**或**第 10.4 款**的规定发出的**通知**，应无效。

10.7 根据下述**第 11.5 款**（b）段的规定，在本条规定辞职或终止的情况下，**争端避免／裁决委员会成员**仍有权获得**争端避免／裁决委员会协议书**规定的、其**争端避免／裁决委员会协议书**终止之日尚未支付的任何费用和／或开支。⊖

10.8 **争端避免／裁决委员会成员**根据本条规定辞职或终止其**争端避免／裁决委员会协议书**后，**争端避免／裁决委员会成员**应：

（a） 仍受上述**第 7.1 款**规定的保密义务约束；（以及）

（b） 在一方书面请求并支付费用的情况下，将其持有的与**争端避免／裁决委员会**活动有关的任何文件的原件，返还给提交此类文件的一方。

10.9 在遵守**争端避免／裁决委员会协议书**适用法律任何强制性要求的前提下，根据本条终止**争端避免／裁决委员会协议书**，除本条另有规定外，双方或**争端避免／裁决委员会成员**（视情况而定）应无须采取任何形式的行动。⊖

11 质疑

11.1 双方不得对**争端避免／裁决委员会成员**提出异议，除非任一方因涉嫌缺乏独立性或公正性或其他原因提出异议，在这种情况下，**争端避免／裁决委员会规则第 10 条**的反对程序和**第 11 条**的质疑程序应适用。⊖

11.2 根据**争端避免／裁决委员会规则第 11 条**做出的决定（一般条件中的"**质疑决定**"）应为最终的且具有决定性的。

11.3 发出**质疑决定**前，被质疑的**争端避免／裁决委员会成员**可随时根据上述**第 10.1 款**辞职，在这种情况下，质疑方应通知**国际商会**（ICC）。但是，**第 10.2 款**不适用于此类辞职，辞职后，**争端避免／裁决委员会成员**的**争端避免／裁决委员会协议书**应立即终止。⊖

11.4 除非被质疑的**争端避免／裁决委员会成员**辞职，或其**争端避免／裁决委员会协议书**已根据上述**第 10.3 款**规定终止，否则**争端避免／裁决委员会成员**和其他成员（如果有）应继续**争端避免／裁决委员会**活动，直到发布有关**质疑决定**为止。⊖

11.5 如果**质疑决定**是成功的质疑：

（a） 被质疑的**争端避免／裁决委员会成员**的任命及其**争端避免／裁决委员会协议书**，应被视为在国际商会发出**质疑决定**通知之日即刻终止；⊖

（b） 被质疑的**争端避免／裁决委员会成员**自国际商会发出**质疑决定**通知之日起，将无权根据其**争端避免／裁决委员会协议书**获得任何费用或开支；⊖

（c） **争端避免／裁决委员会**根据合同条件**第 21.4.3 项**［**争端避免／裁决委员会的决定**］的规定做出的任何决定：

　　（i） 根据**争端避免／裁决委员会规则第 11 条**提出质疑后；（以及）

⊖ 此处中文按照勘误表改正后的英文翻译。——译者注

(ii) before the resignation (if any) of the challenged DAAB Member under Sub-Clause 11.3 above, or his/her DAA Agreement is terminated under sub-paragraph (a) above or under Sub-Clause 10.3 above

shall become void and ineffective. In the case of a sole-member DAAB, all other DAAB's Activities during this period shall also become void and ineffective. In the case of a three-member DAAB, all other DAAB's Activities during this period shall remain unaffected by the Decision on the Challenge except if there has been a challenge to all three members of the DAAB and such challenge is successful;

(d) the successfully challenged DAAB Member shall be removed from the DAAB; and

(e) the Parties shall, without delay, appoint a replacement DAAB Member in accordance with Sub-Clause 21.1 [*Constitution of the DAAB*] of the Conditions of Contract.

12
Disputes under the DAA Agreement

Any dispute arising out of or in connection with the DAA Agreement, or the breach, termination or invalidity thereof, shall be finally settled by arbitration under the Rules of Arbitration of the International Chamber of Commerce 2017 by one arbitrator appointed in accordance with these Rules of Arbitration, and Article 30 and the Expedited Procedure Rules at Appendix VI of these Rules of Arbitration shall apply.

(ii) 根据上述第 11.3 款被质疑的**争端避免 / 裁决委员会**成员辞职前（如果有），或其**争端避免 / 裁决委员会**协议书根据上述（a）段或第 10.3 款的规定终止后。⊖

应作废和无效。如果是唯一成员的**争端避免 / 裁决委员会**，在此期间所有其他**争端避免 / 裁决委员会**活动也将作废和无效。如果是三人成员的**争端避免 / 裁决委员会**，在此期间所有其他**争端避免 / 裁决委员会**活动均不受质疑决定的影响，除非对**争端避免 / 裁决委员会**的所有三名成员都提出了质疑，并且该质疑是成功的；

（d） 被成功质疑的**争端避免 / 裁决委员会**成员应从**争端避免 / 裁决委员会**中除名；（以及）

（e） 双方应立即根据合同条件第 21.1 款［**争端避免 / 裁决委员会**的组成］的规定任命**争端避免 / 裁决委员会**的替代成员。

| 12 争端避免 / 裁决委员会协议书中的争端⊖ | 由**争端避免 / 裁决委员会**协议书或与之有关的，或因对其违反或其终止或无效而引起的任何**争端**，应根据国际商会 2017 年的**仲裁规则**，由按照这些**仲裁规则**任命的一位仲裁员最终解决。此仲裁规则第 30 条以及附录 6 的加急程序规则应适用。⊖ |

⊖ 此处中文按照勘误表改正后的英文翻译。——译者注

Annex DAAB PROCEDURAL RULES

Rule 1
Objectives

1.1 The objectives of these Rules are:

(a) to facilitate the avoidance of Disputes that might otherwise arise between the Parties; and
(b) to achieve the expeditious, efficient and cost effective resolution of any Dispute that arises between the Parties.

1.2 These Rules shall be interpreted, the DAAB's Activities shall be conducted and the DAAB shall use its powers under the Contract and these Rules, in the manner necessary to achieve the above objectives.

Rule 2
Avoidance of Disputes

2.1 Where Sub-Clause 21.3 [*Avoidance of Disputes*] of the Conditions of Contract applies, the DAAB (in the case of a three-member DAAB, all three DAAB Members acting together) may give Informal Assistance during discussions at any meeting with the Parties (whether face-to-face or by telephone or by video conference) or at any Site visit or by an informal written note to the Parties.

Rule 3
Meetings and Site Visits

3.1 The purpose of meetings with the Parties and Site visits by the DAAB is to enable the DAAB to:

(a) become and remain informed about the matters described in sub-paragraphs (d)(i) to (d)(iii) of Sub-Clause 5.1 of the GCs;
(b) become aware of, and remain informed about, any actual or potential issue or disagreement between the Parties; and
(c) give Informal Assistance if and when jointly requested by the Parties.

3.2 As soon as practicable after the DAAB is appointed, the DAAB shall convene a face-to-face meeting with the Parties. At this meeting, the DAAB shall establish a schedule of planned meetings and Site visits in consultation with the Parties, which schedule shall reflect the requirements of Rule 3.3 below and shall be subject to adjustment by the DAAB in consultation with the Parties.

3.3 The DAAB shall hold face-to-face meetings with the Parties, and/or visit the Site, at regular intervals and/or at the written request of either Party. The frequency of such meetings and/or Site visits shall be:

(a) sufficient to achieve the purpose under Rule 3.1 above;
(b) at intervals of not more than 140 days unless otherwise agreed jointly by the Parties and the DAAB; and
(c) at intervals of not less than 70 days, subject to Rules 3.5 and 3.6 below and except as required to conduct a hearing as described under Rule 7 below, unless otherwise agreed jointly by the Parties and the DAAB.

3.4 In addition to the face-to-face meetings referred to in Rules 3.2 and 3.3 above, the DAAB may also hold meetings with the Parties by telephone or video conference as agreed with the Parties (in which case each Party bears the risk of interrupted or faulty telephone or video conference transmission and reception).

附件　争端避免／裁决委员会程序规则

规则 1
目标

1.1 这些**规则**的目标是：

(a) 帮助避免双方之间可能发生的**争端**；(以及)

(b) 迅速、有效且具有成本效益地解决双方之间发生的任何**争端**。

1.2 这些**规则**应得到解释，**争端避免／裁决委员会**活动应得到执行，并且**争端避免／裁决委员会**应以实现上述目标的必要方式，行使其**合同**和这些**规则**规定的权利。

规则 2
争端避免

2.1 在**合同**条件第 21.3 款［*争端避免*］适用的情况下，**争端避免／裁决委员会**（如果是三人成员的**争端避免／裁决委员会**，三名**争端避免／裁决委员会成员**共同行动）可在与双方的任何会议上的讨论中（无论是面对面还是通过电话或视频会议的），或在任何**现场**考察中，或通过给双方的非正式书面说明中，提供非正式协助。

规则 3
会议及现场考察

3.1 与双方会议及**争端避免／裁决委员会现场**考察的目的，是使**争端避免／裁决委员会**能够：

(a) 了解一般条件的第 5.1 款（d）(i) 至（d）(iii) 段中所述的事项；

(b) 知晓并了解双方之间的任何实际或潜在的问题或分歧；(以及)

(c) 如果并当双方共同要求时，则提供非正式协助。

3.2 任命**争端避免／裁决委员会**后，**争端避免／裁决委员会**应在切实可行的范围内，尽快与双方举行面对面会议。会议上，**争端避免／裁决委员会**应与双方协商制定会议计划和**现场**考察时间表，该时间表应反映下述**规则** 3.3 的要求，并应由**争端避免／裁决委员会**与双方协商调整。

3.3 **争端避免／裁决委员会**应定期和／或在任一方的书面要求下，与双方进行面对面的会议，和／或考察**现场**。此类会议和／或**现场**考察的频率应为：

(a) 足以达到上述**规则** 3.1 规定的目的；
(b) 除非双方和**争端避免／裁决委员会**另有商定，间隔不超过 140 天；(以及)

(c) 除非下述**规则** 7 要求进行的意见听取会，在遵守下述**规则** 3.5 和**规则** 3.6 的前提下，间隔不少于 70 天，除非双方和**争端避免／裁决委员会**另有商定。

3.4 除上述**规则** 3.2 和**规则** 3.3 提及的面对面会议外，**争端避免／裁决委员会**还可以按照与双方达成的商定，通过电话或视频会议方式与双方举行会议（在这种情况下，各方承担电话或视频会议传输中断和接收中断或故障的风险）。

3.5 At times of critical construction events (which may include suspension of the Works or termination of the Contract), the DAAB shall visit the Site at the written request of either Party. This request shall describe the critical construction event. If the DAAB becomes aware of an upcoming critical construction event, it may invite the Parties to make such a request.

3.6 Either Party may request an urgent meeting or Site visit by the DAAB. This shall be a written request and shall give reasons for the urgency of the meeting or Site visit. If the DAAB agrees that such a meeting or Site visit is urgent, the DAAB Members shall use all reasonable endeavours to:

(a) hold a meeting with the Parties by telephone or video conference (as agreed with the Parties under Rule 3.4 above) within 3 days after receiving the request; and

(b) if requested and (having given the other Party opportunity at this meeting to respond to or oppose this request) the DAAB agrees that a Site visit is necessary, visit the Site within 14 days after the date of this meeting.

3.7 The time of, and agenda for, each meeting and Site visit shall be set by the DAAB in consultation with the Parties.

3.8 Each meeting and Site visit shall be attended by the Employer, the Contractor and the Engineer.

3.9 Each meeting and Site visit shall be co-ordinated by the Contractor in co-operation with the Employer and the Engineer. The Contractor shall ensure the provision of appropriate:

(a) personal safety equipment, security controls (if necessary) and site transport for each Site visit;

(b) meeting room/conference facilities and secretarial and copying services for each face-to-face meeting; and

(c) telephone conference or video conference facilities for each meeting by telephone or video conference.

3.10 At the conclusion of each meeting and Site visit and, if possible before leaving the venue of the face-to-face meeting or the Site (as the case may be) but in any event within 7 days, the DAAB shall prepare a report on its activities during the meeting or Site visit and shall send copies of this report to the Parties and the Engineer.

Rule 4
Communications and Documentation

4.1 The language to be used:

(a) in all communications to and from the DAAB and the Parties (and, in the case of a three-member DAAB, between the DAAB Members);
(b) in all reports and decisions issued by the DAAB; and
(c) during all Site visits, meetings and hearings relating to the DAAB's Activities

shall be the language for communications defined in Sub-Clause 1.4 [*Law and Language*] of the Conditions of Contract, unless otherwise agreed jointly by the Parties and the DAAB.

4.2 All communications and/or documents sent between the DAAB and a Party shall simultaneously be copied to the other Party. In the case of a three-member DAAB, the sending Party shall send all communications and/or documents to the chairman of the DAAB and simultaneously send copies of these communications and/or documents to the Other Members.

3.5 在重大施工事件（可能包括工程暂停或合同终止）发生时，**争端避免/裁决委员会**应在任一方书面请求下考察**现场**。该请求应就重大施工事件进行说明。如果**争端避免/裁决委员会**意识到即将发生的重大施工事件，可以邀请**双方**提出此类请求。

3.6 任一方均可请求**争端避免/裁决委员会**召开紧急会议或进行**现场**考察。请求应为书面形式，并应说明会议或**现场**考察的紧迫性。如果**争端避免/裁决委员会**同意举行这样的紧急会议或**现场**考察，则**争端避免/裁决委员会成员**应尽一切合理努力：

(a) 在收到请求后 3 天内通过电话或视频会议形式（按上述**规则 3.4** 与双方商定的）与双方举行会议；（以及）

(b) 应请求，并且（在本次会议上给另一方机会回应或反对该请求）**争端避免/裁决委员会**认为**现场**考察有必要，则本次会议召开后 14 天内考察**现场**。

3.7 每次会议和**现场**考察的时间和议程应由**争端避免/裁决委员会**与双方协商确定。

3.8 每次会议和**现场**考察均应由**雇主**、**承包商**和**工程师**参加。

3.9 每次会议和**现场**考察均应由**承包商**与**雇主**和**工程师**进行协商。**承包商**应确保提供适当的：

(a) 每次**现场**考察的个人安全设备、安全控制（如必要）及现场交通；

(b) 每次面对面会议的会议室/会议设施以及秘书和复印服务；（以及）

(c) 每次电话或视频会议的电话会议或视频会议设施。

3.10 在每次会议和**现场**考察结束时，以及在可能的情况下离开面对面会议或**现场**（视情况而定）之前，但无论如何在 7 天内，**争端避免/裁决委员会**应编写一份有关其在会议或**现场**考察期间的活动报告，并应将该报告的副本发送给双方和**工程师**。

规则 4
沟通与文件化

4.1 使用的语言：

(a) **争端避免/裁决委员会**和双方之间的所有通信交流中（如果是三人成员的**争端避免/裁决委员会**，则在两名**争端避免/裁决委员会成员**之间）；
(b) 在**争端避免/裁决委员会**发布的所有报告和决定中；（以及）
(c) 在与**争端避免/裁决委员会**活动有关的所有**现场**考察、会议和意见听取会期间。

应采用**合同条件**第 1.4 款[*法律和语言*]中规定的通信交流语言，除非双方和**争端避免/裁决委员会**另有共同商定。

4.2 **争端避免/裁决委员会**与一方之间进行的所有通信交流和/或文件应同时抄送给另一方。如果是三人成员的**争端避免/裁决委员会**，发送方应将所有通信交流和/或文件发送给**争端避免/裁决委员会**主席，同时将这些通信交流和/或文件的副本发送给**其他成员**。⊖

⊖ 此处中文按照勘误表改正后的英文翻译。——译者注

4.3 The Parties shall provide the DAAB with a copy of all documents which the DAAB may request, including:

(a) the documents forming the Contract;
(b) progress reports under Sub-Clause 4.20 [*Progress Reports*] of the Conditions of Contract;
(c) the initial programme and each revised programme under Sub-Clause 8.3 [*Programme*] of the Conditions of Contract;
(d) relevant instructions given by the Engineer, and Variations under Clause 13.3 [*Variation Procedure*] of the Conditions of Contract;
(e) Statements submitted by the Contractor, and all certificates issued by the Engineer under the Contract;
(f) relevant Notices;
(g) relevant communications between the Parties and between either Party and the Engineer

and any other document relevant to the performance of the Contract and/or necessary to enable the DAAB to become and remain informed about the matters described in sub-paragraphs (d)(i) to (d)(iii) of Sub-Clause 5.1 of the GCs.

Rule 5
Powers of the DAAB

5.1 In addition to the powers granted to the DAAB under the Conditions of Contract, the General Conditions of the DAAB Agreement and elsewhere in these Rules, the Parties empower the DAAB to:

(a) establish the procedure to be applied in making Site visits and/or giving Informal Assistance;
(b) establish the procedure to be applied in giving decisions under the Conditions of Contract;
(c) decide on the DAAB's own jurisdiction, and the scope of any Dispute referred to the DAAB;
(d) appoint one or more experts (including legal and technical expert(s)), with the agreement of the Parties;
(e) decide whether or not there shall be a hearing (or more than one hearing, if necessary) in respect of any Dispute referred to the DAAB;
(f) conduct any meeting with the Parties and/or any hearing as the DAAB thinks fit, not being bound by any rules or procedures for the hearing other than those contained in the Contract and in these Rules;
(g) take the initiative in ascertaining the facts and matters required for a DAAB decision;
(h) make use of a DAAB Member's own specialist knowledge, if any;
(i) decide on the payment of financing charges in accordance with the Contract;
(j) decide on any provisional relief such as interim or conservatory measures;
(k) open up, review and revise any certificate, decision, determination, instruction, opinion or valuation of (or acceptance, agreement, approval, consent, disapproval, No-objection, permission, or similar act by) the Engineer that is relevant to the Dispute; and
(l) proceed with the DAAB's Activities in the absence of a Party who, after receiving a Notification from the DAAB, fails to comply with Sub-Clause 6.3 of the GCs.

5.2 The DAAB shall have discretion to decide whether and to what extent any powers granted to the DAAB, under the Conditions of Contract, the GCs and these Rules, may be exercised.

Rule 6
Disputes

6.1 If any Dispute is referred to the DAAB in accordance with Sub-Clause 21.4.1 [*Reference of a Dispute to the DAAB*] of the Conditions of Contract, the DAAB shall proceed in accordance with Sub-Clause 21.4 [*Obtaining DAAB's Decision*] of the Conditions of Contract and these DAAB Rules, or as otherwise agreed by the Parties in writing.

4.3 双方应向**争端避免/裁决委员会**提供其可能要求的所有文件的副本，包括：

（a） 构成**合同**的文件；
（b） **合同条件第 4.20 款**[*进度报告*]规定的进度报告；
（c） **合同条件第 8.3 款**[*进度计划*]规定的初始进度计划和每个修订的进度计划；
（d） 工程师发出的相关指示以及**合同条件第 13.3 款**[*变更程序*]中规定的变更；
（e） **承包商**提交的**报表**以及工程师按照**合同**签发的所有证书；
（f） 有关**通知**；
（g） 双方之间以及任一方与工程师之间的相关通信交流

以及与履行**合同**有关的，和/或使**争端避免/裁决委员会**能够了解一般条件第 5.1 款（d）（i）至（d）（iii）段所述事项的必要信息的任何其他文件。

规则 5
争端避免/裁决委员会的权利

5.1 除**合同条件**、**争端避免/裁决委员会协议书**一般条件以及本规则其他部分赋予**争端避免/裁决委员会**的权利外，双方还授权**争端避免/裁决委员会**：

（a） 制定应用于**现场**考察和/或提供**非正式协助**的程序；

（b） 制定应用于根据**合同条件**做出决定的程序；

（c） 决定**争端避免/裁决委员会**自身的管辖权，以及提交给**争端避免/裁决委员会**的任何**争端**的范围；
（d） 经双方商定，任命一名或多名专家（包括法律和技术专家）；

（e） 决定是否应就提交给**争端避免/裁决委员会**的任何**争端**举行意见听取会（或必要时，多场意见听取会）；
（f） 在**争端避免/裁决委员会**认为合适的情况下，与双方举行任何会议和/或任何意见听取会，不受除**合同**和本规则规定以外的规则或程序的约束；

（g） 主动查明**争端避免/裁决委员会**决定所需的事实和事项；

（h） 利用**争端避免/裁决委员会**成员自身的专业知识（如果有）；
（i） 根据合同决定融资费用的付款；
（j） 决定任何临时性救济事项，如临时性或保全性措施；
（k） 开启、审核并修改与**争端**相关的工程师的任何证书、决定、确定、指示、意见或估价（或接受、商定、通过、同意、不同意、不反对、允许或类似行为）；（以及）

（l） 在收到**争端避免/裁决委员会**通知后，未能遵守一般条件第 6.3 款规定的一方缺席的情况下，开展**争端避免/裁决委员会**活动。

5.2 **争端避免/裁决委员会**应有权酌情决定是否以及多大程度上，行使**合同条件**、一般条件和本规则赋予**争端避免/裁决委员会**的任何权利。

规则 6
争端

6.1 如果根据**合同条件**第 21.4.1 项[*争端提交给争端避免/裁决委员会*]的规定，将任何**争端**提交给**争端避免/裁决委员会**，则**争端避免/裁决委员会**应按照**合同条件**第 21.4 款[*取得争端避免/裁决委员会的决定*]的规定以及这些**争端避免/裁决委员会**规则进行处理，或以双方另有书面商定的方式处理。

6.2 The DAAB shall act fairly and impartially between the Parties and, taking due regard of the period under Sub-Clause 21.4.3 [*The DAAB's decision*] of the Conditions of Contract and other relevant circumstances, the DAAB shall:

(a) give each Party a reasonable opportunity (consistent with the expedited nature of the DAAB proceeding) of putting forward its case and responding to the other Party's case; and

(b) adopt a procedure in coming to its decision that is suitable to the Dispute, avoiding unnecessary delay and/or expense.

Rule 7
Hearings

7.1 In addition to the powers under Rule 5.1 above, and except as otherwise agreed in writing by the Parties, the DAAB shall have power to:

(a) decide on the date and place for any hearing, in consultation with the Parties;
(b) decide on the duration of any hearing;
(c) request that written documentation and arguments from the Parties be submitted to it prior to the hearing;
(d) adopt an inquisitorial procedure during any hearing:
(e) request the production of documents, and/or oral submissions by the Parties, at any hearing that the DAAB considers may assist in exercising the DAAB's power under sub-paragraph (g) of Rule 5.1 above;
(f) request the attendance of persons at any hearing that the DAAB considers may assist in exercising the DAAB's power under sub-paragraph (g) of Rule 5.1 above;
(g) refuse admission to any hearing, or audience at any hearing, to any persons other than representatives of the Employer, the Contractor and the Engineer;
(h) proceed in the absence of any party who the DAAB is satisfied received timely notice of the hearing;
(i) adjourn any hearing as and when the DAAB considers further investigation by one Party or both Parties would benefit resolution of the Dispute, for such time as the investigation is carried out, and resume the hearing promptly thereafter.

7.2 The DAAB shall not express any opinions during any hearing concerning the merits of any arguments advanced by either Party in respect of the Dispute.

7.3 The DAAB shall not give any Informal Assistance during a hearing, but if the Parties request Informal Assistance during any hearing:

(a) the hearing shall be adjourned for such time as the DAAB is giving Informal Assistance;
(b) if the hearing is so adjourned for longer than 2 days, the period under Sub-Clause 21.4.3 [*The DAAB's decision*] of the Conditions of Contract shall be temporarily suspended until the date that the hearing is resumed; and
(c) the hearing shall be resumed promptly after the DAAB has given such Informal Assistance.

Rule 8
The DAAB's Decision

8.1 The DAAB shall make and give its decision within the time allowed under Sub-Clause 21.4 [*Obtaining DAAB's Decision*] of the Conditions of Contract, or other time as may be proposed by the DAAB and agreed by the Parties in writing.

8.2 In the case of a three-member DAAB:

(a) it shall meet in private (after the hearing, if any) in order to have discussions and to start preparation of its decision;
(b) the DAAB Members shall use all reasonable endeavours to reach a unanimous decision;

6.2 **争端避免/裁决委员会**应在双方之间公平、公正地采取行动，并在适当考虑合同条件第 21.4.3 项［*争端避免/裁决委员会的决定*］规定的时限和其他相关情况后，应：

（a） 给各方提供合理的机会（与**争端避免/裁决委员会**程序的快速性相一致），提出其案件并对另一方的案件做出回应；（以及）

（b） 采取适合**争端**的决定程序，避免不必要的延误和/或费用。

规则 7
意见听取会

7.1 除上述**规则** 5.1 规定的权利外，除非双方另有书面商定，否则**争端避免/裁决委员会**有权：

（a） 与双方商定意见听取会的日期和地点；
（b） 决定意见听取会的时长；
（c） 要求双方在意见听取会前提交书面文件和论据；

（d） 在任何意见听取会上采取研信程序；
（e） 在**争端避免/裁决委员会**认为可能有助于行使上述**规则** 5.1（g）段规定的权利的任何意见听取会上，要求双方出示文件和/或口头陈述；

（f） 在**争端避免/裁决委员会**认为可能有助于行使上述**规则** 5.1（g）段规定的权利的任何意见听取会上，要求人员出席；

（g） 拒绝除**雇主、承包商和工程师代表**以外的任何人参加任何意见听取会；

（h） 在**争端避免/裁决委员会**认为已及时收到意见听取会通知的任一方缺席的情况下继续进行；
（i） 在**争端避免/裁决委员会**认为一方或双方进行的进一步调查有利于解决**争端**时，或在开展调查期间，延期举行意见听取会，并在调查结束后立即复会。

7.2 在意见听取会上，**争端避免/裁决委员会**不得就任一方就**争端**提出的任何论据的是非曲直发表任何意见。

7.3 **争端避免/裁决委员会**不得在意见听取会期间提供任何非正式协助，但如果双方在任何意见听取会期间要求提供非**正式协助**，则：

（a） 意见听取会应在**争端避免/裁决委员会**提供非**正式协助**的时间休会；

（b） 如果因此休会超过 2 天，则合同条件第 21.4.3 项［*争端避免/裁决委员会的决定*］规定的时限应暂时终止，直到意见听取会恢复的日期为止；（以及）

（c） **争端避免/裁决委员会**提供此类非**正式协助**后，应立即恢复意见听取会。

规则 8
争端避免/裁决委员会的决定

8.1 **争端避免/裁决委员会**应在合同条件第 21.4 款［*取得争端避免/裁决委员会的决定*］规定允许的时间内，或**争端避免/裁决委员会**提议并经双方书面商定的其他时间内做出决定。

8.2 如果是三人成员的**争端避免/裁决委员会**：

（a） 为了进行讨论并开始起草其决定，该委员会应举行非公开会议（意见听取会之后，如果有）；
（b） **争端避免/裁决委员会**成员应尽一切合理努力达成一致决定；

(c) if it is not possible for the DAAB Members to reach a unanimous decision, the applicable decision shall be made by a majority of the DAAB Members, who may require the minority DAAB Member to prepare a separate written report (with reasons and supporting particulars) which shall be issued to the Parties; and

(d) if a DAAB Member fails to:

(i) attend a hearing (if any) or a DAAB Members' meeting; or
(ii) fulfil any required function (other than agreeing to a unanimous decision)

the Other Members shall nevertheless proceed to make a decision, unless:

- such failure has been caused by exceptional circumstances, of which the Other Members and the Parties have received a Notification from the DAAB Member;
- the DAAB Member has suspended his services under sub-paragraph (a) of Sub-Clause 9.7 of the GCs; or
- otherwise agreed by the Parties in writing.

8.3 If, after giving a decision, the DAAB finds (and, in the case of a three-member DAAB, they agree unanimously or by majority) that the decision contained any error:

(a) of a typographical or clerical nature; or
(b) of an arithmetical nature

the chairman of the DAAB or the sole DAAB Member (as the case may be) shall, within 14 days after giving this decision, advise the Parties of the error and issue an addendum to its original decision in writing to the Parties.

8.4 If, within 14 days of receiving a decision from the DAAB, either Party finds a typographical, clerical or arithmetical error in the decision, that Party may request the DAAB to correct such error. This shall be a written request and shall clearly identify the error.

8.5 If, within 14 days of receiving a decision from the DAAB, either Party believes that such decision contains an ambiguity, that Party may request clarification from the DAAB. This shall be a written request and shall clearly identify the ambiguity.

8.6 The DAAB shall respond to a request under Rule 8.4 or Rule 8.5 above within 14 days of receiving the request. The DAAB may decline (at its sole discretion and with no requirement to give reasons) any request for clarification under Rule 8.5. If the DAAB agrees (in the case of a three-member DAAB they agree unanimously or by majority) that the decision did contain the error or ambiguity as described in the request, it may correct its decision by issuing an addendum to its original decision in writing to the Parties, in which case this addendum shall be issued together with the DAAB's response under this Rule.

8.7 If the DAAB issues an addendum to its original decision under Rule 8.3 or 8.6 above, such an addendum shall form part of the decision and the period stated in sub-paragraph (c) of Sub-Clause 21.4.4 [*Dissatisfaction with DAAB's decision*] of the Conditions of Contract shall be calculated from the date the Parties receive this addendum.

Rule 9
In the event of Termination of DAAB Agreement

9.1 If, on the date of termination of a DAAB Member's DAAB Agreement arising from resignation or termination under Clause 10 of the GCs, the DAAB is dealing with any Dispute under Sub-Clause 21.4 [*Obtaining DAAB's Decision*] of the Conditions of Contract:

（c） 如果**争端避免/裁决委员会成员**不可能达成一致决定，则适用的决定应由大多数**争端避免/裁决委员会成员**做出，他们可能会要求少数**争端避免/裁决委员会成员**起草单独的书面报告（并附有理由和详细证明资料），该报告应发给双方；（以及）

（d） 如果**争端避免/裁决委员会成员**未能：

（i） 参加意见听取会（如果有）或**争端避免/裁决委员会成员会议**；（或）
（ii） 履行任何必要的职能（达成一致的决定除外）。

其他成员仍应继续做出决定，除非：

- 此类未履职是由特殊情况引起的，**其他成员和双方已收到争端避免/裁决委员会成员的通知**；
- **争端避免/裁决委员会成员**已根据一般条件第9.7款（a）段规定暂停服务；（或）
- 双方另有书面商定。

8.3 如果做出决定后，**争端避免/裁决委员会**发现（如果是三人成员的**争端避免/裁决委员会**，他们一致或多数同意）决定中有任何错误：

（a） 印刷或文书性质的；（或）
（b） 算数性质的。

争端避免/裁决委员会主席或唯一**争端避免/裁决委员会成员**（视情况而定）应在做出此决定后14天内将错误告知双方，并以书面形式将原始决定的附录发送给双方。⊖

8.4 如果任一方在收到**争端避免/裁决委员会**的决定后14天内发现该决定中存在印刷、文书或算数错误，则该方可要求**争端避免/裁决委员会**纠正该错误。应为书面要求，并应明确指出错误。

8.5 如果任一方在收到**争端避免/裁决委员会**的决定后14天内认为该决定含糊不清，则该方可要求**争端避免/裁决委员**做出澄清。应为书面要求，并应明确指出含糊不清之处。

8.6 **争端避免/裁决委员会**应在收到请求后14天内回应上述**规则8.4**或**规则8.5**的请求。**争端避免/裁决委员会**可根据**规则8.5**拒绝（自行决定，且不要求提供理由）任何澄清要求。如果**争端避免/裁决委员会**同意（如果是三人成员的**争端避免/裁决委员会**，他们一致或多数同意），该决定确实包含了请求中所述的错误或含糊不清，则可以通过以书面形式向双方发布原始决定的附录更正其决定。在这种情况下，该附录应与**争端避免/裁决委员会**根据本**规则**做出的答复一起发布。

8.7 如果**争端避免/裁决委员会**根据上述**规则8.3**或**规则8.6**发布了其原始决定的附录，则该附录应构成该决定的一部分，合同条件第21.4.4项［*对争端避免/裁决委员会决定不满意*］（c）段所述的时限，应从双方收到该附录之日起计算。

规则9

争端避免/裁决委员会协议书终止的情况

9.1 如果在一般条件第10条规定的辞职或终止，导致**争端避免/裁决委员会成员**的**争端避免/裁决委员会协议书**终止之日，**争端避免/裁决委员会**正在根据合同条件第21.4款［*取得争端避免/裁决委员会的决定*］的规定处理任何争端，则：

⊖ 此处中文按照勘误表改正后的英文翻译。——译者注

(a) the period under Sub-Clause 21.4.3 [The DAAB's decision] of the Conditions of Contract shall be temporarily suspended; and

(b) when a replacement DAAB Member is appointed in accordance with Sub-Clause 21.1 [*Constitution of the DAAB*] of the Conditions of Contract, the full period under Sub-Clause 21.4.3 [*The DAAB's decision*] of the Conditions of Contract shall apply from the date of this replacement DAAB Member's appointment.

9.2 In the case of a three-member DAAB and if one DAAB Member's DAAB Agreement is terminated as a result of resignation or termination under Clause 10 of the GCs, the Other Members shall continue as members of the DAAB except that they shall not conduct any hearing or make any decision prior to the replacement of the DAAB Member unless otherwise agreed jointly by the Parties and the Other Members.

Rule 10
Objection Procedure

10.1 The following procedure shall apply to any objection against a DAAB Member:

(a) the objecting Party shall, within 7 days of becoming aware of the facts and/or events giving rise to the objection, give a Notification to the DAAB Member of its objection. This Notification shall:

(i) state that it is given under this Rule;
(ii) state the reason(s) for the objection;
(iii) substantiate the objection by setting out the facts, and describing the events, on which the objection is based, with supporting particulars; and
(iv) be simultaneously copied to the other Party and the Other Members;

(b) within 7 days after receiving a notice under sub-paragraph (a) above, the objected DAAB Member shall respond to the objecting Party. This response shall be simultaneously copied to the other Party and the Other Members. If no response is given by the DAAB Member within this period of 7 days, the DAAB Member shall be deemed to have given a response denying the matters on which the objection is based;

(c) within 7 days after receiving the objected DAAB Member's response under sub-paragraph (b) above (or, if there is no such response, after expiry of the period of 7 days stated in sub-paragraph (b) above) the objecting Party may formally challenge a DAAB member in accordance with Rule 11 below;

(d) if the challenge is not referred within the period of 7 days stated in sub-paragraph (c) above, the objecting Party shall be deemed to have agreed to the DAAB Member remaining on the DAAB and shall not be entitled to object and/or challenge him/her thereafter on the basis of any of the facts and/or evidence stated in the notice given under sub-paragraph (a) above;

Rule 11
Challenge Procedure

11.1 If and when the objecting Party challenges a DAAB Member, within 21 days of learning of the facts upon which the challenge is based, the provisions of this Rule shall apply. Any challenge is to be decided by the International Chamber of Commerce (ICC) and administered by the ICC International Centre for ADR.

11.2 The procedure for such challenge and information on associated charges to be paid are set out at http://fidic.org and http://iccwbo.org.

- (a) 合同条件第 21.4.3 项［**争端避免／裁决委员会的决定**］规定的期限应临时暂停；（以及）
- (b) 根据合同条件第 21.1 款［**争端避免／裁决委员会的组成**］的规定任命**争端避免／裁决委员会**替代成员时，合同条件第 21.4.3 项［**争端避免／裁决委员会的决定**］规定的完整期限应从该**争端避免／裁决委员会**替代成员任命之日起生效。

9.2 在三人成员的**争端避免／裁决委员会**的情况下，且如果一名**争端避免／裁决委员会**成员的**争端避免／裁决委员会协议书**，由于一般条件第 10 条规定的辞职或终止而被终止，则**其他成员**应继续履行**争端避免／裁决委员会成员**职责，但在替换**争端避免／裁决委员会**成员之前不得召开意见听取会或做出任何决定，除非双方和**其他成员**另有共同商定。

规则 10
反对程序

10.1 以下程序应适用于对**争端避免／裁决委员会成员**的任何反对：

- (a) 反对方应在获悉引起反对的事实和／或事件之日起 7 天内，**通知**其反对的**争端避免／裁决委员会成员**。该**通知**应：

  - (i) 说明是根据本**规则**提出的；
  - (ii) 说明反对的理由；
  - (iii) 通过陈述事实、用详细证明资料阐述反对的事件，证实反对；（以及）

  - (iv) 同时抄送给另一方和**其他成员**。

- (b) 在收到根据上述（a）段发出的通知后 7 天内，被反对的**争端避免／裁决委员会成员**应回复反对方。该回复同时抄送给另一方和**其他成员**。如果**争端避免／裁决委员会成员**在 7 天内没有做出回复，应被视为已拒绝回复该反对事项；

- (c) 在收到根据上述（b）段被反对的**争端避免／裁决委员会成员**答复后的 7 天内［或，如未有此类回复，则在上述（b）段所述的 7 天期限届满后］，反对方可根据下述规则 11 对**争端避免／裁决委员会成员**提出正式质疑；

- (d) 如果在上述（c）段所述的 7 天内未提出质疑，则反对方应视为已同意该**争端避免／裁决委员会成员**留任在**争端避免／裁决委员会**，并且以后无权基于上述（a）段发出的通知中陈述的任何事实和／或证据对其进行反对和／或质疑。

规则 11
质疑程序

11.1 如果反对方在获知基于质疑的事实后 21 天内，质疑**争端避免／裁决委员会成员**，本**规则**的规定应适用。任何质疑均由**国际商会**决定，并由**国际商会国际中心**负责管理，应用**替代性纠纷解决方法（ADR）**。

11.2 有关此类质疑程序和应付费用的相关信息，请登录 http://fidic.org 和 http://iccwbo.org。

INDEX OF SUB-CLAUSES

	Sub-Clause	Page
Accepted Contract Amount, Sufficiency of the	4.11	72
Access after Taking Over, Right of	11.7	124
Access Route	4.15	74
Access to the Site, Right of	2.1	38
Adjustments for Changes in Cost	13.7	142
Adjustments for Changes in Laws	13.6	140
Advance Payment	14.2	144
Advance Warning	8.4	102
Agreement, Contract	1.6	28
Agreement or Determination	3.7	46
Amicable Settlement	21.5	210
Arbitration	21.6	210
Archaeological and Geological Findings	4.23	80
Assignment	1.7	30
Assistance	2.2	38
Authorities, Delays Caused by	8.6	104
Avoidance of Interference	4.14	74
Care of the Works, Responsibility for	17.1	176
Care of the Works, Liability for	17.2	176
Certificate, Performance	11.9	126
Claims	20.1	194
Claims For Payment and/or EOT	20.2	194
Clearance of Site	11.11	126
Commencement of Works	8.1	98
Completion of Outstanding Work and Remedying Defects	11.1	118
Completion, Statement at	14.10	156
Completion, Time for	8.2	100
Conditions, Unforeseeable Physical	4.12	72
Confidentiality	1.12	34
Contract Termination	1.16	38
Contract Price, The	14.1	144
Contractor to Search	11.8	124
Contractor's Claims	20.1	194
Contractor's Documents	4.4	60
Contractor's Documents, Employer's Use of	1.10	32
Contractor's Equipment	4.17	76
Contractor's General Obligations	4.1	52
Contractor's Obligations after Termination	16.3	176
Contractor's Obligations: Tests on Completion	9.1	110
Contractor's Operations on Site	4.22	80
Contractor's Personnel	6.9	88
Contractor's Records	6.10	90
Contractor's Representative	4.3	58
Contractor's Superintendence	6.8	88
Co-operation	4.6	64

条款索引

	条款	页
中标合同金额的充分性	4.11	73
接收后的进入权	11.7	125
进场通路	4.15	75
现场进入权	2.1	39
因成本改变的调整	13.7	143
因法律改变的调整	13.6	141
预付款	14.2	145
预先警示	8.4	103
合同协议书	1.6	29
商定或确定	3.7	47
友好解决	21.5	211
仲裁	21.6	211
考古和地质发现	4.23	81
权益转让	1.7	31
协助	2.2	39
部门造成的延误	8.6	105
避免干扰	4.14	75
工程照管的职责	17.1	177
工程照管的责任	17.2	177
履约证书	11.9	127
索赔	20.1	195
付款和/或竣工时间延长的索赔	20.2	195
现场清理	11.11	127
工程的开工	8.1	99
完成扫尾工作和修补缺陷	11.1	119
竣工报表	14.10	157
竣工时间	8.2	101
不可预见的物质条件	4.12	73
保密	1.12	35
合同终止	1.16	39
合同价格	14.1	145
承包商调查	11.8	125
索赔	20.1	195
承包商文件	4.4	61
雇主使用承包商文件	1.10	33
承包商设备	4.17	77
承包商的一般义务	4.1	53
终止后承包商的义务	16.3	177
承包商的义务	9.1	111
承包商的现场作业	4.22	81
承包商人员	6.9	89
承包商的记录	6.10	91
承包商代表	4.3	59
承包商的监督	6.8	89
合作	4.6	65

Cost, Adjustments for Changes in	13.7	142
Currencies of Payment	14.15	162
DAAB - see Dispute Avoidance/Adjudication Board		
Daywork	13.5	140
Defective Work off Site, Remedying of	11.5	122
Defects, Failure to Remedy	11.4	120
Defects and Rejection	7.5	96
Defects Notification Period, Extension of	11.3	120
Defects, Remedying of	11.1	118
Definitions	1.1	12
Delay Damages	8.8	106
Delayed Drawings or Instructions	1.9	30
Delays Caused by Authorities	8.6	104
Discharge	14.12	160
Disorderly Conduct	6.11	90
Dispute Avoidance/Adjudication Board, Constitution of the	21.1	202
Dispute Avoidance/Adjudication Board's Decision, Failure to Comply with	21.7	210
Dispute Avoidance/Adjudication Board's Decision, Obtaining	21.4	206
Dispute Avoidance/Adjudication Board, No DAAB In Place	21.8	212
Dispute Avoidance/Adjudication Board Member(s), Failure to Appoint	21.2	204
Disputes, Avoidance of	21.3	204
Documents, Care and Supply of	1.8	30
Documents, Contractor's Use of Employer's	1.11	32
Documents, Employer's Use of Contractor's	1.10	32
Documents, Priority of	1.5	28
Duty to Minimise Delay	18.3	184
Employer-Supplied Materials and Employer's Equipment	2.6	42
Employer's Claims: Currencies of Payment	14.15	162
Employer's Documents, Contractor's Use of	1.11	32
Employer's Financial Arrangements	2.4	40
Employer's Liability, Cessation of	14.14	162
Employer's Personnel and Other Contractors	2.3	40
Engineer, The	3.1	42
Engineer, Delegation by the	3.4	44
Engineer, Replacement of	3.6	46
Engineer's Duties and Authority	3.2	42
Engineer's Instructions	3.5	44
Engineer's Representative	3.3	44
Environment, Protection of the	4.18	76
Exceptional Events	18.1	182
Exceptional Event, Notice of an	18.2	184
Exceptional Event, Consequences of an	18.4	184
Extension of Defects Notification Period	11.3	120
Extension of Time for Completion	8.5	102
Failure to Pass Tests on Completion	9.4	112
FPC, Issue of	14.13	160
Goods, Transport of	4.16	76
Health and Safety Obligations	4.8	66
Health and Safety of Personnel	6.7	88

条目	条款	页码
因成本改变的调整	13.7	143
支付的货币	14.15	163
争端避免/裁决委员会 - 参见争端避免/裁决委员会		
计日工作	13.5	141
现场外缺陷工程的修补	11.5	123
未能修补缺陷	11.4	121
缺陷和拒收	7.5	97
缺陷通知期限的延长	11.3	121
完成扫尾工作和修补缺陷	11.1	119
定义	1.1	13
误期损害赔偿费	8.8	107
延误的图纸或指示	1.9	31
部门造成的延误	8.6	105
结清证明	14.12	161
无序行为	6.11	91
争端避免/裁决委员会的组成	21.1	203
未能遵守争端避免/裁决委员会的决定	21.7	211
取得争端避免/裁决委员会的决定	21.4	207
未设立争端避免/裁决委员会	21.8	213
未能任命争端避免/裁决委员会成员	21.2	205
争端避免	21.3	205
文件的照管和提供	1.8	31
承包商使用雇主文件	1.11	33
雇主使用承包商文件	1.10	33
文件优先次序	1.5	29
将延误减至最小的义务	18.3	185
雇主提供的材料和雇主设备	2.6	43
支付的货币	14.15	163
承包商使用雇主文件	1.11	33
雇主的资金安排	2.4	41
雇主责任的中止	14.14	163
雇主人员和其他承包商	2.3	41
工程师	3.1	43
由工程师付托	3.4	45
工程师的替代	3.6	47
工程师的任务和权利	3.2	43
工程师的指示	3.5	45
工程师代表	3.3	45
环境保护	4.18	77
例外事件	18.1	183
例外事件的通知	18.2	185
例外事件的后果	18.4	185
缺陷通知期限的延长	11.3	121
竣工时间的延长	8.5	103
未能通过竣工试验	9.4	113
最终付款证书的签发	14.13	161
货物运输	4.16	77
健康和安全义务	4.8	67
人员的健康和安全	6.7	89

Indemnities by Contractor	17.4	180
Indemnities by Employer	17.5	182
Indemnities, Shared	17.6	182
Inspection	7.3	92
Insurance to be provided by the Contractor	19.2	190
Insurance: General Requirements	19.1	188
Instructions, Delayed Drawings or	1.9	30
Intellectual and Industrial Property Rights	17.3	178
Interference, Avoidance of	4.14	74
Interference with Tests on Completion	10.3	116
Interim Payment, Application for	14.3	146
Interpretation	1.2	24
Issue of FPC	14.13	160
Issue of IPC	14.6	152
Joint and Several Liability	1.14	36
Labour, Engagement of Staff and	6.1	86
Labour, Facilities for Staff and	6.6	86
Law and Language	1.4	28
Laws, Adjustments for Changes in	13.6	140
Laws, Compliance with	1.13	34
Laws, Labour	6.4	86
Liability, Cessation of Employer's	14.14	162
Liability, Joint and Several	1.14	36
Liability, Limitation of	1.15	36
Manner of Execution	7.1	92
Meetings	3.8	52
Method of Measurement	12.2	130
Nominated Subcontractors	5.2	82
Notice to Correct	15.1	164
Notices and Other Communications	1.3	26
Obligations, Unfulfilled	11.10	126
Obligations, Contractor's General	4.1	52
Omissions	12.4	132
Payment	14.7	154
Payment after Termination by the Contractor	16.4	176
Payment, Currencies of	14.15	162
Payment, Delayed	14.8	156
Payments, Schedule of	14.4	148
Performance Certificate	11.9	126
Performance Security	4.2	56
Personnel, Key	6.12	90
Personnel, Contractor's	6.9	88
Personnel and other Contractors, Employer's	2.3	40
Persons, Recruitment of	6.3	86
Plant and Materials intended for the Works	14.5	150
Plant and Materials, Ownership of	7.7	98
Priority of Documents	1.5	28
Programme	8.3	100
Progress, Rate of	8.7	104
Progress Reports	4.20	78
Provisional Sums	13.4	138

由承包商保障	17.4	181
由雇主保障	17.5	183
保障分担	17.6	183
检验	7.3	93
由承包商提供的保险	19.2	191
保险：一般要求	19.1	189
延误的图纸或指示	1.9	31
知识产权和工业产权	17.3	179
避免干扰	4.14	75
对竣工试验的干扰	10.3	117
期中付款的申请	14.3	147
解释	1.2	25
最终付款证书的签发	14.13	161
期中付款证书的签发	14.6	153
共同的和各自的责任	1.14	37
员工的雇用	6.1	87
为员工提供设施	6.6	87
法律和语言	1.4	29
因法律改变的调整	13.6	141
遵守法律	1.13	35
劳动法	6.4	87
雇主责任的中止	14.14	163
共同的和各自的责任	1.14	37
责任限度	1.15	37
实施方法	7.1	93
会议	3.8	53
测量方法	12.2	131
指定分包商	5.2	83
通知改正	15.1	165
通知和其他通信交流	1.3	27
未履行的义务	11.10	127
承包商的一般义务	4.1	53
删减	12.4	133
付款	14.7	155
由承包商终止后的付款	16.4	177
支付的货币	14.15	163
延误的付款	14.8	157
付款计划表	14.4	149
履约证书	11.9	127
履约担保	4.2	57
关键人员	6.12	91
承包商人员	6.9	89
雇主人员和其他承包商	2.3	41
招聘人员	6.3	87
拟用于工程的生产设备和材料	14.5	151
生产设备和材料的所有权	7.7	99
文件优先次序	1.5	29
进度计划	8.3	101
工程进度	8.7	105
进度报告	4.20	79
暂列金额	13.4	139

Quality Management and Compliance Verification Systems	4.9	68
Records, Contractor's	6.10	90
Release from Performance under the Law	18.6	186
Remedial Work	7.6	96
Remedy Defects, Failure to	11.4	120
Remedying Defects	11.1	118
Remedying Defects, Cost of	11.2	118
Replacement of the Engineer	3.6	46
Representative, Contractor's	4.3	58
Representative, Engineer's	3.3	44
Responsibility for Care of the Works	17.1	176
Resumption of Work	8.13	110
Retention Money, Release of	14.9	156
Retesting after Failure of Tests on Completion	9.3	112
Right to Vary	13.1	132
Rights of Way and Facilities	4.13	74
Royalties	7.8	98
Samples	7.2	92
Schedule of Payments	14.4	148
Search, Contractor to	11.8	124
Security of the Site	4.21	80
Security, Performance	4.2	56
Setting Out	4.7	64
Site, Clearance of	11.11	126
Site, Contractor's Operations on	4.22	80
Site Data, Use of	4.10	70
Site Data and Items of Reference	2.5	40
Site, Right of Access to the	2.1	38
Site, Security of the	4.21	80
Staff and Labour, Engagement of	6.1	86
Staff and Labour, Facilities for	6.6	86
Statement at Completion	14.10	156
Statement, Final	14.11	158
Subcontractors	5.1	82
Subcontractors, nominated	5.2	82
Superintendence, Contractor's	6.8	88
Surfaces Requiring Reinstatement	10.4	118
Suspension, Consequences of Employer's	8.10	108
Suspension by Contractor	16.1	172
Suspension, Employer's	8.9	106
Suspension, Payment for Plant and Materials after Employer's	8.11	108
Suspension, Prolonged	8.12	108
Taking Over Parts	10.2	116
Taking Over the Works and Sections	10.1	114
Taking Over: Surfaces Requiring Reinstatement	10.4	118
Temporary Utilities	4.19	78
Termination, Optional	18.5	186
Termination by Contractor	16.2	172
Termination by Contractor, Payment after	16.4	176
Termination for Contractor's Default	15.2	164
Termination for Contractor's Default, Valuation after	15.3	168
Termination for Contractor's Default, Payment after	15.4	168

条目	条款	页码
质量管理和合规验证体系	4.9	69
承包商的记录	6.10	91
依法解除履约	18.6	187
修补工作	7.6	97
未能修补缺陷	11.4	121
完成扫尾工作和修补缺陷	11.1	119
修补缺陷的费用	11.2	119
工程师的替代	3.6	47
承包商代表	4.3	59
工程师代表	3.3	45
工程照管的职责	17.1	177
复工	8.13	111
保留金的发放	14.9	157
重新试验	9.3	113
变更权	13.1	133
道路通行权和设施	4.13	75
土地（矿区）使用费	7.8	99
样品	7.2	93
付款计划表	14.4	149
承包商调查	11.8	125
现场安保	4.21	81
履约担保	4.2	57
放线	4.7	65
现场清理	11.11	127
承包商的现场作业	4.22	81
现场数据的使用	4.10	71
现场数据和参考事项	2.5	41
现场进入权	2.1	39
现场安保	4.21	81
员工的雇用	6.1	87
为员工提供设施	6.6	87
竣工报表	14.10	157
最终报表	14.11	159
分包商	5.1	83
指定分包商	5.2	83
承包商的监督	6.8	89
需要复原的地面	10.4	119
雇主暂停的后果	8.10	109
由承包商暂停	16.1	173
雇主暂停	8.9	107
雇主暂停后对生产设备和材料的付款	8.11	109
拖长的暂停	8.12	109
部分工程的接收	10.2	117
工程和分项工程的接收	10.1	115
需要复原的地面	10.4	119
临时公用设施	4.19	79
自主选择终止	18.5	187
由承包商终止	16.2	173
由承包商终止后的付款	16.4	177
因承包商违约的终止	15.2	165
因承包商违约终止后的估价	15.3	169
因承包商违约终止后的付款	15.4	169

Termination for Employer's Convenience	15.5	170
Termination for Employer's Convenience, Valuation after	15.6	170
Termination for Employer's Convenience, Payment after	15.7	172
Termination, Contractor's Obligations After	16.3	176
Testing by the Contractor	7.4	94
Tests after Remedying Defects, Further	11.6	122
Tests on Completion, Contractor's Obligations	9.1	110
Tests on Completion, Delayed	9.2	112
Tests on Completion, Failure to Pass	9.4	112
Tests on Completion, Interference with	10.3	116
Time for Completion	8.2	100
Time for Completion, Extension of	8.5	102
Training	4.5	62
Transport of Goods	4.16	76
Unforeseeable Physical Conditions	4.12	72
Unfulfilled Obligations	11.10	126
Valuation of the Works	12.3	130
Value Engineering	13.2	134
Variation Procedure	13.3	136
Wages and Conditions of Labour, Rates of	6.2	86
Working Hours	6.5	86
Works, Valuation of the	12.3	130
Works to be Measured	12.1	128

为雇主便利的终止	15.5	171
为雇主便利终止后的估价	15.6	171
为雇主便利终止后的付款	15.7	173
终止后承包商的义务	16.3	177
由承包商试验	7.4	95
修补缺陷后进一步试验	11.6	123
承包商的义务	9.1	111
延误的试验	9.2	113
未能通过竣工试验	9.4	113
对竣工试验的干扰	10.3	117
竣工时间	8.2	101
竣工时间的延长	8.5	103
培训	4.5	63
货物运输	4.16	77
不可预见的物质条件	4.12	73
未履行的义务	11.10	127
工程的估价	12.3	131
价值工程	13.2	135
变更程序	13.3	137
工资标准和劳动条件	6.2	87
工作时间	6.5	87
工程的估价	12.3	131
需测量的工程	12.1	129

施工合同条件
Conditions of Contract for
CONSTRUCTION

用于由雇主设计的建筑和工程
FOR BULDING AND ENGINEERING WORKS DESIGED BY THE EMPLOYER

专用条件编写指南
Guidance for the Preparation of Particular Conditions

通用条件
GENERAL CONDITIONS

专用条件编写指南和附件：
担保函格式
GUIDANCE FOR THE PREPARATION OF PARTICULAR CONDITIONS AND ANNEXES: FORMS OF SECURITIES

投标函、中标函、合同协议书
和争端避免/裁决委员会
协议书格式
FORMS OF LETTER OF TENDER, LETTER OF ACCEPTANCE, CONTRACT AGREEMENT AND DAAB AGREEMENT

The official and authentic text of this publication is the version in the English language. Further to the Guidance for Preparation of Particular Conditions included in this publication, no alteration has been made to the General Conditions to take account of the bilingual nature of that document. Consequently, for the second paragraph of Sub-Clause 1.4 [Law and Language] to operate, it is essential that the ruling language is stated in the Contract Data.

本出版物正式的和权威性的文本为英文版本。除本出版物中包含的专用条件编写指南外，考虑到该文件的双语性质，未对通用条件进行任何修改。因此，为使第1.4款[法律和语言]的第2段易于操作，必须在合同数据中规定主导语言。

Guidance
for the Preparation of Particular Conditions

CONTENTS

INTRODUCTORY GUIDANCE NOTES		258
CONTRACT DATA		260
INTRODUCTION		270
NOTES ON THE PREPARATION OF TENDER DOCUMENTS		274
NOTES ON THE PREPARATION OF SPECIAL PROVISIONS		278
1	General Provisions	280
2	The Employer	290
3	The Engineer	292
4	The Contractor	296
5	Subcontracting	308
6	Staff and Labour	312
7	Plant, Materials and Workmanship	318
8	Commencement, Delays and Suspension	320
9	Tests on Completion	324
10	Employer's Taking Over	324
11	Defects after Taking Over	324
12	Measurement and Valuation	326
13	Variations and Adjustments	326
14	Contract Price and Payment	330
15	Termination by Employer	344
16	Suspension and Termination by Contractor	344
17	Care of the Works and Indemnities	346
18	Exceptional Events	348
19	Insurance	348
20	Employer's and Contractor's Claims	348
21	Disputes and Arbitration	348

ADVISORY NOTES TO USERS OF FIDIC CONTRACTS WHERE THE PROJECT IS TO INCLUDE BUILDING INFORMATION MODELLING SYSTEMS 362

ANNEXES   FORMS OF SECURITIES 368

专用条件编写指南

目录

介绍性指南说明	259
合同数据	261
引言	271
编写招标文件注意事项	275
编写特别规定注意事项	279
第1条　一般规定	281
第2条　雇主	291
第3条　工程师	293
第4条　承包商	297
第5条　分包	309
第6条　员工	313
第7条　生产设备、材料和工艺	319
第8条　开工、延误和暂停	321
第9条　竣工试验	325
第10条　雇主的接收	325
第11条　接收后的缺陷	325
第12条　测量和估价	327
第13条　变更和调整	327
第14条　合同价格和付款	331
第15条　由雇主终止	345
第16条　由承包商暂停和终止	345
第17条　工程照管和保障	347
第18条　例外事件	349
第19条　保险	349
第20条　雇主和承包商的索赔	349
第21条　争端和仲裁	349
项目使用建筑信息模型系统的 FIDIC 合同用户的建议说明	363
附件　　担保函格式	369

Particular Conditions Part A - Contract Data

INTRODUCTORY GUIDANCE NOTES

These Introductory Guidance Notes are for use in completing the Contract Data and shall not form part of the Contract Data.

Certain Sub-Clauses in the General Conditions require that specific information is provided in the Contract Data.

The Employer should amend as appropriate and complete all data and should insert "Not Applicable" in the space next to any Sub-Clause which the Employer does not wish to use.

The Employer should insert "Tenderer to Complete" in the space next to any Sub-Clause which the Employer wishes Contract Data to be completed by the tenderers. Except where indicated "Tenderer to Complete" tenderers shall not amend the Contract Data as provided by the Employer.

All italicised text and any enclosing square brackets are for use in preparing the Contract Data and should be deleted from the final version of the Contract Data.

Failure by the Employer to provide the information and details required in the Contract Data could mean either that the documents forming the Contract are incomplete with vital information missing, or that the fall-back provisions to be found in some of the Sub-Clauses in the General Conditions will automatically take effect.

专用条件 A 部分——合同数据[一]

介绍性指南说明

这些**介绍性指南说明**仅用于填写**合同数据**,并不构成**合同数据**的一部分。

通用条件中的某些**条款**要求在**合同数据**中提供特定信息。

雇主应适当修改并填写所有数据,并在**雇主**不希望使用的任何**条款**旁边的空白处插入"**不适用**"。

雇主应在**雇主**希望由投标人填写**合同数据**的任何**条款**旁边的空白处插入"**由投标人填写**"。除注明"**由投标人填写**"外,投标人不得修改**雇主**提供的**合同数据**。

所有斜体文本及任何所附方括号均用于编写**合同数据**,应从**合同数据**的最终版本中删除。

雇主未提供**合同数据**中要求的信息和细节,可能意味着构成合同的文件不完整,缺少重要信息,或者意味着**通用条件**的某些**条款**中的兜底条款将自动生效。

[一] 此处按英文版翻译。——译者注

CONTRACT DATA

Sub-Clause	Data to be given	Data
1.1.20	where the Contract allows for Cost Plus Profit, percentage profit to be added to the Cost:	_____%
1.1.27	Defects Notification Period (DNP):	_____days
1.1.31	Employer's name and address:	
1.1.35	Engineer's name and address:	
1.1.84	Time for Completion:	_____days
1.3 (a)(ii)	agreed methods of electronic transmission:	
1.3(d)	address of Employer for communications:	
1.3(d)	address of Engineer for communications:	
1.3(d)	address of Contractor for communications:	
1.4	Contract shall be governed by the law of:	
1.4	ruling language:	
1.4	language for communications:	
1.8	number of additional paper copies of Contractor's Documents	
1.15	total liability of the Contractor to the Employer under or in connection with the Contract:	_____(sum)
2.1	after receiving the Letter of Acceptance, the Contractor shall be given right of access to all or part of the Site within:	_____Days

合同数据

条款	需提供的数据	数据
1.1.20	如果合同允许成本加利润，百分比的利润应加到成本中：	........%
1.1.27	缺陷通知期限（DNP）：	........天
1.1.31	雇主的姓名和地址：	
1.1.35	工程师的姓名和地址：	
1.1.84	竣工时间：	........天
1.3（a）（ii）	商定的电子传输方式：	
1.3（d）	雇主的通信地址：	
1.3（d）	工程师的通信地址：	
1.3（d）	承包商的通信地址：	
1.4	合同应由 ...... 法律管辖：	
1.4	主导语言：	
1.4	交流语言：	
1.8	承包商文件的附加纸质副本的数量：	
1.15	合同规定的或与合同相关的承包商对雇主的全部责任：	........（金额）
2.1	收到中标函后，承包商应在 ...... 天内有权进入全部或部分现场：	........天

Sub-Clause	Data to be given	Data
2.4	Employer's financial arrangements:	
4.2	Performance Security (as percentages of the Accepted Contract Amount in Currencies):	
	percent:	........%
	currency:	
	percent:	........%
	currency:	
4.7.2	period for notification of errors in the items of reference.	........days
4.19	period of payment for temporary utilities	........days
4.20	number of additional paper copies of progress reports..	
5.1(a)	maximum allowable accumulated value of work subcontracted (as a percentage of the Accepted Contract Amount)	........%
5.1(b)	parts of the Works for which subcontracting is not permitted:	
6.5	normal working hours on the Site	
8.3	number of additional paper copies of programmes	
8.8	Delay Damages payable for each day of delay	
8.8	maximum amount of Delay Damages	
12.2	method of measurement:	
12.3	percentage profit:	as stated under 1.1.19 above
13.4(b)(ii)	percentage rate to be applied to Provisional Sums for overhead charges and profit	........%
14.2	total amount of Advance Payment (as a percentage of Accepted Contract Amount)	........%
14.2	currency or currencies of Advance Payment	

条款	需提供的数据	数据
2.4	雇主的资金安排：	
4.2	履约保函（占货币中标合同金额的百分比）：	
	百分比：	_____%
	货币：	
	百分比：	_____%
	货币：	
4.7.2	参考事项中的差错通知期限	_____天
4.19	临时公用设施的付款期限	_____天
4.20	进度报告附加纸质副本的数量	
5.1（a）	所允许的分包工作的最大累积价值（占中标合同金额的百分比）	_____%
5.1（b）	不准分包工程的部分：	
6.5	在现场的正常工作时间	
8.3	进度计划附加纸质副本的数量	
8.8	每日延误可支付的误期损害赔偿费	
8.8	误期损害赔偿费的最高金额	
12.2	测量方法：	
12.3	利润比例：	按上述第1.1.19项规定
13.4（b）（ii）	管理费和利润的暂列金额使用的百分比	_____%
14.2	预付款总额（占中标合同金额的比例）	_____%
14.2	预付款的币种	

Sub-Clause	Data to be given	Data
14.2.3	percentage deductions for the repayment of the Advance Payment	........%
14.3	period of payment	
14.3(b)	number of additional paper copies of Statements	
14.3(iii)	percentage of retention	........%
14.3(iii)	limit of Retention Money (as a percentage of Accepted Contract Amount)	........%
14.5(b)(i)	Plant and Materials for payment when shipped	
14.5(c)(i)	Plant and Materials for payment when delivered to the Site	
14.6.2	minimum amount of Interim Payment Certificate (IPC)	
14.7(a)	period for payment of Advance Payment to the Contractor	........days
14.7(b)(i)	period for the Employer to make interim payments to the Contractor under Sub-Clause 14.6 [*Interim Payment*]	........days
14.7(b)(ii)	period for the Employer to make interim payments to the Contractor under Sub-Clause 14.13 [*Final Payment*]	........days
14.7(c)	period for the Employer to make final payment to the Contractor	........days
14.8	financing charges for delayed payment (percentage points above the average bank short–term lending rate as referred to under sub-paragraph (a))	........%
14.11.1(b)	number of additional paper copies of draft Final Statement	
14.15	currencies for payment of Contract Price:	
14.15(a)(i)	proportions or amounts of Local and Foreign Currencies are: Local Foreign	

条款	需提供的数据	数据
14.2.3	预付款再支付的扣减比例	_____%
14.3	付款期限	_____
14.3（b）	报表附加纸质副本的数量	_____
14.3（iii）	保留金比例	_____%
14.3（iii）	保留金限额（占中标合同金额的比例）	_____%
14.5（b）（i）	运输时应支付的生产设备和材料	_____ _____
14.5（c）（i）	发运到现场时应支付的生产设备和材料	_____ _____
14.6.2	期中付款证书（IPC）的最小金额	_____
14.7（a）	给承包商预付款的支付期限	_____天
14.7（b）（i）	按照第14.6款［期中付款］的规定，雇主给承包商期中付款的期限	_____天
14.7（b）（ii）	按照第14.13款［最终付款］的规定，雇主给承包商期中付款的期限	_____天
14.7（c）	雇主给承包商最终付款的期限	_____天
14.8	延误付款的融资费用［高于（a）段所指的平均银行短期贷款利率的百分点］	_____%
14.11.1（b）	最终报表草案附加纸质副本的数量	_____
14.15	合同价格支付的币种：	_____ _____ _____
14.15（a）（i）	当地和外国货币的比例或金额是：	
	当地	_____
	外国	_____

Sub-Clause	Data to be given	Data
14.15(c)	currencies and proportions for payment of Delay Damages	
14.15(f)	rates of exchange	
17.2(d)	forces of nature, the risks of which are allocated to the Contractor	
19.1	permitted deductible limits:	
	insurance required for the Works	
	insurance required for Goods	
	insurance required for liability for breach of professional duty	
	insurance required against liability for fitness for purpose (if any is required)	
	insurance required for injury to persons and damage to property	
	insurance required for injury to employees	
	other insurances required by Laws and by local practice:	
19.2(1)(b)	additional amount to be insured (as a percentage of the replacement value, if less or more than 15%)	_____%
19.2(1)(iv)	list of Exceptional Risks which shall not be excluded from the insurance cover for the Works	
19.2.2	extent of insurance required for Goods	
	amount of insurance required for Goods	
19.2.3(a)	amount of insurance required for liability for breach of professional duty	
19.2.3(b)	insurance required against liability for fitness for purpose	yes/no (*delete as appropriate*)
19.2.3	period of insurance required for liability for breach of professional duty	
19.2.4	amount of insurance required for injury to persons and damage to property	

条款	需提供的数据	数据
14.15（c）	**延误损害赔偿费**的支付货币和比例............................	
14.15（f）	汇率............................	
17.2（d）	自然力的风险分配给**承包商**............................	
19.1	允许的免赔额度：	
	工程要求的保险............................	
	货物要求的保险............................	
	违反职业职责的责任要求的保险............................	
	满足使用功能（如有任何需要）的责任要求的保险...	
	人身伤害和财产损害要求的保险............................	
	雇员的人身伤害要求的保险............................	
	法律和当地惯例要求的其他保险：	
	....	
	....	
	....	
19.2（1）（b）.	投保的附加金额（如果低于或不超过 15%，按重置价值的比例）............................	............%
19.2（1）（iv）	不应排除在**工程**保险范围以外的**例外风险**清单............	
19.2.2	**货物**要求的保险范围............................	
	货物要求的保险金额............................	
19.2.3（a）	违反职业职责的责任要求的保险金额............................	
19.2.3（b）	满足使用功能责任要求的保险............................	是 / 否 （适用的情况下可删除）
19.2.3	违反职业职责的责任要求的保险期限............................	
19.2.4	人身伤害和财产损害要求的保险金额............................	

Sub-Clause	Data to be given	Data
19.2.6	other insurances required by Laws and by local practice (give details)	
21.1	time for appointment of DAAB	
21.1	the DAAB shall comprise	............members
21.1	list of proposed members of DAAB - proposed by Employer	1. 2. 3.
	- proposed by Contractor	1. 2. 3.
21.2	Appointing entity (official) for DAAB members	

(*unless otherwise stated here, it shall be the President of FIDIC or a person appointed by the president*)

Definition of Sections (if any):

Description of parts of the Works that shall be designated a section for the purposes of the Contract (Sub-Clause 1.1.73)	Value: Percentage* of Accepted contract Amount (Sub-Clause 14.9)	Time for Completion (Sub-Clause 1.1.84)	Delay Damages (Sub-Clause 8.8)

* These percentages shall also be applied to each half of the Retention Money under Sub-Clause 14.9.

条款	需提供的数据	数据
19.2.6	**法律**和当地惯例要求的其他保险（详细情况）	....
21.1	任命**争端避免/裁决委员会**的时间	....
21.1	**争端避免/裁决委员会**应由组成	...........成员
21.1	拟任**争端避免/裁决委员会**成员名单	
	- 由**雇主**提议	1. .... 2. .... 3. ....
	- 由**承包商**提议	1. .... 2. .... 3. ....
21.2	**争端避免/裁决委员会**成员的任命实体（官方）	....
		（除非在此另有规定，否则这将是**菲迪克主席**或其任命的人员）

分项工程的定义（如果有）：

为满足合同目的，应指定为某区段的工程部分的说明（第1.1.73项）	价值：中标合同金额的比例 *（第14.9款）	竣工时间（第1.1.84项）	误期损害赔偿费（第8.8款）

* 这些比例也应适用于**第14.9款**规定的一半的**保留金**。

Particular Conditions Part B —— Special Provisions

INTRODUCTION

The terms of the Conditions of Contract for Construction have been prepared by the Fédération Internationale des Ingénieurs-Conseils (FIDIC) and are recommended for general use for the construction of building or engineering works, where tenders are invited on an international basis.

Modifications to the General Conditions may well be required to account for local legal requirements, particularly if they are to be used on domestic contracts.

Under the usual arrangements for this type of contract, the Contractor is responsible for the construction, in accordance with the design of the Employer, of building and/or engineering works. These Conditions allow for the possibility that the Contractor may be required to design a small proportion or a minor element of the Permanent Works, but they are not intended for use where significant design input by the Contractor is required or the Contractor is required to design a large proportion or any major elements of the Permanent Works. In this latter case, it is recommended that the Employer consider using FIDIC's Conditions of Contract for Plant and Design-Build, Second Edition 2017 (or, alternatively and if suitable for the circumstances of the project, FIDIC's Conditions of Contract for EPC/Turnkey Projects, Second Edition 2017).

The guidance hereafter is intended to assist drafters of the Special Provisions (Particular Conditions – Part B) by giving options for various sub-clauses where appropriate. In some cases example wording is included between lines, while in other instances only an aide-memoire is given.

FIDIC strongly recommends that the Employer, the Contractor and all drafters of the Special Provisions take all due regard of the five FIDIC Golden Principles:

GP1: **The duties, rights, obligations, roles and responsibilities of all the Contract Participants must be generally as implied in the General Conditions, and appropriate to the requirements of the project.**

GP2: **The Particular Conditions must be drafted clearly and unambiguously.**

GP3: **The Particular Conditions must not change the balance of risk/reward allocation provided for in the General Conditions.**

GP4: **All time periods specified in the Contract for Contract Participants to perform their obligations must be of reasonable duration.**

GP5: **All formal disputes must be referred to a Dispute Avoidance/Adjudication Board (or a Dispute Adjudication Board, if applicable) for a provisionally binding decision as a condition precedent to arbitration.**

These FIDIC golden principles are described and explained in the publication FIDIC's Golden Principles (http://http://fidic.org/bookshop[†]), and are necessary to ensure that modifications to the General Conditions:

- are limited to those necessary for the particular features of the Site and the project, and necessary to comply with the applicable law;

- do not change the essential fair and balanced character of a FIDIC contract; and

  the Contract remains recognisable as a FIDIC contract.

专用条件 B 部分 —— 特别规定

引言

国际咨询工程师联合会（FIDIC，菲迪克）已编制了《施工合同条件》的条款，推荐用于进行国际招标的建筑或工程。

为考虑当地法律要求，特别是用于国内合同时，可能需要对**通用条件**做些修改。

根据此类合同的通常安排，**承包商**应根据**雇主**的设计负责建筑和／或工程的施工。这些**条件**容许**承包商**进行少部分非主要部分**永久工程**的设计，但本条件不是为要求**承包商**进行大量设计投入或需要**承包商**设计大部分或主要部分的**永久工程**情况下使用的。在后一种情况下，建议**雇主**考虑使用菲迪克（FIDIC）2017 年第 2 版《生产设备和设计 - 施工合同条件》[或，如果适用于项目情况，使用菲迪克（FIDIC）2017 年第 2 版《设计采购施工（EPC）／交钥匙工程合同条件》]。

下述指南旨在通过适当给出各类备选条款，为**特别规定（专用条件——B 部分）**的编写人提供帮助。在某些情况下，包括一些文字间的范例措辞，但在其他情况下只给出备忘要点。

菲迪克（FIDIC）强烈建议雇主、承包商和特别规定的所有编写人都要充分考虑菲迪克（FIDIC）的五项黄金原则：

1. 合同各方的职责、权利、义务、角色和责任必须在通用条件中予以明确，并符合项目的要求。

2. 专用条件必须清楚明确。

3. 专用条件不得改变通用条件中规定的风险／报酬分配平衡。

4. 合同中规定的合同参与方履行其义务的所有时间期限必须合理。

5. 除非与合同的管辖法律有冲突，所有的正式争端都必须提交给争端避免／裁决委员会（或争端裁决委员会）进行有约束力的临时裁决，作为仲裁的先决条件。[一]

这些黄金原则在出版物《菲迪克（FIDIC）的黄金原则》（http:// http://fidic.org/bookshop[†]）中进行了说明和解释，是修改通用条件的必要保证：

- 仅限于现场和项目的特定功能所必需的，并符合适用法律的；

- 不得改变菲迪克（FIDIC）合同的基本公平和平衡特性；（以及）

- 合同仍被认为是菲迪克（FIDIC）合同。

[一] 此处中文按照勘误表改正后的英文翻译。——译者注

Before incorporating any new or changed sub-clauses, the wording must be carefully checked to ensure that it is wholly suitable for the particular circumstances. Unless it is considered suitable, example wording should be amended before use.

Where any amendments or additions are made to the General Conditions, great care must be taken to ensure that the wording does not unintentionally alter the meaning of other clauses in the Conditions of Contract, does not inadvertently change the obligations assigned to the Parties or the balance of risks shared between them and/or does not create any ambiguity or misunderstanding in the rest of the Contract documents.

Each time period stated in the General Conditions is what FIDIC believes is reasonable, realistic and achievable in the context of the obligation to which it refers, and reflects the appropriate balance between the interests of the Party required to perform the obligation, and the interests of the other Party whose rights are dependent on the performance of that obligation. If consideration is given to changing any such stated time period in the Special Provisions (Particular Conditions – Part B), care should be taken to ensure that the amended time period remains reasonable, realistic and achievable in the particular circumstances.

There are a number of Sub-Clauses in the General Conditions which require data to be provided by the Employer and/or the Contractor and inserted into the Contract Data (Particular Conditions – Part A). However, there are no Sub-Clauses in the General Conditions which require data or information to be included in the Special Provisions (Particular Conditions – Part B).

Provisions found in the Contract documents under Special Provisions (Particular Conditions – Part B) indicate that the General Conditions have been amended or supplemented.

In describing the Conditions of Contract in the tender documents, the following text can be used:

> "The Conditions of Contract comprise the "General Conditions", which form part of the "Conditions of Contract for Construction" Second Edition 2017 published by the Fédération Internationale des Ingénieurs-Conseils (FIDIC), the Contract Data (Particular Conditions – Part A) and the following "Special Provisions" (Particular Conditions – Part B), which include amendments and additions to such General Conditions."

The provisions of the Special Provisions (Particular Conditions – Part B) will always over-rule and supersede the equivalent provisions in the General Conditions, and it is important that the changes are easily identifiable by using the same clause numbers and titles as appear in the General Conditions. Furthermore, it is necessary to add a statement in the tender documents for a contract that:

> "The provisions to be found in the Special Provisions (Particular Conditions – Part B) take precedence over the equivalent provisions found under the same Sub-Clause number(s) in the General Conditions, and the provisions of the Contract Data (Particular Conditions – Part A) take precedence over the Special Provisions (Particular Conditions – Part B)."

† Please Note: all web links referred to in these guidance notes are up-to-date as of the date of this publication but it is recommended that users of these guidance notes check online, at the time that they wish to reference the relevant document, for the most up-to-date version of the document.

在使用任何新的或修改的条款前，必须认真核实，以确保其完全适用于特定的情况。除非认为是合适的，使用范例措辞前应进行修改。

对**通用条件**进行任何修改或补充时，必须注意确保其措辞不会无意改变**合同条件**其他条款的含义，不会无意改变分配给双方的义务或其间风险的平衡分担，和/或不会对**合同**文件的其他部分产生任何歧义或误解。

通用条件中规定的每个时间期限，都是菲迪克（FIDIC）认为在其所指义务的范围内是合理、现实和可实现的，并能反映履行该义务所需的一方利益之间的适当平衡，另一方权利和义务的履行依赖于另一方的利益。如果考虑修改**特别规定**（专用条件——B部分）中规定的任何时间期限，应注意确保修改后的时间期限在特定情况下仍然合理、现实和可实现。

很多**通用条件的条款**要求**雇主**和/或**承包商**提供数据，并将数据插入合同数据中（专用条件——A部分）。但是，**通用条件**中没有**条款**要求将数据或信息包括在**特别规定**（专用条件——B部分）中。

特别规定（专用条件——B部分）合同文件的规定表明，**通用条件**已修改或补充。

在说明招标文件中的**合同条件**时，可以使用以下文本：

> "包含'**通用条件**'的'**合同条件**'，构成**国际咨询工程师联合会**（FIDIC，菲迪克）出版的2017年第2版《施工合同条件》的一部分、合同数据（专用条件——A部分），以及包括对此类**通用条件**进行的修改和补充的下述'**特别规定**'（专用条件——B部分）。"

特别规定（专用条件——B部分）的规定，将始终优先于并取代**通用条件**的同等条款。重要的是，通过使用**通用条件**中出现的相同条款编号和标题，可以很容易识别出修改之处。此外，有必要在合同的招标文件中增加以下说明：

> "在**特别规定**（专用条件——B部分）中的规定优先于**通用条件**中相同条款编号的同等条款的情况下，合同数据（专用条件——A部分）优先于**特别规定**（专用条件——B部分）。"

注意：这些指南说明中提到的所有网络链接自出版之日起均为最新的，建议指南说明的用户在希望参考相关文件时，进行在线检索，以获取最新版本的文本。

NOTES ON THE PREPARATION OF TENDER DOCUMENTS

When preparing the tender documents and planning the tender process, Employers should read the publication FIDIC Procurement Procedures Guide 1st edition 2011 (http://fidic.org/books/fidic-procurement-procedures-guide-1st-ed-2011) which presents a systematic approach to the procurement of engineering and building works for projects of all sizes and complexity, and gives invaluable help and advice on the contents of the tender documents, and the procedures for receiving and evaluating tenders. This publication provides internationally acceptable, comprehensive, best practice procedures designed to increase the probability of receiving responsive, clear and competitive tenders using FIDIC forms of contract. FIDIC intends to update the FIDIC Procurement Procedures Guide (planned for publication at a later date) to make specific reference to these Conditions of Contract for Construction, Second Edition 2017.

The tender documents should be prepared by suitably qualified engineers who are familiar with the technical aspects of the required works and the particular requirements and contractual provisions of a construction project. Furthermore, a review by suitably-qualified lawyers is advisable.

The tender documents issued to tenderers should normally include the following:

- Letter of invitation to tender
- Instructions to Tenderers (including advice on any matters which the Employer wishes tenderers to include in their Tenders but which do not form part of the Specification and/or Drawings)
- Form of Letter of Tender and required appendices (if any)
- Conditions of Contract: General and Particular
- General information and data
- Technical information and data (including the data referred to in Sub-Clause 2.5 [*Site Data and Items of Reference*] of the General Conditions)
- the Specification
- the Drawings
- Schedules from the Employer
- details of schedules and other information required from tenderers
- required forms of agreement, securities and guarantees.

The publication FIDIC Procurement Procedures Guide referred to above provides useful guidance as to the content and format of each of the above.

For this type of contract where the Works are normally valued by measurement, the Bill of Quantities will usually be the most important Schedule. A Daywork Schedule may also be necessary to cover minor works to be executed at cost. In addition, each of the tenderers should receive the data referred to in Sub-Clause 2.5 [*Site Data and Items of Reference*], and the Instructions to Tenderers should advise them of any special matters which the Employer wishes them to take into account when pricing the Works but which are not to form part of the Contract.

When the Employer accepts the Letter of Tender, the Contract (which then comes into full force and effect) includes these completed Schedules.

The following Sub-Clauses make express reference to matters to be stated in the Specification and/or Drawings. However, it may also be necessary under other Sub-Clauses for the Employer to give specific information in the Specification (for example, under Sub-Clause 7.2 [*Samples*]).

1.8	Care and Supply of Documents
1.13	Compliance with Laws
2.1	Right of Access to the Site
2.5	Site Data and Items of Reference
2.6	Employer-Supplied Materials and Employer's Equipment
4.1	Contractor's General Obligations
4.4	Contractor's Documents

编写招标文件注意事项

编写招标文件并设计招标程序时，**雇主**应阅读 2011 年第 1 版《FIDIC 采购程序指南》（http：//fidic.org/books/fidic-procurement-procedures-guide-1st-ed-2011）。该指南为各种规模和复杂性的项目进行工程和建筑的采购提供了一种系统方法，并就招标文件的内容以及接受和评估投标的程序提供了有价值的帮助和建议。该出版物提供了国际上可接受的、全面的最佳实践程序，旨在提高采用 FIDIC 合同格式接收响应性、明确性和竞争性投标的可能性。菲迪克（FIDIC）打算参照 2017 年第 2 版《**施工合同条件**》，更新《FIDIC 采购程序指南》（计划以后出版）。

招标文件应由具备适当资质、熟悉要建工程技术情况、特别要求以及施工项目合同条款的工程师编写。此外，建议由具有适当资质的律师进行审核。

发给投标人的招标文件通常应包括以下内容：

- 招标书；
- **投标人须知**（包括**雇主**希望投标人将任何事项纳入**标书**，而又不构成**规范要求**和 / 或**图纸**一部分的建议）；

- **标书格式**及规定的附录（如有）；
- **合同条件**：**通用条件**和**专用条件**；
- 一般信息和数据；
- 技术信息和数据（包括**通用条件**第 2.5 款 [*现场数据和参考事项*] 中提到的数据）；

- **规范要求**；
- **图纸**；
- **雇主的资料表**；
- 招标人要求的详细资料表和其他信息；
- 所需的协议书、担保和保函格式。

上述《FIDIC 采购程序指南》提供了以上有关内容和格式的指南。

对此类合同，工程一般通过测量来估价，**工程量清单**通常是最重要的**资料表**。计日工作表也可能是必需的，以便包括需要估价的次要工作。此外，每位投标人都应收到在**第 2.5 款** [*现场数据和参考事项*] 中提到的资料，且**投标人须知**应告诉他们**雇主**希望其在给**工程**定价时要考虑的任何特定事项，但这些事项不构成**合同**的一部分。

当**雇主**接受**投标函**时，合同（此时全面实施和生效）包括此类填写好的**资料表**。

下列**条款**参考了**规范要求**和 / 或**图纸**中规定的事项。但是，**雇主**也有必要根据其他**条款**，在**规范要求**中提供具体信息（如，在**第 7.2 款** [*样品*] 中）。

    1.8   文件的照管和提供
    1.13   遵守法律
    2.1   现场进入权
    2.5   现场数据和参考事项
    2.6   雇主提供的材料和雇主设备
    4.1   承包商的一般义务
    4.4   承包商文件

4.5	Training
4.6	Co-operation
4.8	Health and Safety Obligations
4.9	Quality Management and Compliance Verification Systems
4.16	Transport of Goods
4.18	Protection of the Environment
4.19	Temporary Utilities
4.20	Progress Reports
5.2	Nominated Subcontractors
6.1	Engagement of Staff and Labour
6.6	Facilities for Staff and Labour
6.7	Health and Safety of Personnel
6.12	Key Personnel
7.3	Inspection
7.4	Testing by the Contractor
7.8	Royalties
8.3	Programme
9.1	Contractor's Obligations
10.2	Taking Over Parts
11.11	Clearance of Site.

Many Sub-Clauses in the General Conditions make reference to data being contained in the Contract Data (Particular Conditions – Part A). This data must be provided in the tender documents, and these Conditions of Contract assume that all such data will be provided by the Employer, except as expressly noted in the example form of Contract Data included in this publication. If the Employer requires tenderers to provide any of the other information required in the Contract Data, the tender documents must make this clear.

If the Employer requires tenderers to provide additional data or information, a convenient way of doing this is to provide a suitably worded questionnaire with the tender documents.

The Instructions to Tenderers may need to specify any constraints on the completion of the Contract Data and/or Schedules, and/or specify the extent of other information which each tenderer is to include with his/her Tender. If each tenderer is to produce a tender security and/ or a parent company guarantee, these requirements should be included in the Instructions to Tenderers: example forms are included at the end of this publication.

The Instructions to Tenderers may require the tenderer to provide information on the matters referred to in some or all of the following Sub-Clauses:

4.3	Contractor's Representative
6.12	Key Personnel
19	Insurance

It is important for the Parties to understand which of the documents included in the tender dossier, and which of the documents submitted by tenderers, will form part of the Contract and therefore have continuing effect. For example, the Instructions to Tenderers are not, by definition, a part of the Contract. They are simply instructions and information on the preparation and submission of the tender, and they should not contain anything of a binding or contractual nature.

Finally, when planning the overall programme for the project, Employers must remember to allow a realistic time for:

- tenderers to prepare and submit a responsive Tender (avoiding time that is either too short which can reduce competition and result in inadequate submittals, or too long which can be wasteful to all parties involved); and
- the review and evaluation of tenders and the award of the Contract to the successful tenderer. This will be the minimum time which tenderers should be asked to hold their tenders valid and open to acceptance.

4.5	培训
4.6	合作
4.8	健康和安全义务
4.9	质量管理和合规验证体系
4.16	货物运输
4.18	环境保护
4.19	临时公用设施
4.20	进度报告
5.2	指定分包商
6.1	员工的雇用
6.6	为员工提供设施
6.7	人员的健康和安全
6.12	关键人员
7.3	检验
7.4	由承包商试验
7.8	土地（矿区）使用费
8.3	进度计划
9.1	承包商的义务
10.2	部分工程的接收
11.11	现场清理

很多**通用条件**中的**条款**都引用**合同数据**（**专用条件——A 部分**）中的数据。招标文件必须提供该数据，并且合同条件认为所有此类数据将由**雇主**提供，本出版物中包含的**合同数据**范例格式中明确说明的除外。如果**雇主**要求投标人提供**合同数据**需要的任何其他信息，则投标文件必须对此予以明确。

如果**雇主**要求投标人提供其他数据或信息，一种简便的方法是在提供招标文件时，也同时提供适当措辞的问卷。

投标人须知可能需要规定对填写**合同数据**和 / 或**资料表**的任何限制条件，和 / 或规定每个投标人应在**标书**中包含的其他信息。如果每个投标人要提供投标保证金和 / 或母公司担保，则这些要求应包含在**投标人须知**中：本出版物最后附有范例格式。

投标人须知可能会要求投标人提供下述某些或所有**条款**规定事项的信息：

4.3	承包商代表
6.12	关键人员
19	保险

对双方而言，重要的是要了解招标文件中包含哪些文件，以及招标人提交的哪些文件将构成合同的一部分，并因此而持续有效。例如，根据定义，**投标人须知**不构成合同的一部分。它们只是有关投标准备和提交的须知和信息，不应包含任何具有约束力或合同性质的内容。

最后，在设计项目的总体进度计划时，**雇主**必须记住留出实际时间用于：

- 投标人应编写并提交响应性**投标书**（避免时间太短，这可能会降低竞争，导致提交文件不足，或时间过长会对所有相关方造成浪费）；（以及）

- 对投标书进行评审，并将**合同**授予中标人。这将是要求投标人保持其投标书有效并接受的最短时间。

NOTES ON THE PREPARATION OF SPECIAL PROVISIONS

It is very important that Employers (and all drafters of the Special Provisions) work with their professional advisers to review specific terminology in the General Conditions for compliance and consistency with accepted practice in the legal jurisdiction they are operating in.

For example:

- under a number of legal systems (notably in some common law jurisdictions) the term "gross negligence" has no clear definition and, as such, is often avoided in legal documents; and
- under English law, the term "indemnity" or "indemnify" has a specific meaning, entitling the indemnified party to recover certain losses that may not otherwise be recoverable at law. For example, a Contractor's indemnity may be construed to permit the Employer to recover losses or damages that might otherwise be considered indirect or consequential, whereas for some events which may arise both Parties' losses or damages are intended to be limited by Sub-Clause 1.15 [Limitation of Liability] to direct damages only.

The following references and examples show some of the Sub-Clauses in the General Conditions which may need amending to suit the needs of the project or the requirements of the Employer. The selected Sub-Clauses and the example wording are included as examples only. They also include, as an aide-memoire, references to other documents such as the Specification and/or Drawings and the Contract Data, where particular issues may need to be addressed.

The selected Sub-Clauses do not necessarily require changing and the example wording may not suit the needs of a particular project or Employer. It is the responsibility of the drafter of the Special Provisions to ensure that the selection of the Sub-Clauses and the choice of wording is appropriate to the project concerned.

Furthermore, there may be other Sub-Clauses, not mentioned below, which need to be amended. Great care must be taken when amending the wording of Sub-Clauses from the General Conditions, or adding new provisions, to ensure that the balance of obligations and rights of the Parties are not unintentionally compromised.

编写特别规定注意事项

雇主（以及**特别规定**的所有编写人）与其专业顾问一起审核**通用条件**中的专用术语，以确保与其业务所在地区的法律管辖所公认的惯例保持符合性和一致性，这一点非常重要。

例如：

- 在很多法律体系中（特别是在一些英美法系国家），"严重过失"这一术语没有明确的定义，因此在法律文件中经常避免使用；（以及）

- 根据英国法律，词语"赔偿"或"补偿"具有特定含义，使受害方有权追偿某些本来无法追回的损失。例如，**承包商**赔偿可解释为，允许**雇主**追偿可能被认为是间接的或后果性的损失或损害，而对于某些可能引起双方损失或损害的事件，仅限于**第 1.15 款**[*责任限度*]规定的直接损害赔偿。

以下参考和示例展示了**通用条件**中一些需要修改的**条款**，以满足项目需要或**雇主**要求。选定的**条款**和范例措辞仅作为示例提供。作为辅助备忘录，它们还包括对其他文件如**规范要求**和/或**图纸**和**合同数据**的引用，其中可能需要解决特定问题。

选定的**条款**不一定需要修改，并且范例措辞可能不适合特定项目或**雇主**的需求。**特别规定**的编写人有责任保证，**条款**的选定和措辞的选择适合于相关项目。

此外，可能还有下述未提到其他的**条款**需要修改。在修改**通用条件**中**条款**措辞或增加新的规定时，必须格外谨慎，以确保双方的义务和权利的平衡不会被无意损害。

Clause 1   General Provisions

Sub-Clause 1.1   **Definitions**

The opening words of Sub-Clause 1.1 mean that the definitions apply, not only to the Conditions of Contract, but to **all** the documents of the Contract. Therefore, the Employer should take care (particularly when drafting the Specification) and the Contractor should take care (particularly when drafting the documents of the Tender) to use terms in accordance with those as defined in the definitions.

Regarding the definition of "Base Date" under Sub-Clause 1.1.4, if the Employer makes significant data/information available to tenderers 28 days or less before the due date for submission of tenders, then it is recommended that the Employer consider extending that date so that such information is provided before the Base Date for the purposes of the Contract conditions (including Sub-Clause 2.5 [*Site Data and Items of Reference*] and Sub-Clause 4.12 [*Unforeseeable Physical Conditions*]).

The defined term "Site", under Sub-Clause 1.1.74, is used for a number of different purposes in the General Conditions (for example: insurances and security) and so, if it is anticipated that any item(s) of equipment, Plant, Materials and Temporary Works are to be located or stored in places other than where the Permanent Works are to be executed, it is recommended that consideration be given to specifying such places in the Contract (by the Employer in the Specification and/or by the Contractor in the Tender).

In general, changes should not be made to the definitions as this could have serious consequences on the interpretation of the documents of the Contract, particularly the Conditions of Contract.

However, there are limited circumstances where it may be desirable to amend some of the definitions. For example:

    1.1.4   the Base Date could be defined as a particular calendar date
    1.1.42 one particular Foreign Currency may be required
    1.1.52 a different currency may be required to be the contract Local Currency

Also, some of the defined terms in the General Conditions may not be appropriate and may need changing or developing.

For example, if the extent of the Site crosses the border between two countries, it is recommended that consideration be given to the following changes:

EXAMPLE	1.1.20
	" "Country" means either xxxxx or yyyyy depending on the location to which the reference will apply."
EXAMPLE	1.1.52
	" "Local Currency" means the currency of (*insert name of Country*) or (*insert name of Country*)."

条款[一]

第 1 条　　一般规定

第 1.1 款　　定义

第 1.1 款的开场词表示该定义不仅适用于**合同条件**，而且适用于**合同**的**所有**文件。因此，**雇主**（特别是在编写**规范要求**时）和**承包商**（尤其是在起草**投标**文件时）均应注意根据定义中的界定使用术语。

关于**第 1.1.4 项**中"**基准日期**"的定义，如果在递交投标书的截止日期前 28 天或更短时间内，**雇主**向投标人提供了重要的数据 / 信息，则建议雇主考虑延长这一日期，以便为满足**合同条件**（包括**第 2.5 款**[*现场数据和参考事项*]和**第 4.12 款**[*不可预见的物质条件*]）规定的**基准日期**前，提供此类信息。

第 1.1.74 项定义的"**现场**"一词，在**通用条件**下用于多种不同目的（例如，保险和担保），因此，如果预计有设备、生产设备、材料和临时工程等任何物品放置或存放在要执行**永久工程**的场地以外的地方，建议考虑在**合同**中（由**规范要求**中的**雇主**和 / 或由**投标书**中的**承包商**）明确这些地方。

通常情况下，不应修改定义，因为这可能对**合同**文件，尤其是**合同条件**的解释产生严重影响。

但是，在少数情况下，可能需要修改某些定义。例如：

  1.1.4 **基准日期**可规定为某一特定日历日期
  1.1.42 可能要求某种特定**外国货币**
  1.1.52 合同**当地货币**可能要求另一种货币

同样，**通用条件**中某些定义的术语可能不适用，可能需要修改或延展。

例如，如果**现场**范围跨越两国边界，建议考虑做以下修改：

范例   1.1.20

    "'工程所在国'既可以是指 X 国又可以是指 Y 国，取决于引用适用的处所。"

范例   1.1.52

    "'当地货币'是指（插入工程所在国名称）或（插入工程所在国名称）的货币。"

[一] 译者增补。——译者注

Sub-Clause 1.1.82 "Tests after Completion"

If the Contract specifies that the Contractor shall design any part of the Permanent Works and if, after such part has been taken over, the Employer wishes to carry out Tests after Completion on that part, detailed requirements regarding these tests should be stated in the Specification.

The Specification should describe the tests the Employer will carry out to verify that the part designed by the Contractor fulfils the Employer's performance requirements, which requirements should also be stated in the Specification.

Further, it is recommended that consideration be given to including provisions in the Special Provisions that are based on those of Clause 12 in FIDIC's Conditions of Contract for Plant & Design-Build Projects, Second Edition 2017.

If it is necessary to introduce new terminology into the text of the Special Provisions, each term should be carefully and properly defined, using a clear Sub-Clause numbering system for the new/additional defined terms. It is recommended, that the numbering of new definitions does not interfere with the numbering as originally included in the General Conditions. For example:

EXAMPLE   New definition B 1.1.x

" "Safety Regulations" means the Employer's safety regulations existing on the Site which the Contractor is required to follow."

In some jurisdictions, it may be advisable to add additional definitions to clarify certain terms/expressions which appear in the General Conditions but are not defined. A typical example might be:

EXAMPLE   New definition B 1.1.'n'

" "Gross Negligence" means any act or omission of a party which is contrary to the most elementary rules of diligence which a conscientious employer or contractor would have observed in similar circumstances, and/or which show serious reckless disregard for the consequences of such an act or omission. It involves materially more want of care than mere inadvertence or simple negligence."

Sub-Clause 1.2 Interpretation

In relation to the meaning of "consent" under sub-paragraph (g), it should be noted that this does not mean "approve" or "approval" which, under some legal jurisdictions, may be interpreted as accepting or acceptance that the requested matter is wholly satisfactory - following which the requesting party may no longer have any responsibility or liability for it.

Sub-Clause 1.3 Notices and Other Communications

If Notices are only to be given in paper format by post, then consideration may be given to increasing the particular timescales for Notices in the Conditions of Contract to allow extra days for the Notice to be delivered to the recipient.

Sub-Clause 1.5 Priority of Documents

An order of precedence is usually necessary, in case a conflict is subsequently found between the documents of the Contract.

If the Specification is to comprise a number of documents, it may be useful to amend sub-paragraph (g) of this Sub-Clause 1.5 by adding an order of precedence between the documents of the Specification.

第 1.1.82 项　　　　"竣工后试验"

如果合同规定，**承包商**应设计**永久工程**的任何部分，并且在该部分接收后，**雇主**希望该部分**竣工后试验**，则应在**规范要求**中说明有关这些试验的详细要求。

规范要求应说明**雇主**将进行的试验，验证**承包商**设计的部分满足**雇主**的性能要求，这些要求也应在**规范要求**中加以说明。

此外，建议根据菲迪克（FIDIC）《生产设备和设计 - 施工合同条件》（2017 年第 2 版）**第 12 条**的规定，考虑将**特别规定**包括在内。

如果有必要把新术语引入**特别规定**的文本，应使用明确的**条款**编号系统对新的 / 附加的术语进行仔细、适当的定义。建议新定义的编号不能扰乱原**通用条件**使用的编号。例如：

范例　　　　新定义 B 1.1.x

"'安全守则'系指**承包商**必须遵守的**雇主**制定的**现场**安全规程。"

在某些法律辖区，建议增加附加定义，以澄清**通用条件**中出现但又未定义的某些术语 / 短语。典型范例如下：

范例　　　　新定义 B 1.1. 'n'

"'重大过失'系指任一方的作为或不作为，有悖于勤勉认真的**雇主**或**承包商**在类似情况下应遵守的最基本的尽职规则，和 / 或完全无视对这种作为或不作为造成的后果。它涉及的不仅仅是纯粹的疏忽或简单的过失，更多的是缺乏照管。"

第 1.2 款　　　　解释

关于（g）段中"同意"的含义，应当注意，这并非是指"同意"或"批准"，在某些法律管辖区，这可以解释为接受或认可所请求的事项完全令人满意——此后，请求方可能对此不再承担任何责任。

第 1.3 款　　　　通知和其他通信交流

如果仅通过邮寄方式发出纸质**通知**，则可以考虑增加**合同条件**中**通知**的特定期限，以增加**通知**送达收件人的额外天数。

第 1.5 条　　　　文件优先次序

如果随后发现**合同**文件之间存在冲突，通常需要优先顺序。

如果**规范要求**包括多个文件，可以在本**规范要求**的文件之间增加优先顺序，修改**第 1.5 款**的（g）段。

If no order of precedence of any of the documents of the Contract is to be prescribed, this Sub-Clause may be varied:

EXAMPLE   Delete Sub-Clause 1.5 and substitute:

"The documents forming the Contract are to be taken as mutually explanatory of one another. If an ambiguity or discrepancy is found, the priority shall be such as may be accorded by the governing law. The Engineer has authority to issue any instruction which he/she considers necessary to resolve an ambiguity or discrepancy."

Sub-Clause 1.6   Contract Agreement

The form of the Contract Agreement should be included in the tender documents as an annex to the Special Provisions: an example form is included at the end of this publication (in the section "Sample Forms").

If lengthy tender negotiations were necessary, it is recommended that consideration be given to recording in the Contract Agreement: the Accepted Contract Amount, the Base Date and/or the Commencement Date.

Entry into a formal Contract Agreement may be mandatory under the applicable law. In such a case, the words "unless they agree otherwise" should be deleted from the text.

Sub-Clause 1.10   Employer's Use of Contractor's Documents

Additional provisions may be required if all rights to particular items of computer software (for example) are to be assigned to the Employer. These provisions should take account of the applicable law.

Sub-Clause 1.13   Compliance with Laws

In respect of sub-paragraphs (b) and (c) of this Sub-Clause, if (in addition to the provisions of Sub-Clause 4.18 [*Protection of the Environment*]) the applicable law requires the Contractor to apply for and/or comply with particular environmental permits, these permits should be stated in the Specification together with the Contractor's obligations associated with each permit.

Sub-Clause 1.14   Joint and Several Liability

If it is likely that one or more of the tenderers will be a Joint Venture, detailed requirements for the JV may need to be specified in addition to those listed in the definition of "JV Undertaking" under Sub-Clause 1.1.47. For example, it may be desirable for each member of the JV to produce a parent company guarantee: an example form is included at the end of this publication (in the section "Sample Forms").

These requirements should be included in the Instructions to Tenderers. Normally the Employer will wish the leader of the JV to be appointed at an early stage, providing a single point of contact thereafter, and will not wish to be involved in a dispute between the members of a JV.

The Employer should scrutinise the JV Undertaking carefully and, where relevant, check if the project's financing institution(s) has/have to give consent.

Sub-Clause 1.15   Limitation of Liability

If it is required that the limitation of each Party's liability to the other Party is to include certain "indirect or consequential loss or damage" and is also to take into account liabilities which are to be insured under Clause 19 [*Insurance*], the Contract Data and this Sub-Clause may be varied:

如未规定**合同**文件的优先顺序，则**本款**可修改为：

范例　　　　　　删除**第 1.5 款**，代之以：

"组成**合同**的各项文件将被认为是互作说明的。如出现含糊或歧义时，应按管辖地法律确定先后顺序。**工程师**有权发出其认为必要的任何指示，解决此模糊或歧义。"

第 1.6 款　　　　　　合同协议书

合同协议书的格式应作为**特别规定**的附件包含在招标文件中：在本文本的后面有其范例格式（在"**格式**"一节）。⊖

如投标谈判需要较长时间，建议考虑在**合同协议书**中写入：**中标合同金额、基准日期**和/或**开工日期**。

根据适用法律，签订正式**合同协议书**可能是强制性的。在这种情况下，"除非同意"的词句应该从文本中删除。

第 1.10 款　　　　　　雇主使用承包商文件

如果要将计算机软件（比如说）的特定项目的所有权利都分配给**雇主**，可能需要附加规定。这些规定应考虑适用法律。

第 1.13 款　　　　　　遵守法律

关于**本款**第（b）和（c）段，如果（除**第 4.18 款**［*环境保护*］的规定外）适用法律要求**承包商**申请和/或遵守特定的环境许可证，这些许可证应在**规范要求**中加以明确，并规定**承包商**与每种许可证有关的义务。

第 1.14　　　　　　共同的和各自的责任

如果一个或多个投标人可能是一家**联营体**，则除**第 1.1.47 项**"**联营体承诺书**"的定义中列出的要求外，可能还需要对**联营体**规定一些具体要求。如可能希望每个成员提交一份母公司保函：本文本最后附有范例格式（见"**格式**"部分）。⊖

这些要求应包含在**投标人须知**中。一般情况下，**雇主**将希望早期指定**联营体**的负责方，以便此后有一个单独的联系方，避免卷入**联营体**成员间的争端。

雇主应仔细审核**联营体承诺书**，并在相关情况下检查项目的融资机构是否必须同意。

第 1.15 款　　　　　　责任限度

如果要求一方对另一方的责任限度包括某些"间接或重大损失或损害"，并考虑到**第 19 条**［*保险*］规定的将投保的责任，则**合同数据**和**本款**可以修改：

⊖ 此处中文按照勘误表改正后的英文翻译。——译者注

EXAMPLE In the Contract Data, add the following:

" 1.15 Amount of Sum A: _____
Amount of Sum B: _____
Amount of Sum C: _____ "

Delete Sub-Clause 1.15 and substitute:

"Except as stated in this Sub-Clause, neither Party shall be liable to the other Party for loss of use of any Works, loss of profit, loss of any contract or for any indirect or consequential loss or damage which may be suffered by the other Party in connection with the Contract.

The Contractor's liability to the Employer under or in connection with the Contract shall be limited as follows:

(a) for failure to comply with Sub-Clause 8.2 [*Time for Completion*], the limit shall be the maximum amount of Delay Damages stated in the Contract Data;

(b) for loss of profit, loss of any contract or loss of use of any part of the Permanent Works (after taking over under the Contract), caused by:

   (i) defects attributable to the Contractor, the limit shall be the sum A stated in the Contract Data;

   (ii) damage to the Works caused by the Contractor, the limit shall be the sum B stated in the Contract Data;

   (iii) damage caused by the Contractor to the Employer's property other than the Works, the limit shall be fifty percent (50%) of the sum B stated in the Contract Data; and

   (iv) any other matter attributable to the Contractor, the limit shall be fifty percent (50%) of the sum A stated in the Contract Data;

(c) for damage caused by the Contractor to the Works, the limit shall be the required value of cover of the insurance under Sub-Clause 19.2.1 [*The Works*];

(d) for damage caused by the Contractor to the Employer's property other than the Works, the limit shall be the required value of cover of the insurance under Sub-Clause 19.2.4 [*Injury to persons and damage to property*];

(e) for death or injury to the Employer's personnel or the Engineer's personnel by a cause attributable to the Contractor, the limit shall be the required value of cover of the insurance under Sub-Clause 19.2.5 [*Injury to employees*];

(f) for the Contractor's indemnity to the Employer for third party claims under the first paragraph of Sub-Clause 17.4 [*Indemnities by Contractor*], there shall be no limit; and

| 范例 | 在**合同数据**中增加下述内容： |

"第1.15款　金额 A：_____
　　　　　　金额 B：_____
　　　　　　金额 C：_____"

删除**第1.15款**，并代之以：

"除非本**款**另有规定，任一方均不对另一方因使用任何**工程**造成的损失、利润损失、任何合同损失，或另一方遭受的与**合同**相关的任何间接或重大损失或损害承担责任。"

合同规定的或与**合同**相关的**承包商**对**雇主**的责任应限定在以下方面：

（a）对于未能遵守**第8.2款**[*竣工时间*]的规定，限额应为**合同数据**中规定的最高延误损失赔偿费；

（b）对于以下原因造成的利润损失、任何合同损失或**永久工程**的任何部分的使用损失（根据合同接收后）：

（i）由**承包商**造成的缺陷，限额应为**合同数据**中规定的金额 A；

（ii）由**承包商**对**工程**造成的损害，限额应为**合同数据**中规定的金额 B；

（iii）由**承包商**对**工程**以外的**雇主**财产造成的损害，限额应为**合同数据**中规定的金额 B 的百分之五十（50%）；（以及）

（iv）由**承包商**造成的任何其他事项，限额应为**合同数据**中规定的金额 A 的百分之五十（50%）。

（c）由**承包商**对**工程**造成的损害，限额应为**第19.2.1项**[*工程*]规定的保额；

（d）由**承包商**对工程以外的**雇主**财产造成的损害，限额应为**第19.2.4项**[*人身伤害和财产损失*]规定的保额；

（e）由**承包商**造成**雇主**人员或**工程师**人员的死亡或伤害，限额应为**第19.2.5项**[*雇员的人身伤害*]规定的保额；

（f）对于**第17.4款**[*由承包商保障*]第1段规定的**承包商**对**雇主**的第三方赔偿要求，应无限额；（以及）

(g) for the Contractor's indemnity to the Employer under the second paragraph of Sub-Clause 17.4 [*Indemnities by Contractor*], the limit shall be the required value of cover of the insurance under Sub-Clause 19.2.3 [*Liability for breach of professional duty*];

(h) for all matters other than those described in sub-paragraphs (b) to (g) above and other than those under Sub-Clause 17.3 [*Intellectual and Industrial Property Rights*], the limit shall be the sum C stated in the Contract Data.

The Employer's liability to the Contractor under or in connection with the Contract shall be limited as follows:

(i) for loss of profit, loss of any contract or any other direct loss caused by termination of the contract under Sub-Clause 15.5 [*Termination for Employer's Convenience*] and Sub-Clause 16.2 [*Termination by Contractor*], the limit shall be twenty percent (20%) of the Accepted Contract Amount;

(ii) for damage caused by the Employer or Employer's Personnel to the Temporary Works or Plant and Materials not included in the Permanent Works, the limit shall be thirty percent (30%) of the required value of cover of the insurance under Sub-Clause 19.2.1 [*The Works*];

(iii) for damage caused by the Employer or Employer's Personnel to the Contractor's Equipment, Materials, Plant and/or Temporary Works, the limit shall be the required value of cover of the insurance under Sub-Clause 19.2.2 [*Goods*];

(iv) for death or injury to the Contractor's Personnel by a cause attributable to the Employer the limit shall be the required value of cover of the insurance under Sub-Clause 19.2.5 [*Injury to employees*];

(v) for the Employer's indemnity to the Contractor for third party claims under Sub-Clause 17.5 [*Indemnities by Employer*], there shall be no limit; and

(vi) for all matters other than those described in sub-paragraph (i) to (v) above and other than under Sub-Clause 17.3 [*Intellectual and Industrial Property Rights*], the limit shall be in accordance with Sub-Clause 14.14 [*Cessation of Employer's Liability*] as may be amended under Sub-Clause 21.4 [*Obtaining DAAB's Decision*], Sub-Clause 21.5 [*Amicable Settlement*] or Sub-Clause 21.6 [*Arbitration*].

This Sub-Clause shall not limit liability in any case of fraud, gross negligence, deliberate default or reckless misconduct by the defaulting Party."

(g) 对于第 17.4 款 [*由承包商保障*] 第 2 段规定的**承包商**对**雇主**的赔偿，限额应为第 19.2.3 项 [*违反职业职责的责任*] 所要求的保额；

(h) 对于除上述（b）至（g）段所述的，以及第 17.3 款 [*知识产权和工业产权*] 所述外的所有事项，限额应为**合同数据**中规定的金额 C。

合同规定的或与**合同**相关的**雇主**对**承包商**的责任应限定在以下情况：

(i) 对于因第 15.5 款 [*为雇主便利的终止*] 和第 16.2 款 [*由承包商终止*] 导致的合同终止而造成的利润损失、任何合同损失或任何其他直接损失，限额应为**中标合同金额**的百分之二十（20%）；

(ii) 对于**雇主**或**雇主人员**对**永久工程**以外的**临时工程**或**生产设备**和**材料**造成的损害，限额应为第 19.2.1 项 [*工程*] 规定的保额价值的百分之三十（30%）；

(iii) 对于**雇主**或**雇主人员**对**承包商**的设备、材料、生产设备和/或临时工程造成的损害，限额应为第 19.2.2 项 [*货物*] 规定的保额价值；

(iv) 对于**雇主**对**承包商人员**造成的死亡或伤害，限额应为第 19.2.5 项 [*雇员的人身伤害*] 规定的保额；

(v) 对于第 17.5 款 [*由雇主保障*] 规定的**雇主**对**承包商**第三方索赔的赔偿，应没有限额；（以及）

(vi) 除上述第（i）至（v）段所述的事项以及第 17.3 款 [*知识产权和工业产权*] 规定以外的所有事项，限额应符合第 14.14 款 [*雇主责任的中止*] 的规定，可以根据第 21.4 款 [*取得争端避免/裁决委员会的决定*]、第 21.5 款 [*友好解决*] 或第 21.6 款 [*仲裁*] 的规定进行修改。

如果违约**方**有任何欺诈、严重过失、故意违约或轻率不当行为的任何情况，本**款**均不限制责任。

Clause 2   The Employer

Sub-Clause 2.1   Right of Access to the Site

It may be necessary for the Contractor to have early access to the Site for the purposes of survey and sub-surface investigations. If so, the details of such early right of access should be stated in the Specification, including any restrictions on such access and whether or not it shall be exclusive to the Contractor.

If right of access to, and possession of, the Site cannot be granted by the Employer in the normal way, details should be stated in the Specification and appropriate amendments made to the first paragraph of this Sub-Clause and, if necessary, to the second paragraph.

Sub-Clause 2.3   Employer's Personnel and Other Contractors

The provisions concerning co-operation between contractors should be reflected in the Employer's contracts with any other contractors working on or near the Site.

Sub-Clause 2.6   Employer-Supplied Materials and Employer's Equipment

If Employer-Supplied Materials are listed in the Specification for the Contractor's use in the execution of the Works, the following provisions may be added:

EXAMPLE   After the last paragraph of Sub-Clause 2.6, add:

"The Employer shall supply to the Contractor the Employer-Supplied Materials listed in the Specification, at the time(s) stated in the Specification (if not stated, within the times that shall be required to enable the Contractor to proceed with execution of the Works in accordance with the Programme).

When made available by the Employer, the Contractor shall visually inspect the Employer-Supplied Materials and shall promptly give a Notice to the Engineer of any shortage, defect or default in them. Thereafter, the Contractor shall rectify such shortage, defect or default to the extent instructed by the Engineer. Such instruction shall be deemed to have been given under Sub-Clause 13.3.1 [*Variation by Instruction*].

After this visual inspection, the Employer-Supplied Materials shall come under the care, custody and control of the Contractor. The Contractor's obligations of inspection, care, custody, and control shall not relieve the Employer of liability for any shortage, defect or default not apparent from a visual inspection."

In the last paragraph of Sub-Clause 17.1 [*Responsibility for Care of the Works*] after the two instances of "Goods" add

", Employer-Supplied Materials,".

If Employer's Equipment is listed in the Specification for the Contractor's use in the execution of the Works, the following provisions may be added:

EXAMPLE   After the last paragraph of Sub-Clause 2.6, add:

"The Employer shall make the Employer's Equipment listed in the Specification available to the Contractor at the time(s) stated in the Specification (if not stated, within the times that shall be required to enable the Contractor to proceed with execution of the Works in accordance with the Programme).

第 2 条　　　　雇主

第 2.1 款　　　现场进入权

承包商可能需要早期进入**现场**进行勘测和地下调查。如果是这样，应在**规范要求**中规定早期进入**现场**的详情，包括对此类进入**现场**的任何限制，以及是否对**承包商**无限制等。

如果**雇主**不能以正常方式赋予**现场**进入权和占有权，应在**规范要求**中列出详情，并对本**款**第一段；（以及）（如有必要）第二段进行适当修改。

第 2.3 款　　　雇主人员和其他承包商

承包商之间合作的有关规定，应反映在**雇主**与在**现场**或附近工作的其他承包商的合同中。

第 2.6 条　　　雇主提供的材料和雇主设备

如果**规范要求**中列出的**雇主提供的材料**供**承包商**在**工程**实施中使用，则可以增加以下规定：

范例　　　　在**第 2.6 款**最后一段，增加：

"**雇主**应在**规范要求**中规定的时间（如未规定，应在使**承包商**能够按照**进度计划**继续实施**工程**所需的时间内），向**承包商**提供**规范要求**所列的**雇主提供的材料**。

雇主提供后，**承包商**应目检**雇主提供的材料**，并应立即将任何短缺、缺陷或违约的情况**通知工程师**。随后，**承包商**应按照工程师的指示纠正这种短缺、缺陷或违约。此类指示应视为已根据**第 13.3.1 项**［*指示变更*］的规定发出。

目检完毕后，**承包商**应照管、保管和控制**雇主提供的材料**。**承包商**的检查、照管、保管和控制义务不应免除**雇主**对目检未发现的任何短缺、缺陷或违约的责任。"

在**第 17.1 款**［*工程照管的责任*］的最后一段中，在"**货物**"的两个实例之后增加

"，**雇主提供的材料**，"。

如果**雇主设备**在**规范要求**中规定是由**承包商**实施**工程**使用的，应增加以下规定：

范例　　　　**第 2.6 款**最后一段后，增加：

"**雇主**应在**规范要求**中规定的时间（如未规定，应在使**承包商**能够按照**进度计划**继续实施**工程**所需的时间内），向**承包商**提供**规范要求**中所列的**雇主设备**。

Unless expressly stated otherwise in the Specification, the Employer's Equipment shall be provided for the exclusive use of the Contractor.

When made available by the Employer, the Contractor shall visually inspect the Employer's Equipment and shall promptly give a Notice to the Engineer of any shortage, defect or default in them. Thereafter, the Contractor shall rectify such shortage, defect or default to the extent instructed by the Engineer. Such instruction shall be deemed to have been given under Sub-Clause 13.3.1 [*Variation by Instruction*].

The Contractor shall be responsible for the Employer's Equipment while it is under the Contractor's control and/or any of the Contractor's Personnel is operating it, driving it, directing it, using it, or in control of it.

The Contractor shall not remove from the Site any items of the Employer's Equipment without the consent of the Employer. However, consent shall not be required for vehicles transporting Goods or Contractor's Personnel to or from the Site."

In the last paragraph of Sub-Clause 17.1 [*Responsibility for Care of the Works*] after the two instances of "Goods" add:

", Employer's Equipment",

Clause 3 The Engineer

Sub-Clause 3.1 The Engineer

It is recommended that the engineer who was appointed by the Employer to carry out the Employer's design of the Works is retained by the Employer to act as the Engineer under the Contract for services during construction of the Works.

In the appointment of the Engineer, it is recommended that the Employer and the Engineer use the form of agreement FIDIC Clients/Consultants Model Services Agreement fifth Edition, 2017 **(http://fidic.org/books/clientconsultant-model-services-agreement-5th-ed-2017-white-book)**.

Sub-Clause 3.2 Engineer's Duties and Authority

In performing the Engineer's duties and in exercising his/her authority under the Contract, the Engineer should take due regard of:

a) FIDIC's Code of Ethics for consulting engineers: http://fidic.org/about-fidic/fidic-policies/fidic-code-ethics[†];

and

b) the duty to prevent corruption and bribery as described in the publications:

- FIDIC Guidelines for Integrity Management (FIM) in the consulting industry, Part 1- Policies and principles first edition, 2011 (http://fidic.org/books/integrity-management-system-fims-guidelines-1st-ed-2011-part1[†]) and

- FIDIC Guidelines for Integrity Management System in the Consulting Industry 1st Ed (2015) Part 2, FIMS Procedures

除非**规范要求**中另有明确规定，应提供**承包商**专用的**雇主设备**。

雇主提供时，**承包商**应目检**雇主设备**，并应立即将其短缺、缺陷或违约的情况**通知工程师**。随后，**承包商**应根据**工程师**指示纠正这种短缺、缺陷或违约。此类指示应视为已根据**第 13.3.1 项**[*指示变更*]的规定发出。

承包商应在其控制和/或任何**承包商人员**操作、驾驶、指挥、使用或控制的情况下，对**雇主设备**负责。

未经**雇主**同意，**承包商**不得从**现场**运走**雇主设备**的任何物品。但是，将**货物**或**承包商人员**运入或运离**现场**的车辆无须征得同意。"

在**第 17.1 款**[*工程照管的责任*]的最后一段中，在"**货物**"的两个实例之后增加：

"，**雇主设备**，"。

第 3 条　　　　　　　　工程师

第 3.1 款　　　工程师

建议**雇主**聘请由其任命进行工程设计的**工程师**，为工程施工期间服务**合同**的**工程师**。

在任命**工程师**时，建议**雇主**和**工程师**使用 2017 年第 5 版《菲迪克（FIDIC）客户/咨询工程师（咨询单位）服务协议范本》的协议格式（请参见 http://fidic.org/books/clientconsultant-model-services-agreement-5th-ed-2017-white-book）。

第 3.2 款　　　工程师的任务和权利

在履行**工程师**的任务和行使其**合同**规定的权利时，**工程师**应适当考虑：

a) 菲迪克（FIDIC）咨询工程师职业道德规范：请参见 http://fidic.org/about-fidic/fidic-policies/fidic-code-ethics[†]；

（以及）

b) 文本规定的防止腐败和贿赂的职责：

- 2011 年第 1 版《菲迪克（FIDIC）工程咨询业廉洁管理指南（FIM）》第 1 部分——政策和原则（请参见 http://fidic.org/books/integrity-management-system-fims-guidelines-1st-ed-2011-part1[†]）；（和）

- 2015 年第 1 版《菲迪克（FIIDC）工程咨询业廉洁管理体系指南》，第 2 部分 FIMS 程序（请参见 http://fidic.org/books/guidelines-integri-

(http://fidic.org/books/guidelines-integrity-management-system-consulting-industry-1st-ed-2015-part-2[†]).

If there are any requirements for Employer's consent, this Sub-Clause requires that these should be set out in the Special Provisions and so the following provisions should be added to this Sub-Clause:

EXAMPLE	"The Engineer shall obtain the consent of the Employer before taking action under the following Sub-Clauses of these Conditions:
	(a) Sub-Clause _____ *
	(b) Sub-Clause _____ * "
	* *[Insert Sub-Clause number (not Sub-Clause 3.7 [Agreement or Determination] – please see below) and describe the specific procedure to be followed by the Engineer, where appropriate].*

This list should be extended or reduced as necessary. If the obligation to obtain the Employer's consent only applies beyond certain limits, financial or otherwise, the example wording should be varied accordingly.

It should be noted that any such requirement shall not be applied to any action by the Engineer under Sub-Clause 3.7 [*Agreement or Determination*], as stated in Sub-Clause 3.2 [*Engineer's Duties and Authority*] of the General Conditions.

Sub-Clause 3.4   **Delegation by the Engineer**

The Engineer's assistants may include: design engineers, other construction professionals, technicians, inspectors and/or specialist independent engineers and/or inspectors appointed to monitor and review the execution of the Works (including inspecting and/or testing items of Plant and/or Materials).

If it is anticipated that the Engineer's assistants may not all be fluent in the language for communications defined in Sub-Clause 1.4 [*Law and Language*], consideration should be given to amending this Sub-Clause:

EXAMPLE	In the second paragraph of Sub-Clause 3.4, after "Sub-Clause 1.4 [*Law and Language*]" add the following:
	"If any assistants are not fluent in this language the Engineer shall make competent interpreters available during all working hours, in a number sufficient for those assistants to properly perform their assigned duties and/or exercise their delegated authority."

Sub-Clause 3.5   **Engineer's Instructions**

If the applicable law prevents the Contractor from complying with any Engineer's instruction that may have an adverse effect on the health and safety of the Contractor's Personnel, sub-paragraph (b) of this Sub-Clause may be amended:

EXAMPLE	At the end of sub-paragraph (b) of Sub-Clause 3.5, add the following:
	"or will adversely affect the health and safety of the Contractor's Personnel".

If the Conditions of Contract are to allow for the giving of oral instructions by the Engineer, this Sub-Clause may be amended:

ty-management-system-consulting-industry-1st-ed-2015-part-2[†])。

如有需要**雇主**同意的任何要求，**本款**规定这些要求应在**特别规定**中列出，因此应在本款增加以下规定：

范例　　　　"**工程师**在根据**本条件**的以下各**款**采取措施前，应征得**雇主**的同意：

（a）　第 _____ 款 *

（b）　第 _____ 款 * "

*　*插入各款款号（不含第 3.7 款［商定或确定］——请见下述说明）并在适用情况下，说明工程师应遵守的特别程序。*

该列表应根据需要进行扩展或缩减。如果取得**雇主**同意的义务只适用于超出某些限制的范围，无论是财务上的还是其他方面的，则应相应地改变范例措辞。

应当指出，任何此类要求均不适用于**工程师**根据**通用条件**第 3.2 款［*工程师的任务和权利*］，以及第 3.7 款［*商定或确定*］的规定采取的任何行动。

第 3.4 款　　　　由工程师付托

工程师的助手可以包括设计工程师、其他施工专业人员、技术人员、检验人员和 / 或指定的监督和评估**工程**实施的独立工程师和 / 或监理人员（包括**生产设备**和 / 或**材料**的检验和 / 或试验）。

如果不希望**工程师**的助手都能够熟练使用**第 1.4 款**［*法律和语言*］规定的交流语言，应考虑修改**此款**：

范例　　　　在**第 3.4 款**第二段，"**第 1.4 款**［*法律和语言*］"之后，增加以下内容：

"如果助手不能流利使用本语言交流，**工程师**应派足够数量胜任的翻译在所有工作时间随时在场，为助手们完成分配的任务和 / 或行使授予他们的权利而提供帮助。"

第 3.5 款　　　　工程师的指示

如果适用法律使**承包商**不能执行可能对**承包商人员**的健康和安全具有不良影响的**工程师**指示，本**款**（b）段则可修改为：

范例　　　　在**第 3.5 款**（b）段之后，增加以下词句：

"或将影响**承包商人员**的健康和安全。"

如果**合同条件**允许**工程师**发出口头指示，**该款**可修改为：

EXAMPLE   At the end of Sub-Clause 3.5, add the following:

"If the Engineer or a delegated assistant:

(a) gives an oral instruction;

(b) receives a written confirmation of the instruction, from the Contractor, within two working days after giving the oral instruction; and

(c) does not reply by issuing a written rejection and/or instruction within two working days after receiving the confirmation

then the Contractor's confirmation shall constitute the written instruction of the Engineer or delegated assistant (as the case may be)."

Sub-Clause 3.7   **Agreement or Determination**

The Engineer shall not delegate any of his/her duties under this Sub-Clause to the Engineer's Representative (if appointed) or to any assistant, as stated under Sub-Clause 3.4 [*Delegation by the Engineer*].

The opening words of this Sub-Clause state that the Engineer "shall act neutrally between the Parties" when carrying out his/her duties under this Sub-Clause and "shall not be deemed to act for the Employer". By these statements it is intended that, although the Engineer is appointed by the Employer and acts for the Employer in most other respects under the Contract, when acting under this Sub-Clause the Engineer treats both Parties even-handedly, in a fair-minded and unbiased manner.

Sub-Clause 3.8   **Meetings**

It may be useful to give information in the Specification of a planned timetable of meetings such as management meetings, site meetings, technical meetings, and progress meetings.

Clause 4   The Contractor

Sub-Clause 4.1   **Contractor's General Obligations**

In some projects there may be an item or items of Temporary Works for which the Contractor will not be fully responsible. For example, an Employer-designed temporary diversion of an existing road or waterway may be necessary to facilitate the Works. In this case, and in order for the second sentence of the third paragraph to apply ("Except to the extent specified in the Contract ..."), the item(s) of Temporary Works and the extent of the Employer's responsibility should be clearly stated in the Contract (by the Employer in the Specification, or by the Contractor in the Tender).

Sub-Clause 4.2   **Performance Security**

The acceptable form(s) of Performance Security should be included in the tender documents. Example forms are included at the end of this publication (in the section "Sample Forms"). They incorporate two sets of Uniform Rules published by the International Chamber of Commerce (the "ICC", which is based at 33-43 Avenue du Président Wilson, 75116 Paris, France), which also publishes guides to these Uniform Rules.

These example forms and the wording of the Sub-Clause may have to be amended to comply with applicable law.

EXAMPLE   At the end of the first paragraph of Sub-Clause 4.2.1 [*Contractor's Obligations*], insert:

范例	在**第 3.5 款**后,增加下述内容:

"如果**工程师**或其授权的助手:

(a) 发出口头指示;

(b) 在发出口头指示两个工作日内收到**承包商**的书面确认;(以及)

(c) 收到确认后的两个工作日内没有签发书面反对和 / 或指示。

那么,**承包商**的确认将构成**工程师**或其授权代表(视情况而定)的书面指示。"

第 3.7 款　　商定或确定

根据第 3.4 款[*由工程师付托*]的规定,**工程师**不得将本款规定其的任务付托给**工程师代表**(如任命)或任何助手。

本款的开头语规定,**工程师**在履行**本款**规定的任务时"应在双方之间保持中立",并且"不应被视为代**雇主**行事"。这些表述意思是说,尽管**工程师**是由**雇主**任命的,并按**合同**规定在大多数情况下代表**雇主**,但根据**本款**规定,**工程师**应以公正的态度公平地对待**双方**。

第 3.8 款　　会议

规范要求中提供会议时间计划表,如管理会议、现场会议、技术会议和进度会议等可能会很有益处。

第 4 条　　承包商

第 4.1 款　　承包商的一般义务

在某些项目中,**临时工程**的某一项或多项可能不是由**承包商**完全负责的。例如,为便于**工程**实施,可能需要对现有道路或水路进行由**雇主**设计的临时改道。在这种情况下,为了使第三段的第二句适用("除**合同**规定的情况外……"),应在**合同**中(由**雇主**在**规范要求**中或由**投标人**在**投标书**中)明确说明**临时工程**的事项和**雇主**职责的范围。

第 4.2 款　　履约担保

招标文件中应包括可接受的**履约担保**格式。范例格式包括在本文本后("**格式**"一节)。它们体现**国际商会**(总部位于 33-43 Avenue du Président Wilson, 75116 Paris, France)出版的两套**统一规则**,**国际商会**还出版了这些**统一规则**的指南。㊀

这些范例格式及**条款**的措辞可能需要进行修改,以符合适用法律。

范例	在**第 4.2.1 项**[*承包商义务*]第一段末尾插入:

㊀ 此处中文按照勘误表改正后的英文翻译。——译者注

"If the Performance Security is in the form of a bank guarantee, it shall be issued either (i) by a bank located in the Country, or (ii) directly by a foreign bank to which the Employer gives consent. If the Performance Security is not in the form of a bank guarantee, it shall be issued by a financial entity registered, or licensed to do business, in the Country."

The example forms of Performance Security that are included at the end of this publication (in the section "Forms") provide for the option to reduce the amount of the Performance Security following issue of the Taking-Over Certificate for the whole of the Works under Clause 10. If neither of the example forms is to be used, consideration should be given to adding provisions for this option to Sub-Clause 4.2.1.

Sub-clause 4.3   Contractor's Representative

If the Contractor's Representative is known at the time of submission of the tender, the tenderer may propose that person in his/her tender. The tenderer may wish to propose alternatives, especially if the contract award seems likely to be delayed.

The "main engineering discipline applicable to the Works" in which the Contractor's Representative is required to be qualified, experienced and competent should be the engineering discipline of the Works which is of highest value proportionate to the value of the Works. If it is necessary to stipulate that the Contractor's Representative shall be qualified, experienced and competent in a particular engineering discipline in relation to the Works, the following amendment will need to be made to this Sub-Clause:

EXAMPLE   In the second paragraph of Sub-Clause 4.3 delete "the main engineering discipline applicable to the Works" and replace with:

[Insert description of engineering discipline],

If it is necessary that the Contractor's Representative shall also be fluent in a particular language other than the language for communications defined in Sub-Clause 1.4 [*Law and Language*], the following amendment may be made:

EXAMPLE   At the end of the second paragraph of Sub-Clause 4.3, add:

"The Contractor's Representative shall also be fluent in _____ (*insert name of language*)."

If it is permissible that the Contractor's Representative is not fluent in the language for communications defined in Sub-Clause 1.4 [*Law and Language*], consideration should be given to amending this Sub-Clause:

EXAMPLE   Insert at the end of the second paragraph of Sub-Clause 4.3:

"If the Contractor's Representative is not fluent in this language the Contractor shall make competent interpreter(s) available during all working hours, sufficient for the Contractor's Representative to properly perform his/her duties and exercise his/her authority under the Contract."

If it is permissible that the Contractor's Representative's delegates are not all fluent in the language for communications defined in Sub-Clause 1.4 [*Law and Language*], consideration should be given to amending this Sub-Clause:

EXAMPLE   Insert at the end of the last paragraph of Sub-Clause 4.3:

> "如果**履约担保**是银行保函的形式,它应(i)由**工程所在国**的银行;(或)(ii)直接由**雇主**同意的外国银行出具。如**履约担保**不是银行保函的形式,应由在**工程所在国**注册或取得营业执照的金融实体提供。"

本文本末尾("**格式**"一节)包括的**履约担保**的范例格式,提供了一种选项,可以按**第 10 条**规定签发整个**工程**的**接收证书**后,减少**履约担保**的数额。如果都不使用范例格式,应考虑在**第 4.2.1 项**中增加对此选项的规定。

第 4.3 款   承包商代表

如果在递交投标书时已确定**承包商代表**,投标人可在投标书中提出人选建议。投标人可能希望提出替代人选,特别是看来要推迟授予合同时。

要求**承包商代表**有资格、有经验和有能力的"适用于**工程**的主要工程学科",应是占**工程**价值最高比例**工程**的工程学科。如果有必要规定**承包商代表**应具有与**工程**有关的特定工程学科的资格、经验和能力,则需要对**本款**做以下修改:

范例   在**第 4.3 款**的第二段删除"适用于**工程**的主要工程学科",并代之以:

[插入对工程学科的说明],

如果需要**承包商代表**能流利使用**第 1.4 款**[*法律和语言*]中规定的通信交流语言以外的某种语言,可以进行以下修改:

范例   在**第 4.3 款**第二段末尾增加:

"**承包商代表**还应能流利使用_____(填入语言名称)。"

如果允许**承包商代表**可不用**第 1.4 款**[*法律和语言*]规定的语言流利交流,则应考虑修改**此款**:

范例   在**第 4.3 款**第二段末尾插入:

"如果**承包商代表**不能流利使用本语言交流,**承包商**应派胜任的翻译在所有工作时间随时在场,以便使**承包商代表**能够适当履行其**合同**规定的义务,行使其权利。"

如果允许**承包商代表**可不流利使用**第 1.4 款**[*法律和语言*]规定的语言,则应考虑修改**此款**:

范例   在**第 4.3 款**最后一段末尾,插入:

"If any of these persons is not fluent in this language the Contractor shall make competent interpreters available during all working hours, in a number sufficient for those persons to properly perform their delegated powers, functions and/or authority."

Sub-clause 4.4   **Contractor's Documents**

Sub-Clause 4.4.1   **Preparation and Review**

It is important that the Specification should clearly specify which Contractor's Documents the Employer requires the Contractor to prepare, which may not necessarily include (for example) all the technical documents which the Contractor's Personnel will need in order to execute the Works.

Further, if there are Contractor's Documents that the Contractor must submit to the Engineer for review, it is important that they are clearly identified in the Specification.

If a different Review Period than that stated under Sub-Clause 4.4.1 is considered necessary, or different Review Periods are considered necessary for different (types of) Contractor's Documents, such different Review Period(s) should be clearly stated in the Specification.

Sub-Clause 4.4.3   **Operation and Maintenance Manuals**

If the Contractor is required to supply spare parts under the Contract, these should be detailed in the Specification, which should also specify the required guarantee period for these spare parts.

Sub-Clause 4.5   **Training**

If the Works include Plant that comprises (in whole or in part) new or innovative technology in the Country or at the Employer's location, it is recommended that the Parties consider including in the Contract provision for training (by the Employer in the Specification and/or by the Contractor in the Tender) of the Employer's Personnel at the location of another plant facility, or at a number of other plant facilities, that are similar in nature to the Works.

Sub-Clause 4.8   **Health and Safety Obligations**

If the Contractor is sharing occupation of the Site with others, it may not be appropriate for him/her to provide some of the listed items. In these circumstances:

- this Sub-Clause should be amended to specify exactly what health and safety obligations are the Contractor's under the Contract, and

- the health and safety obligations which are to be fulfilled by the Employer and/or others should be stated in the Specification

so that it is clear which Party is responsible for what in respect of the health and safety obligations for the Site and for the Works.

Sub-Clause 4.9.1   **Quality Management System**

In the performance of the Contractor's obligations under this Sub-Clause, it is recommended that the Contractor and the Engineer take due regard of the FIDIC publication Improving the Quality of Construction - a guide for actions, 2004 (http://fidic.org/books/improving-quality-construction-2004[†]).

If the Employer requires the Contractor to have a Quality Manager employed on the Site, such position should be stated in the Specification as one of the positions of Key Personnel.

"如果这些人中没有人可以流利使用本语言，**承包商**应派足够数量胜任的翻译在所有工作时间随时在场，以便使这些人能够适当行使委托给他们的权利、职能及 / 或授权。"

第 4.4 款　　　　　　承包商文件

第 4.4.1 项　　　　　　编制和审核

规范要求中应明确规定**雇主**要求**承包商**编制的**承包商文件**，不一定包括（例如）**承包商人员**执行工程所需的所有技术文件。

此外，如果有些**承包商文件**需提交**工程师**进行审核，应在**规范要求**中明确说明。

如果认为有必要在**第 4.4.1 项**所述之外规定不同的**审核期**，或认为有必要对不同（类型）的**承包商文件**设定不同的**审核期**，则应在**规范要求**中明确说明这些不同的**审核期**。

第 4.4.3 项　　　　　　操作和维护手册

如需**承包商**提供**合同**规定的零部件，则应在**规范要求**中详细说明，同时还应明确这些零部件备件所需的保证期。

第 4.5 款　　　　　　培训

如果**工程**包括构成**工程**所在国或**雇主**所在地的（全部或部分）新技术或创新技术的**生产设备**，建议双方在合同中考虑包括另一生产设备设施或具有类似**工程**性质的多种生产设备设施所在地，（由**规范要求**中的**雇主**和 / 或由**投标书**中的**承包商**）培训**雇主**人员的条款。

第 4.8 款　　　　　　健康和安全义务

如果**承包商**与他人共同占用**场地**，其可能不合适提供所列的一些事项。在这些情况下：

- 应当修改本款，以确切说明**合同**规定的**承包商**的健康和安全义务；（以及）

- **规范要求**中应说明**雇主**和 / 或其他人应履行的健康和安全义务。

明确各方应履行的**现场**和**工程**的健康和安全义务。

第 4.9.1 项　　　　　　质量管理体系

在根据**本款**规定履行**承包商**义务时，建议**承包商**和**工程师**充分考虑菲迪克（FIDIC）2004 年出版的《提高施工质量——行动指南》（http://fidic.org/books/improving-quality-construction-2004†）。

如果**雇主**要求**承包商**聘用**现场质量经理**，则该职位应在**规范要求**中规定为关键人员职位之一。

If the Employer requires the Contractor to interface with others and/or with the Employer, the Specification should clearly describe all such interfaces and should specify the extent in which the Contractor shall allow for such interfaces in the QM System.

Sub-Clause 4.12   **Unforeseeable Physical Conditions**

In the case of a construction project which involves significant sub-surface works, it is recommended that the Employer consider the allocation of the risk of encountering adverse sub-surface conditions when the tender documents are being prepared.

If this risk is to be shared between the Parties, this Sub-Clause will need to be amended:

EXAMPLE   In Sub-Clause 4.12.4 [*Delay and/or Cost*], delete the words "payment of such cost" and substitute with

"payment of _____ percent (_____%) of such Cost (the balance of _____ percent (_____%) shall be borne by the Contractor)."

In order to assist the Engineer in performing his duties under Sub-Clause 4.12.5 [*Agreement or Determination of Delay and/or Cost*] in the event that Unforeseeable physical conditions are encountered by the Contractor, the Employer may consider including the physical/geological/sub-surface conditions that are known at the Base Date in the Contract - in the form of a '"Baseline Report".

Sub-Clause 4.16   **Transport of Goods**

In some cases, the Contractor may be required to get permission prior to delivery of Goods to the Site. In such cases those Goods for which permission is required should be stated in the Specification, and the following wording may be added to this Sub-Clause:

EXAMPLE   Insert at the end of Sub-Clause 4.16:

"The Contractor shall obtain the Engineer's permission prior to delivering to the Site any item of Goods which is identified in the Specification as requiring such permission. No such Goods shall be delivered without this permission, which shall not relieve the Contractor from any duty, obligation or responsibility under or in connection with the Contract."

Sub-Clause 4.17   **Contractor's Equipment**

If the Contractor is not to provide all the Contractor's Equipment necessary to execute the Works, the Employer's obligations should be specified: please see Sub-Clause 2.6 [*Employer-Supplied Materials and Employer's Equipment*] and the guidance notes for Sub-Clause 2.6 above.

If vesting of Contractor's Equipment to the Employer is required, and such vesting is consistent with the Laws of the Country, further paragraphs may be added to this Sub-Clause:

EXAMPLE   At the end of Sub-Clause 4.17, add the following paragraphs:

"Each item of Contractor's Equipment shall become the property of the Employer (free from liens and other encumbrances) when it arrives on the Site.

This vesting of property from the Contractor to the Employer shall not:

(a)   affect the responsibility or liability of the Employer under the Contract;

如果**雇主**要求**承包商**与他人和／或**雇主**衔接，则**规范要求**应明确说明所有此类衔接，并应规定**承包商**应在**质量管理体系**中允许此类衔接的范围。

第 4.12 款   **不可预见的物质条件**

对于涉及重大地下工程的施工项目，建议**雇主**在编写招标文件时考虑对遇到的不利地下条件的风险进行分配。

如果双方分担这种风险，则该款需修改为：

范例	删除**第 4.12.4 项**［**延误和／或费用**］中的"此类费用的支付"并代之以
	"此类费用百分之_____（_____%）的支付［剩余的百分之_____（_____%）应由**承包商**承担］。"

承包商遇到**不可预见**的**物质条件**时，为帮助**工程师**履行**第 4.12.5 项**［**延误和／或费用的商定或确定**］规定的职责，**雇主**可考虑将在**合同基准日期**已知的**物质／地质条件**包括在"**基线报告**"格式中。

第 4.16 款   **货物运输**

某些情况下，**承包商**把**货物**运到**现场**之前可能需要事先获得许可。在这种情况下，应在**规范要求**中说明需要许可的那些**货物**，并可以在此**款**中增加以下措辞：

范例	在**第 4.16 款**末尾插入：
	"**规范要求**要求需要获得许可的任何**货物**在运到**现场**之前，**承包商**均应获得**工程师**的事先许可。没有这样的许可，不得发运，这些均不应减轻**合同**规定的或与合同相关的**承包商**的职责、义务或责任。"

第 4.17 款   **承包商设备**

如果**承包商**不提供实施**工程**所需的所有**承包商设备**，应规定雇主义务：请参阅**第 2.6 款**［**雇主提供的材料和雇主设备**］以及上述**第 2.6 款**的指南说明。

如果需要将**承包商设备**归属**雇主**，并且这种归属符合**工程所在国**的法律，则可以在此**款**中增加以下段落：

范例	在**第 4.17 款**末尾，增加以下段落：
	"每件**承包商设备**到达**现场**后，均应归雇主所有（无留置权和其他产权负担）。
	从**承包商**到**雇主**的这种财产归属不得：
	（a）影响**合同**规定的**雇主**职责或责任；

(b) prejudice the Contractor's right to exclusive use of all items of Contractor's Equipment for the purpose of the Works; and/or

(c) relieve the Contractor from any duty, obligation or responsibility to operate and maintain all items of Contractor's Equipment.

The property in each item of Contractor's Equipment shall be deemed to re-vest in the Contractor (free from liens and other encumbrances) when he/she is entitled to remove it from the Site or to receive the Taking-Over Certificate for the Works, whichever is the earlier."

Sub-Clause 4.18   Protection of the Environment

If the applicable environmental law requires the Contractor to prepare project-specific environmental management plans, for review by the Engineer and/or the Employer and approval by regulatory authorities, these plans should be clearly described in detail in the Specification together with the review/approval process associated with each plan.

Sub-Clause 4.19   Temporary Utilities

If services are to be available on the Site for the Contractor to use, details of such services should be stated in the Specification, including a description of each utility, the capacity of each utility that is available for the Contractor's use, its location and the price per unit of consumption.

Sub-Clause 4.21   Security of the Site

If the Contractor is sharing occupation of the Site with others, it may not be appropriate for the Contractor to be responsible for its security. In these circumstances, it is recommended that:

a) this Sub-Clause should be amended to specify exactly what security obligations are the Contractor's under the Contract,

and

b) the security obligations which are to be fulfilled by the Employer and/or others should be stated in the Specification

so that it is clear which Party is responsible for what in respect of the security of the Site.

Sub-Clause 4.22   Contractor's Operations on Site

If the Contractor is sharing occupation of the Site with others, it is recommended that this Sub-Clause is amended by identifying and allocating responsibility for clearance and removal from the Site of any wreckage, rubbish, hazardous waste, temporary works and surplus material.

Additional Sub-Clause Milestones

If the Employer wishes to have certain parts of the Works completed within certain times but does not wish to take over such parts when completed (as distinct from the parts of the Works which the Employer wishes to take over after completion, which should be defined as Sections in the Contract Data), such parts of the Works should be clearly described in the Specification as 'Milestones' and it is recommended that the following provisions are added to the Contract Data and to the Conditions of Contract:

(b) 损害**承包商**使用**工程**所需的**承包商设备**所有事项的专有权利；（和 / 或）

(c) 免除**承包商**操作和维护**承包商设备**所有事项的任何任务、义务或职责。

当**承包商**有权将**承包商设备**运离现场，或收到**工程接收证书**时（以较早者为准），应将每项**承包商设备**中的财产视为重新归属**承包商**（无留置权和其他产权负担）。"

第 4.18 款　　　　保护环境

如果适用环境法律要求**承包商**编写具体项目的环境管理计划，供**工程师**和 / 或**雇主**审核并获得监管机构的批准，则这些计划应在**规范要求**中加以详细说明，并附有每个计划相关的审核 / 批准程序。

第 4.19 款　　　　临时公用设施

如果**现场**提供可供**承包商**使用的服务，则应在**规范要求**中明确此类服务的详细信息，包括对每个公用设施的说明、每个可供**承包商**使用的设施的容量、位置以及单位消费价格。

第 4.21 款　　　　现场安保

如果**承包商**与他人共同占用**现场**，其安保工作应不宜由**承包商**负责。在这种情况下，建议：

a)　　　应修改本**款**，明确规定**承包商**所应承担的**合同**规定的安保义务；

（以及）

b)　　　**规范要求**应规定**雇主**和 / 或其他人应履行的安保义务。

以明确各**方现场**安保的责任。

第 4.22 款　　　　承包商的现场作业

如果**承包商**与他人共同占用**现场**，则建议对本**款**进行修改，明确并分配从**现场**清理和清除残骸、垃圾、危险废弃物、临时工程和多余的材料的责任。

附加条款　　　　里程碑

如果**雇主**希望在特定时间内完成**工程**的某些部分，但又不希望在竣工时接收这些部分（与**雇主**希望在竣工后接收的工程部分不同，应规定为**合同数据**中的**分项工程**），**规范要求**应将此类部分工程明确规定为"里程碑（译注：重要阶段）"，建议在**合同数据**和**合同条件**中增加以下规定：

EXAMPLE PROVISIONS FOR MILESTONES

In the Contract Data, add the following:

"4.24.......... Definition of Milestones:

Description of a part of the Plant or of the Works that shall be designated a Milestone for the purposes of the Contract	Time for Completion	Delay Damages (as a percentage of final Contract Price per day of delay)
	.... days	.... %
	.... days	.... %
	.... days	.... %
	.... days	.... %

Maximum amount of Delay Damages for Milestones (percent of final Contract Price) ....%. "

Insert two new definitions under Sub-Clause 1.1:

" "**Milestone**" means a part of the Plant and/or a part of the Works stated in the Contract Data (if any), and described in detail in the Specification as a Milestone, which is to be completed by the time for completion stated in Sub-Clause 4.24 [*Milestone Works*] but is not to be taken over by the Employer after completion.

"**Milestone Certificate**" means the certificate issued by the Engineer under Sub-Clause 4.24 [*Milestones*]. "

Add new Sub-Clause 4.24 [*Milestones*].

"If no Milestones are specified in the Contract Data, this Sub-Clause shall not apply.

The Contractor shall complete the works of each Milestone (including all work which is stated in the Specification as being required for the Milestone to be considered complete) within the time for completion of the Milestone, as stated in the Contract Data, calculated from the Commencement Date.

The Contractor shall include, in the initial programme and each revised programme, under sub-paragraph (a) of Sub-Clause 8.3 [*Programme*], the time for completion for each Milestone.

Sub-paragraph (d) of Sub-Clause 8.4 [*Advance Warning*] and Sub-Clause 8.5 [*Extension of Time for Completion*] shall apply to each Milestone, such that "Time for Completion" under Sub-Clause 8.5 shall be read as the time for completion of a Milestone under this Sub-Clause.

The Contractor may apply, by Notice to the Engineer, for a Milestone Certificate not earlier than 14 days before the works of a Milestone will, in the Contractor's opinion, be complete. The Engineer shall, within 28 days after receiving the Contractor's Notice:

里程碑的范例条款

在**合同数据**中增加以下内容:

"第 4.24 款 .... 里程碑的定义:

对应设定合同里程碑的生产设备或工程的某部分的说明	竣工时间	误期损害赔偿费（按每天延误最终合同价格的百分比计算）
	........天	___%
	........天	___%
	........天	___%
	........天	___%

里程碑的误期损害赔偿费的最大数额（最终合同价格的百分比）……___%。"

在**第 1.1 款**中插入两个新定义:

"'**里程碑**'系指**合同数据**（如果有）中规定的**生产设备**的某部分和/或**工程**的某部分，并在**规范要求**中被详细描述为里程碑，该里程碑应在**第 4.24 款**[*里程碑*]规定的竣工时间之前完成，但该**生产设备**或**工程**竣工后**雇主**不接收。

'**里程碑证书**'是指**工程师**根据**第 4.24 款**[*里程碑*]规定签发的证书。"

增加**第 4.24 款**[*里程碑*]。

"如果**合同数据**未规定**里程碑**，此款将不适用。

承包商应按照**合同数据**规定的、从**开工日期**开始的**里程碑**完成时间内，完成每个**里程碑**的工作（包括**规范要求**中规定的视为完成**里程碑**所需的所有工作）。

承包商应按照**第 8.3 款**[*进度计划*]（a）段的规定，在初步进度计划和每项修改的进度计划中，包含每个**里程碑**的完成时间。

第 8.4 款[*预先警示*]（d）段和**第 8.5 款**[*竣工时间的延长*]应适用于每个**里程碑**，因此，**第 8.5 款**中的"竣工时间"应解读为该款规定的**里程碑**的完成时间。

承包商可在其认为**里程碑**工程完成前 14 天内，向**工程师**发出**通知**，申请**里程碑证书**。**工程师**应在收到**承包商**通知后 28 天内:

(a) issue the Milestone Certificate to the Contractor, stating the date on which the works of the Milestone were completed in accordance with the Contract, except for any minor outstanding work and defects (as shall be listed in the Milestone Certificate); or

(b) reject the application, giving reasons and specifying the work required to be done and defects required to be remedied by the Contractor to enable the Milestone Certificate to be issued.

The Contractor shall then complete the work referred to in sub-paragraph (b) of this Sub-Clause before issuing a further Notice of application under this Sub-Clause.

If the Engineer fails either to issue the Milestone Certificate or to reject the Contractor's application within the above period of 28 days, and if the works of a Milestone are complete in accordance with the Contract, the Milestone Certificate shall be deemed to have been issued on the date which is 14 days after the date stated in the Contractor's Notice of application.

If Delay Damages for a Milestone are stated in the Contract Data, and if the Contractor fails to complete the works of the Milestone within the time for completion of the Milestone (with any extension under this Sub-Clause):

(i) the Contractor shall, subject to Sub-Clause 20.1 [*Claims*], pay Delay Damages to the Employer for this default;

(ii) such Delay Damages shall be the amount stated in the Contract Data, for every day which shall elapse between the time for completion for the Milestone (with any extension under this Sub-Clause) and the date stated in the Milestone Certificate;

(iii) these Delay Damages shall be the only damages due from the Contractor for such default; and

(iv) the total amount of Delay Damages for all Milestones shall not exceed the maximum amount stated in the Contract Data (this shall not limit the Contractor's liability for Delay Damages in any case of fraud, gross negligence, deliberate default or reckless misconduct by the Contractor)."

If certain payment(s) to the Contractor is/are to be made on completion of each Milestone, such payment(s) should be specified in a Schedule of Payments in the Contract and consideration should be given to amending Sub-Clause 14.4 [*Schedule of Payments*] to make express reference to the Milestone payments.

Clause 5   Subcontracting

Sub-Clause 5.1   Subcontractors

It may be appropriate and/or desirable, taking account of the circumstances and locality of the project, to encourage the Contractor to employ local contractors in the execution of the Work:

(a) 给**承包商**签发**里程碑证书**，注明**里程碑**的工程已按照合同要求完成的日期，除了任何少量扫尾工作和缺陷（应在**里程碑证书**中列出）；（或）

(b) 拒绝申请，说明原因并指出**承包商**取得**里程碑证书**所需完成的工作和需要修补的缺陷。

承包商随后应在本款规定的进一步申请**通知**前，完成本款（b）项规定的工作。

如果**工程师**在上述 28 天内未能签发**里程碑证书**或拒绝**承包商**的申请，并且**里程碑**的工作已按合同要求完成，则**里程碑证书**将被视为已在**承包商**申请**通知**所述日期后 14 天的日期签发。

如果**合同数据**规定了**里程碑**的**误期损害赔偿费**，且**承包商**未能在**里程碑**竣工时间内完成**里程碑**工作（根据本款规定的任何延长），则：

(i) **承包商**应根据**第 20.1 款**［*索赔*］的规定，就此违约向**雇主**支付**误期损害赔偿费**；

(ii) 此类**误期损害赔偿费**应为**合同数据**中规定的金额，应在**里程碑**完成时间（根据本款规定的任何延长）至**里程碑证书**规定的日期之间按天计算；

(iii) 这些**误期损害赔偿费**应是**承包商**违约而应承担的唯一损害赔偿费；（以及）

(iv) 所有**里程碑**的**误期损害赔偿费**总额不得超过**合同数据**中规定的最高金额（不应减少**承包商**在任何欺诈、重大过失、故意违约或轻率不当行为的情况下对**延误损害赔偿**的责任）。"

如果在每个**里程碑**完成后要向**承包商**支付某些款项，则应在**合同**的**付款计划表**中规定这些款项，并应考虑修改**第 14.4 款**［*付款计划表*］，以明确说明**里程碑**的付款。

第 5 条　　　　分包

第 5.1 款　　　　分包商

考虑到项目的情况和地点，鼓励**承包商**雇用当地承包商来实施**工程**，可能是适当和 / 或可取的：

EXAMPLE	At the end of the second paragraph of Sub-Clause 5.1, add:
	"The Contractor shall give reasonable opportunity to contractors from the Country to tender for subcontracts for the Works, and shall use reasonable endeavours to employ such contractors as Subcontractors."

If no consent by the Engineer to Subcontractors is required, the last three paragraphs of this Sub-Clause may be deleted. If less consent by the Engineer to Subcontractors is preferable, this Sub-Clause may be amended:

EXAMPLE	After sub-paragraph (ii) of Sub-Clause 5.1, add:
	"or
	(iii) a Subcontractor where the value of his/her subcontract is less than one percent (1%) of the Accepted Contract Amount."

If the Engineer's consent is required to the suppliers of certain Materials, this Sub-Clause may be amended:

EXAMPLE	At the end of the third paragraph of Sub-Clause 5.1, add:
	"However, the Engineer's prior consent shall be required for the suppliers of the following Materials:
	_____
	_____ "
	(*insert details: for example, specific manufactured or prefabricated items of Materials*)

If the Employer requires that all the Contractor's subcontracts should provide for assignment of the subcontract to the Employer in the event that Sub-Clause 15.2 [*Termination for Contractor's Default*] applies, the following amendment will need to be made to this Sub-Clause:

EXAMPLE	Insert at the end of the last paragraph of Sub-Clause 5.1:
	"All subcontracts relating to the Works shall include provisions which entitle the Employer to require the subcontract to be assigned to the Employer under sub-paragraph (a) of Sub-Clause 15.2.3 [*After Termination*]."

If the Employer requires that the Contractor assigns the benefit of the relevant subcontract in the event that a Subcontractor's obligations continue after expiry of the DNP relating to that Subcontractor's work, the following amendment will need to be made to this Sub-Clause:

EXAMPLE	Insert at the end of the last paragraph of Sub-Clause 5.1:
	"If a Subcontractor's obligations to the Contractor extend beyond the expiry date of the DNP which is applicable to the Subcontractor's work and if the Contractor receives an instruction from the Engineer to do so not less than 7 days before this expiry date, the Contractor shall assign the benefit of such obligations to the Employer. Unless otherwise stated in the assignment, the Contractor shall have no liability to the Employer for work carried out by the Subcontractor after the assignment takes effect."

范例	在第 5.1 款的第二段末尾，增加：
	"**承包商**应给工程所在国的承包商提供投标分包工程的合理机会，并应用合理方式雇用此类承包商为**分包商**。"

如果无须**工程师**同意**分包商**，则可以删除本款的最后三段。如果**工程师**不太同意分包商，则本款可修改为：

范例	在第 5.1 款第（ii）段后，增加：
	"或
	（iii）分包价值低于**中标合同金额**的 1% 的**分包商**。"

如果要求**工程师**同意某些**材料**的供应商，**该款**可修改为：

范例	在第 5.1 款的第三段末尾，加上：
	"但是，**工程师**应事先同意以下**材料**的供应商：

	--- "
	（插入详细说明：如，特别制造或预制的**材料细项**）

在第 15.2 款 [*因承包商违约的终止*] 适用的情况下，如果**雇主**要求所有**承包商**的分包合同都应规定将分包合同转让给**雇主**，则需要对此款做出以下修改：

范例	在第 5.1 款最后一段的末尾，插入：
	"与**工程**相关的所有分包合同都应包括授权**雇主**要求把分包合同根据第 15.2.3 项 [*终止后*]（a）段转让给**雇主**的条款。"

在与**分包商**工作相关的**缺陷通知期限**到期后**分包商**的义务续存的情况下，如果**雇主**要求**承包商**转让相关**分包商**的收益，则需对该款做出以下修改：

范例	在第 5.1 款最后一段的末尾，插入：
	"如果**分包商**对**承包商**的义务超过适用于**分包商**工作的**缺陷通知期限**的有效期，且如果**承包商**收到**工程师**在到期前 7 天内发出的这样做的指示，**承包商**应把这些义务的收益转让给**雇主**。除非在转让中另有规定，**承包商**应在转让生效后对**分包商**执行的工作不再对**雇主**负责。"

Sub-Clause 5.2   Nominated Subcontractors

Normally the Contractor will select and employ subcontractor(s), (subject to any constraints stated in the Contract), but this Sub-Clause provides for the situation where the Employer may wish to select a particular subcontractor or subcontractors to be employed by the Contractor in the execution of the Works.

If this is the case, it is recommended that the Employer names the particular subcontractor(s) in the Specification so that tenderers are aware of this requirement before submitting their tenders - although this Sub-Clause also makes provision for the Engineer to instruct the Contractor to employ a subcontractor or subcontractors selected by the Employer after contract award.

It should be noted that:

a) once the Contractor has employed a subcontractor who has been nominated by the Employer, the second paragraph of Sub-Clause 5.1 [*Subcontractors*] applies; and

b) sub-paragraph (b) of Sub-Clause 13.4 [*Provisional Sums*] should provide for payment to the Contractor for works or services to be purchased by the Contractor from a nominated Subcontractor.

If the Employer anticipates that a Subcontractor is to be instructed under Sub-Clause 13.3 [*Variation Procedure*] but is not to be a nominated Subcontractor, this Sub-Clause should be amended describing the particular circumstances.

Clause 6   Staff and Labour

Sub-Clause 6.3   Recruitment of Persons

It may be necessary to review and revise this Sub-Clause (or delete it in its entirety) if the relevant labour Laws applicable to the Contractor's Personnel and/or the Employer's Personnel do not permit any restriction on the right of any worker to seek other positions.

Sub-Clause 6.6   Facilities for Staff and Labour

If the Employer plans to make any facilities and/or accommodation (for example, office accommodation) available for the Contractor's occupation and/or use, the Employer's obligations to do so and details of such facilities and/or accommodation should be stated in the Specification. See also the guidance under Clause 17 [*Care of the Works and Indemnities*] below for the suggested additional sub-clause in respect of responsibility for care of such facilities and/or accommodation.

Sub-Clause 6.8   Contractor's Superintendence

If it is permissible that the Contractor's superintending staff are not all fluent in the language for communications defined in Sub-Clause 1.4 [*Law and Language*], then consideration should be given to amending this Sub-Clause:

EXAMPLE   Insert at the end of sub-paragraph (a) of Sub-Clause 6.8:

"or, if not, the Contractor shall make competent interpreters available during all working hours, in a number sufficient for those persons to properly perform their superintendence duties."

If it is necessary to stipulate that the Contractor's superintending staff shall be fluent in a particular language, the following sentence should be added.

第 5.2 款　　　　　指定分包商

一般情况下，**承包商**将选择和雇用**分包商**（受合同规定的任何限制），但本**款**规定了**雇主**可能希望选择**承包商**在实施**工程**中雇用的一个特定分包商或多个分包商的情况。

如果是这种情况，建议**雇主**在**规范要求**中指定特定**分包商**，以便投标人在提交其投标书之前了解此要求——尽管该**款**还规定**工程师**应指示**承包商**在授予合同后雇用由**雇主**选择的一个或多个分包商。

应注意的是：

a)　　　一旦**承包商**雇用了由**雇主**提名的分包商，则第 5.1 款［*分包商*］的第二段适用；（以及）

b)　　　第 13.4 款［*暂列金额*］（b）段应规定，**承包商**向从指定**分包商**购买的工程或服务付款。

如果**雇主**希望**分包商**，但其不是指定**分包商**，根据第 13.3 款［*变更程序*］的规定得到指示，则应对该**款**进行修改，说明此特殊情况。

第 6 条　　　　　　　　　员工

第 6.3 款　　　　　招聘人员

如果适用于**承包商人员**和 / 或**雇主人员**的相关劳动**法律**不允许对任何工人寻求其他职位的权利进行限制，可能有必要审核并修改本**款**（或全部删除）。

第 6.6 款　　　　　为员工提供设施

如果**雇主**计划向**承包商**提供可占有和 / 或使用的任何设施和 / 或住所（例如，办公室住所），**规范要求**应明确**雇主**这样做的义务，以及此类设施和 / 或住所的详细信息。有关此类设施和 / 或住所照管责任建议的附加条款，请参见下文第 17 条［*工程照管和保障*］中的指南。

第 6.8 款　　　　　承包商的监督

如果允许**承包商**的监督人员可以不全部流利使用第 1.4 款［*法律和语言*］规定的交流语言，则应考虑修改**此款**：

范例　　　　　　　在第 6.8 款（a）段的末尾，插入：

"或，如果不是这样，**承包商**应派足够数量胜任的翻译、在所有工作时间随时在场的，以便那些人能够适当履行其监督职责。"

如果有必要规定**承包商**的监督人员应流利使用某种特定语言，则应增加以下词句。

EXAMPLE   Insert at the end of Sub-Clause 6.8:

"A reasonable proportion of the Contractor's superintending staff shall have a working knowledge of

[*Insert name of language*],

or the Contractor shall make interpreters available on Site during all working hours in a number deemed sufficient by the Engineer."

Sub-Clause 6.12   Key Personnel

If it is permissible that all Key Personnel are not fluent in the language for communications defined in Sub-Clause 1.4 [*Law and Language*], then consideration should be given to amending this Sub-Clause:

EXAMPLE   Insert at the end of the last paragraph of Sub-Clause 6.12:

"If any of the Key Personnel are not fluent in this language the Contractor shall make competent interpreter(s) available during all working hours, sufficient for that person to properly perform his/her duties under the Contract."

Additional Sub-Clauses

It may be necessary, appropriate and/or desirable to include additional sub-clauses to take account of the circumstances and locality of the Site:

EXAMPLE SUB-CLAUSE FOR FOREIGN PERSONNEL

"The Contractor may bring into the Country any foreign personnel who are necessary for the execution of the Works to the extent allowed by the applicable Laws. The Contractor shall ensure that these personnel are provided with the required residence visas and work permits. The Employer shall, if requested by the Contractor, use all reasonable endeavours in a timely and expeditious manner to assist the Contractor in obtaining any local, state, national, or government permission required for bringing in the Contractor's personnel.

The Contractor shall be responsible for the return of these personnel to the place where they were recruited or to their domicile. In the event of the death in the Country of any of these personnel or members of their families, the Contractor shall similarly be responsible for making the appropriate arrangements for their return or burial."

EXAMPLE SUB-CLAUSE FOR SUPPLY OF FOODSTUFFS

"The Contractor shall arrange for the provision of a sufficient supply of suitable food as may be stated in the Specification at reasonable prices for the Contractor's Personnel for the purposes of or in connection with the Contract."

EXAMPLE SUB-CLAUSE FOR SUPPLY OF WATER

"The Contractor shall, having regard to local conditions, provide on the Site an adequate supply of drinking and other water for the use of the Contractor's Personnel."

范例	在**第 6.8 款**末尾，插入：
	"合理比例的**承包商**监督人员应具备
	[*插入语言名称*] 的工作知识，
	或，**承包商**应派**工程师**认为足够数量的翻译在所有工作时间在**现场**。"

第 6.12 款　　　　关键人员

如果允许所有**关键人员**可不流利使用**第 1.4 款**[*法律和语言*]规定的交流语言，则应考虑修改此款为：

范例	在**第 6.12 款**最后一段的末尾，插入：
	"如果任何**关键人员**不能流利使用本语言，**承包商**应派足够数量胜任的翻译在所有工作时间随时在场，以便其能够适当履行**合同**规定的任务。"

附加条款

考虑到**现场**的情况和位置，设定一些附加条款可能是有必要、适当和/或希望的：

范例条款 外国人员

"**承包商**可以在适用**法律**允许的范围内，将实施**工程**所需的任何外国人员带入**工程所在国**。**承包商**应确保向这些人员提供所需的居住签证和工作许可证。应**承包商**要求，**雇主**应及时、迅速地采取一切合理的措施，协助**承包商**获得引进**承包商**人员所需的任何地方、州、国家或政府的许可。

承包商应负责这些人员返回到他们被招聘的地方或定居地。如果这些人员或其家庭成员在**工程所在国**死亡，**承包商**应同样负责适当安排其回国或埋葬。"

范例条款 食物供应

"**承包商**应为履行**合同**或与**合同**相关的人员，以合理的价格提供**规范要求**中规定的充足食物。"

范例条款 供水

"**承包商**应根据当地情况，在**现场**提供足够的饮用水和其他水，以供**承包商**的人员使用。"

EXAMPLE SUB-CLAUSE FOR MEASURES AGAINST INSECT AND PEST NUISANCE

"The Contractor shall at all times take the necessary precautions to protect the Contractor's Personnel employed on the Site from insect and pest nuisance, and to reduce the danger to their health. The Contractor shall comply with all the regulations of the local health authorities, including use of appropriate insecticide."

EXAMPLE SUB-CLAUSE FOR ALCOHOLIC LIQUOR OR DRUGS

"The Contractor shall not, other than in accordance with the Laws of the Country, import, sell, give, barter or otherwise dispose of any alcoholic liquor or drugs, or permit or allow importation, sale, gift, barter or disposal thereto by the Contractor's Personnel."

EXAMPLE SUB-CLAUSE FOR ARMS AND AMMUNITION

"The Contractor shall not give, barter, or otherwise dispose of, to any person, any arms or ammunition of any kind, or allow the Contractor's Personnel to do so."

EXAMPLE SUB-CLAUSE FOR FESTIVALS AND RELIGIOUS CUSTOMS

"The Contractor shall respect the Country's recognised festivals, days of rest and religious or other customs."

EXAMPLE SUB-CLAUSE FOR FUNERAL ARRANGEMENTS

"The Contractor shall be responsible, to the extent required by local regulations, for making any funeral arrangements for any of the Contractor's local employees who may die while engaged upon the Works."

EXAMPLE SUB-CLAUSE FOR FORCED LABOUR

"The Contractor shall not employ forced labour, which consists of any work or service, not voluntarily performed, that is exacted from an individual under threat of force or penalty, and includes any kind of involuntary or compulsory labour, such as indentured labour, bonded labour or similar labour-contracting arrangements."

EXAMPLE SUB-CLAUSE FOR CHILD LABOUR

"The Contractor shall not employ children (any natural persons under the age of eighteen years) in a manner that is economically exploitative, or is likely to be hazardous, or to interfere with the child's education, or to be harmful to the child's health or physical, mental, spiritual, moral, or social development. Where the relevant labour Laws of the Country have provisions for employment of minors, the Contractor shall follow those Laws applicable to the Contractor."

EXAMPLE SUB-CLAUSE FOR EMPLOYMENT RECORDS OF WORKERS

"The Contractor shall keep complete and accurate records of the employment of labour at the Site. The records shall include the names, ages, genders, hours worked and wages paid to all workers. These records shall be summarised on a monthly basis and submitted to the Engineer. These records shall be included in the details to be submitted by the Contractor under Sub-Clause 6.10 [*Contractor's Records*]."

范例条款 防虫害措施

"**承包商**应始终采取必要的预防措施,以保护**现场**雇用的**承包商人员**免受虫害的滋扰,并减少对其健康的危害。**承包商**应遵守当地卫生部门的所有法规,包括使用适当的杀虫剂。"

范例条款 酒类或毒品

"**承包商**应遵守**工程所在国**的**法律**,不得进口、出售、赠送、易货或以其他方式处置任何酒类或毒品,或由**承包商人员**进口、出售、赠送、易货或处置。"

范例条款 武器和弹药

"**承包商**不得将任何种类的任何武器或弹药交给任何人,以物易物或以其他方式处置,或允许**承包商人员**这样做。"

范例条款 节日和宗教习俗

"**承包商**应尊重**工程所在国**公认的节日、休息日和宗教或其他习俗。"

范例条款 葬礼安排

"**承包商**应在当地法规要求的范围内,负责为从事该**工程**时可能死亡的**承包商**的当地雇员安排丧葬。"

范例条款 强迫劳动

"**承包商**不得雇用强迫劳动,其中包括非自愿执行的任何工作或服务,是从个人受到武力或罚款威胁的情况下进行的,包括任何形式的非自愿或强制性劳动,例如包身工、债役劳动或类似的劳动合同安排。"

范例条款 童工

"**承包商**不得以经济上剥削,或可能有害的,或干扰儿童的教育,或有害于儿童健康或身体、智力、精神、道德或社会发展的方式雇用儿童(未满 18 岁的自然人)。如果**工程所在国**有关劳动**法**中有未成年人雇用的规定,**承包商**应遵循这些适用于**承包商**的**法律**。"

范例条款 雇用劳工记录

"**承包商**应保留完整和准确的**现场**劳工雇用记录。记录应包括姓名、年龄、性别、工作时间和支付给所有工人的工资等。这些记录应每月汇总,并提交给**工程师**。这些记录应包括在**承包商**根据第 6.10 款 [**承包商的记录**] 的规定提交的详细信息中。"

EXAMPLE SUB-CLAUSE FOR WORKERS' ORGANISATIONS

"In countries where the relevant labour Laws recognise workers' rights to form and to join workers' organisations of their choosing without interference and to bargain collectively, the Contractor shall comply with such Laws. Where the relevant labour Laws substantially restrict workers' organisations, the Contractor shall enable alternative means for the Contractor's Personnel to express their grievances and protect their rights regarding working conditions and terms of employment. In either case described above, and where the relevant labour Laws are silent, the Contractor shall not discourage the Contractor's Personnel from forming or joining workers' organisations of their choosing or from bargaining collectively, and shall not discriminate or retaliate against the Contractor's Personnel who participate, or seek to participate, in such organisations and bargain collectively. The Contractor shall engage with such workers' representatives. Workers' organisations are expected to fairly represent the workers in the workforce."

EXAMPLE SUB-CLAUSE FOR NON-DISCRIMINATION AND EQUAL OPPORTUNITY

"The Contractor shall not make employment decisions on the basis of personal characteristics unrelated to inherent job requirements. The Contractor shall base the employment relationship on the principle of equal opportunity and fair treatment, and shall not discriminate with respect to aspects of the employment relationship, including recruitment and hiring, compensation (including wages and benefits), working conditions and terms of employment, access to training, promotion, termination of employment or retirement, and discipline. In countries where the relevant labour Laws provide for non-discrimination in employment, the Contractor shall comply with such Laws. When the relevant labour Laws are silent on non-discrimination in employment, the Contractor shall meet the requirements under this Sub-Clause. Special measures of protection or assistance to remedy past discrimination or selection for a particular job based on the inherent requirements of the job shall not be deemed discrimination."

Clause 7   Plant, Materials and Workmanship

Sub-Clause 7.7   **Ownership of Plant and Materials**

If the Contractor is to provide high-value items of Plant and/or Materials under the Contract, consideration may be given to amending this Sub-Clause:

EXAMPLE   Insert at the end of Sub-Clause 7.7:

"No Plant and/or Materials that is the property of the Employer shall be removed from the Site. If it becomes necessary to:

(i) remove any item of such Plant from the Site for the purposes of repair, the Contractor shall give a Notice, with reasons, to the Engineer requesting consent to remove the defective or damaged item off the Site. This Notice shall clearly identify the item of defective or damaged Plant, and shall give details of: the defect or damage to be repaired; the place to which defective or damaged item of Plant is to be taken for repair; the transportation to be used (and insurance cover for such transportation); the proposed inspections and testing off the Site; and the planned duration required before

范例条款 劳工组织

"在相关劳动**法律**承认工人有权自由选择成立和参加其选择的劳工组织并集体谈判的国家，**承包商**应遵守这些**法律**。如果相关劳动**法律**在很大程度上限制了劳工组织，**承包商**应为**承包商人员**提供替代手段，以表达他们的不满，并保护他们在工作条件和雇用条件方面的权利。在上述两种情况下，以及在有关劳动**法律**未做规定的情况下，**承包商**均不得劝阻**承包商人员**组成或加入其选择的劳工组织，或进行集体谈判，也不得歧视或报复参与或寻求参与此类组织并进行集体谈判的**承包商人员**。**承包商**应与此类工人的代表进行接触，期望劳工组织在劳动力中公平地代表劳工。"

范例条款 非歧视和平等机会

"**承包商**不得根据与固有工作要求无关的个人特征做出雇用决定。**承包商**应基于平等机会和公平待遇的原则建立雇佣关系，并且不得在雇佣关系的各个方面进行歧视，包括招聘和雇用、薪酬（包括工资和福利）、工作条件和雇佣条件、获得培训、升职、终止雇佣关系或退休，以及纪律的途径。在有关劳动**法律**规定就业不受歧视的国家，**承包商**应遵守这些**法律**。当相关劳动**法律**对雇用方面的非歧视没有规定时，**承包商**应遵守本款要求。为补救过去的歧视或根据工作的固有要求，对某一特定工作进行选择而采取的特别保护或协助措施，不应视为歧视。"

第 7 条   生产设备、材料和工艺

第 7.7 款   生产设备和材料的所有权

如果**承包商**需根据合同提供高价值的**生产设备**和／或**材料**，则可以考虑把此款修改为：

范例   在**第 7.7 款**末尾，插入：

"不得从**现场**移走属于**雇主**财产的**生产设备**和／或**材料**。如果下述情况有必要：

(i) 为修理目的而从**现场**移除该**生产设备**的任何部件，**承包商**应有理由通知工程师，要求工程师同意将有缺陷或损坏的部件从**现场**移除。该通知应明确说明有缺陷或损坏**生产设备**的部件，并应提供以下详细信息：待修复的缺陷或损坏；在何地修理有缺陷或损坏的**生产设备**部件；所使用的交通工具（以及这种交通工具的保险）；建议在**现场**外进行的

the repaired item of Plant shall be returned to the Site. The Contractor shall also provide any further details that the Employer may reasonably require; or

(ii) replace any item(s) of such Plant and/or Materials, the Contractor shall give a Notice, with reasons, to the Engineer clearly identifying the item(s) of Plant and/or Materials to be replaced, and giving details of the due date of delivery to the Site of the replacement item(s).

Where any item of Plant and/or Materials has become the property of the Employer under this Sub-Clause before it has been delivered to the Site, the Contractor shall ensure that such an item is not moved except for its delivery to the Site.

The Contractor shall indemnify and hold the Employer harmless against and from the consequences of any defect in title or encumbrance or charge (except any reasonable restriction arising from the intellectual property rights of the manufacturer or producer) on any item of Plant and/or Materials that has become the property of the Employer under this Sub-Clause."

Additional Sub-Clause

If the Contract is being financed by an institution whose rules or policies require a restriction on the use of its funds, a further sub-clause may be added:

EXAMPLE SUB-CLAUSE FOR GOODS FROM ELIGIBLE SOURCE COUNTRIES

"All Goods shall have their origin in eligible source countries as defined in

[*Insert name of published guidelines for procurement*].

Goods shall be transported by carriers from these eligible source countries, unless exempted by the Employer in writing on the basis of potential excessive costs or delays. Surety, insurance and banking services shall be provided by insurers and bankers from the eligible source countries."

Clause 8   Commencement, Delays and Suspension

Sub-Clause 8.2   Time for Completion

If the Works are to be completed and taken over in stages it is important that each stage is defined as a Section, and the Time for Completion of each Section is stated, in the Contract Data (please also see the guidance below under Sub-Clause 10.1 [*Taking Over the Works and Sections*].

Sub-Clause 8.3   Programme

It is strongly recommended that the programming software that is preferable to the Engineer (in his/her monitoring of the Contractor's progress in executing the Works) be clearly identified in the Specification, and that such software is drawn to the attention of tenderers in the Instructions to Tender.

For less complex projects, the Employer may consider simplifying the requirements for the Contractor's programme as listed in sub-paragraphs (a) to (k) in this Sub-Clause (for example, by replacing sub-paragraphs (a) to (k) in this Sub-Clause with sub-paragraphs (a) to (d) as they appear under Sub-Clause 8.3 [*Programme*] of FIDIC's Conditions of Contract for Construction, First Edition 1999).

检测和试验，以及**生产设备**部件修理后运回**现场**之前所需的计划工期。**承包商**还应提供**雇主**可能合理要求的任何进一步细节；（或）

(ii) 更换此类**生产设备**和／或**材料**的任何部件时，**承包商**应在有理由的情况下向**工程师**发出**通知**，明确说明要更换的**生产设备**和／或**材料**部件，并提供把更换部件运到**现场**的预计日期的详细情况。

如果**生产设备**和／或**材料**部件在运到**现场**前已成为本款规定的**雇主**财产，**承包商**应确保该部件除了运到**现场**外，不得移动。

对于本款规定的、已成为**雇主**财产的任何**生产设备**和／或**材料**部件，**承包商**应保障并使**雇主**免受任何所有权缺陷，或产权负担，或费用（制造商或生产商的知识产权引起的任何合理限制除外）造成的后果。"

附加条款

如果**合同**由规则或政策要求限制其资金使用的机构提供资金，可以增加以下条款：

范例条款 来自合格来源国的货物

"所有**货物**应来源于

［*插入已发布的采购指南名称*］规定的合格国家。

除**雇主**基于潜在的超额费用或延误而书面豁免外，**货物**应由合格来源国的承运人运输。保证金、保险和银行服务应由合格来源国的保险公司和银行提供。"

第 8 条　　　　　　　　开工、延误和暂停

第 8.2 款　　　　　　竣工时间

如果工程要分阶段完成和接收，则有必要把每一阶段规定为一项**分项工程**，并在**合同数据**中注明每一**分项工程**的**竣工时间**（请参见下述**第 10.1 款**［*工程和分项工程的接收*］的指南）。

第 8.3 款　　　　　　进度计划

强烈建议在**规范要求**中明确规定有利于**工程师**的编程软件（在其监督**承包商**实施工程进度时），并在**投标须知**中提请投标人注意此类软件。

对于不太复杂的项目，**雇主**可以考虑简化本款（a）至（k）段所述的**承包商**进度计划的要求［例如，通过用 1999 年第 1 版《菲迪克（FIDIC）施工合同条件》第 8.3 款［*进度计划*］中的（a）至（d）段替换本款的（a）至（k）段］。

Sub-Clause 8.5 Extension of Time for Completion

Sub-paragraph (c) of this Sub-Clause entitles the Contractor to an extension of the Time for Completion for "exceptionally adverse climatic conditions". It may be preferable to set out in the Specification what constitutes an "exceptionally adverse" event - for example, by reference to available weather statistics and return periods. It may also be appropriate to compare the adverse climatic conditions that have been encountered with the frequency with which events of similar adversity have previously occurred at or near the Site. For example, an exceptional degree of adversity might be regarded as one which has a probability of occurrence of once every four or five times the Time for Completion of the Works (for example, once every eight to ten years for a two-year contract).

The final paragraph of this Sub-Clause provides that the rules and procedures for assessing the Contractor's entitlement to an EOT where there is concurrency between delays attributable to both Parties shall be stated in the Special Provisions. This provision has been drafted by FIDIC in this manner because there is no one standard set of rules/procedures in use internationally (though, for example the approach given in the Delay and Disruption Protocol published by the Society of Construction Law (UK):

https://www.scl.org.uk/sites/default/files/SCL_Delay_Protocol_2nd_Edition_Final.pdf[†] is increasingly being adopted internationally and different rules/procedures may apply in different legal jurisdictions.)

In preparing the Special Provisions, therefore, it is strongly recommended that the Employer be advised by a professional with extensive experience in construction programming, analysis of delays and assessment of extension of time in the context of the governing law of the Contract.

Sub-Clause 8.8 Delay Damages

Under the laws of many countries or other jurisdictions, the amount of these pre-defined damages must be a reasonable estimate of the anticipated or actual loss caused to the Employer by the delay. Therefore if the Delay Damages are fixed at an unreasonably large amount, they may be unenforceable in common law jurisdictions or subject to downward adjustment in civil law jurisdictions.

If the Accepted Contract Amount is to be quoted as the sum of figures in more than one currency, it may be preferable to define these damages (per day) as a percentage to be applied to each of these figures. If the Accepted Contract Amount is expressed in the Local Currency, the damages per day may either be defined as a percentage or be defined as a figure in Local Currency: see Sub-Clause 14.15(c).

Additional Sub-Clause

Incentives for early completion may be included (in addition to sub-paragraph (a) of Sub-Clause 13.2 [*Value Engineering*] which refers to accelerated completion):

EXAMPLE SUB-CLAUSE FOR INCENTIVES FOR EARLY COMPLETION

> "The Contractor shall be entitled to a bonus payment if the Works and/or each Section is completed earlier than the Time for Completion for the Works or Section (as the case may be). The amount of bonus for early completion of the Works and/or each Section is stated in the Specification.
>
> For the purposes of calculating any bonus payment, the applicable Time for Completion stated in the Contract Data is fixed and no adjustments of this time by reason of granting an EOT will be allowed."

第 8.5 款　　竣工时间的延长

本款（c）段规定，**承包商**有权因"特别不利的气候条件"而延长**竣工时间**。最好在**规范要求**中列出构成"特别不利"事件的要素——例如，通过参考可得的天气统计数据和返回周期。对曾经发生的不利气候条件和以前在该**现场**或附近发生的类似不利情况的频率进行比较，可能也是恰当的。例如，特别不利的程度可能被认为是**工程竣工时间** 4 到 5 倍的发生概率（例如，两年期合同每 8 至 10 年发生一次）。

本款的最后一段规定，如果双方都有延误，**特别规定**应规定评估**承包商**应享有的**竣工时间的延长**的规则和程序。菲迪克（**FIDIC**）以这种方式起草该规定，因为国际上没有一套可用的标准的规则 / 程序（尽管**英国建筑法学会**发布的《误期与中断协议》：

http: //www.scl.org.uk/sites/default/files/SCL_Delay_Protocol_2nd_Edition_Final.pdf† 规定的方法，在国际上越来越多地被采用，并且不同的规则 / 程序可能适用于不同的法律辖区。

因此，在编写**特别规定**时，强烈建议在**合同管辖法律**的范围内，由在施工规划、延误分析以及时间延长评估方面具有丰富经验的专业人员为**雇主**提供建议。

第 8.8 款　　误期损害赔偿费

根据许多国家或其他司法管辖区的法律，这些事先规定的损害赔偿费的数额，必须是对延误给**雇主**造成的预期或实际损失的合理估算。因此，如果将**误期损害赔偿费**定得过高，其在普通法管辖区可能无法执行，或在民法管辖区可能会向下调整。

如果**中标合同金额**的报价为多种货币的数字总和，最好将这些损害赔偿费（按天）确定为适用于每个货币数字的百分比。如果**中标合同金额**以**当地货币**表示，每天的损害赔偿费既可确定为百分比，也可确定为**当地货币**中的数字：请参阅**第 14.15 款**（c）段。

附加条款

可包括提前竣工的激励措施（除了**第 13.2 款**［*价值工程*］（a）段提到的加速竣工的奖励）：

范例条款 提前竣工的激励措施

"如果**工程**和 / 或每项**分项工程**早于**工程**或**分项工程**的**竣工时间**（视情况而定）完成，**承包商**有权获得奖金。**规范要求**中规定了**工程**和 / 或**分项工程**提前竣工的奖金数额。

为计算任何奖金的支付，**合同数据**规定的适用**竣工时间**是固定的，并且由于授予**竣工时间延长**而不允许对此时间进行任何调整。"

Clause 9   Tests on Completion

Sub-Clause 9.1   Contractor's Obligations

The Specification should describe the tests which the Contractor is to carry out before being entitled to a Taking-Over Certificate. It may also be appropriate for the Contractor's Proposal to include detailed arrangements, instrumentation, etc.

If the Works are to be tested and taken over in Sections, it is strongly recommended that consideration is given to testing requirements in the Specification that take due account of the effect of other parts of the Works being incomplete.

For less complex projects, the Employer may consider simplifying the testing requirements under this Sub-Clause.

Clause 10   Employer's Taking Over

Sub-Clause 10.1   Taking Over the Works and Sections

If the Works are to be completed and taken over in stages it is important that each stage is defined as a Section in the Contract Data. Precise geographical definitions are advisable, and the Contract Data should include a table to also define for each Section:

- the relevant percentage for release of Retention Money,
- the Time for Completion, and
- the applicable Delay Damages.

An example form of such table is shown in the example form of Contract Data included in this publication.

Clause 11   Defects after Taking Over

Sub-Clause 11.3   Extension of Defects Notification Period

Depending on the complexity of the project and the nature of the Works, the Employer may consider amending the period of "two years" stated at the end of the first paragraph of this Sub-Clause to be a longer or a shorter period.

Sub-Clause 11.10   Unfulfilled Obligations

It may be necessary to review the effect of this Sub-Clause in relation to the period of liability imposed by the applicable law.

In particular, the second paragraph of this Sub-Clause may require amending to take account of the applicable law.

Additional Sub-Clause

If the Works include the Plant that comprises (in whole or in part) new or innovative technology in the Country or at the Employer's location, consideration may be given to including in the Specification a requirement for the Contractor to provide supervisory assistance to the Employer's permanent operating personnel in the operation and maintenance of the Plant during the DNP of the Works.

第 9 条　　　　竣工试验

第 9.1 款　　　承包商的义务

规范要求应规定**承包商**在有权取得**接收证书**之前应进行的试验。**投标书**还应包括详细安排、仪器仪表等。[注]

如果**工程**按**分项工程**进行试验和接收，强烈建议考虑**规范要求**中的试验要求，并适当考虑**工程**其他部分不完整的影响。

对于不太复杂的项目，**雇主**可以考虑简化本**款**中的试验要求。

第 10 条　　　　雇主的接收

第 10.1 款　　　工程和分项工程的接收

如果**工程**要分阶段完成和接收，把每个阶段规定为**合同数据**中的一项**分项工程**非常重要。精确的地理定义是可取的，并且**合同数据**应包括一个图表，明确每项**分项工程**：

- 保留金发放的相关百分比；
- 竣工时间；（以及）
- 适用的**误期损害赔偿费**。

此图表的范例格式，请见本文件**合同数据**的范例格式。

第 11 条　　　　接收后的缺陷

第 11.3 款　　　缺陷通知期限的延长

根据项目的复杂性和**工程**的性质，**雇主**可以考虑将本**款**第一段末尾所述的"2 年"期限修改为更长或更短的期限。

第 11.10 款　　　未履行的义务

可能有必要审核本**款**对适用法律规定的责任期限的影响。

特别是，本**款**第二段可能需要在考虑适用法律的基础上进行修改。

附加条款

如果**工程**包括（全部或部分）构成**工程**所在国或**雇主**所在地的新型创新技术的**生产设备**，可考虑在**规范要求**中包括要求**承包商**向**雇主**的永久操作人员，提供**工程缺陷通知期限生产设备**操作与维护的监理协助。

[注]此处中文按照勘误表改正后的英文翻译。——译者注

EXAMPLE SUB-CLAUSE FOR SUPERVISORY ASSISTANCE DURING DNP

"The Contractor shall provide supervisory assistance to the Employer during the DNP for the Works. Such supervisory assistance shall be as described in the Specification for the purpose of supporting the Employer's operation and maintenance of the Plant for the period of [*insert number of months*] after the Date of Completion."

Clause 12   Measurement and Valuation

Sub-Clause 12.1   Works to be Measured

If any part of the Permanent Works is to be measured according to records of its construction, this part should be identified, and the records to be made should be stated in the Specification.

If the Contractor is to be responsible for any such records, this should be specified in the Contract (by the Employer in the Specification or by the Contractor in the Tender).

Clause 13   Variations and Adjustments

Variations can be initiated by any of three ways:

a) the Contractor may initiate his own proposals under Sub-Clause 13.2 [*Value Engineering*], which are intended to benefit both Parties;

b) the Engineer may instruct Variations under Sub-Clause 13.3.1 [*Variation by Instruction*]; or

c) the Engineer may request a proposal from the Contractor under Sub-Clause 13.3.2 [*Variation by Request for Proposal*].

Sub-Clause 13.4   Provisional Sums

Provisional Sums may be required for parts of the Works which are not required to be priced at the risk of the Contractor. For example, a Provisional Sum may be necessary to cover goods or services which the Employer wishes to select after award of the Contract, or to deal with a major uncertainty regarding sub-surface conditions. It is essential to define the scope of each Provisional Sum (it is recommended that this be included in a Schedule prepared by the Employer), since the amount of each Provisional Sum corresponding to the defined scope will then be excluded from the other elements of the Accepted Contract Amount.

Sub-Clause 13.7   Adjustments for Changes in Cost

The provision for adjustments under this Sub-Clause may be required if it would be unreasonable for the Contractor to bear the risk of escalating costs due to inflation.

If the amounts payable to the Contractor are to be adjusted for rises or falls in the cost of labour, Goods and other inputs to the Works, for this Sub-Clause to apply it is important that a Schedule (or Schedules) of cost indexation is (are) included in the Contract and that such Schedule(s) include a formula or formulae for calculation of the applicable adjustment under this Sub-Clause.

It is recommended that the Employer be advised by a professional with experience in construction costs and the inflationary effect on construction costs when preparing the contents of the Schedule(s) of cost indexation.

范例条款 缺陷通知期限的监理协助

> "**承包商**应在**工程**缺陷通知期限为**雇主**提供监理协助。此类监理协助应按照**规范要求**规定提供，以在**竣工日期**后的［*插入月数*］内支持**雇主**对**生产设备**的操作和维护。"

第 12 条　　测量和估价

第 12.1 款　　需测量的工程

如需根据**永久工程**的施工记录对其任何部分进行测量，应确定该部分，并在**规范要求**中规定要进行的记录。

如果**承包商**负责任何此类记录，应在**合同**中（由**规范要求**中的**雇主**或由**投标书**中的**承包商**）加以明确。

第 13 条　　变更和调整

可以通过以下三种方式之一实施变更：

a)　　**承包商**可根据第 13.2 款［*价值工程*］的规定提出对双方有利的建议书；

b)　　**工程师**可根据第 13.3.1 项［*指示变更*］的规定指示**变更**；（或）

c)　　**工程师**可根据第 13.3.2 项［*建议书要求的变更*］的规定要求**承包商**提交建议书。

第 13.4 款　　暂列金额

工程的某些部分可能需要**暂列金额**，**暂列金额**需**承包商**承担定价风险。例如，**暂列金额**对涵盖授予**合同**后**雇主**希望选择的货物或服务，或应对地下条件的重大不确定性，可能是必要的。规定每项**暂列金额**的范围是非常重要的（建议将其包括在由**雇主**编制的**资料表**中），因为与规定范围相关的每项**暂定金额**的数额，将从**中标合同金额**的其他要素中排除。

第 13.7 款　　因成本改变的调整

如果**承包商**不承担因通货膨胀而导致的成本上升的风险，可能需要按照本款规定而设立调整条款。

如果因本款适用的劳务、**货物**和**工程**的投入成本的上升或下降而需要调整应付给**承包商**的金额，则有必要在本**合同**中设立成本指数表的一个（或多个）**资料表**，并且这些**资料表**应包括一个或多个公式，用于计算本款规定的适用调整。

建议**雇主**在编制成本指数表的**资料表**时，由具有施工成本和通货膨胀影响经验的专业人员提供咨询。

SCHEDULE OF COST INDEXATION: EXAMPLE FORMULA FOR ADJUSTMENT FOR CHANGE IN COST

$$P_n = a + b\frac{L_n}{L_o} + c\frac{E_n}{E_o} + d\frac{M_n}{M_o} + \ldots$$

where:

"Pn" is the adjustment multiplier to be applied to the estimated contract value in the relevant currency of the work carried out in period "n", this period being a month;

"a" is a fixed coefficient, representing the non-adjustable portion in contractual payments;

"b", "c", "d", ... are coefficients representing the estimated proportion of each cost element related to the execution of the Works, such tabulated cost elements may be indicative of resources such as labour, equipment and materials;

(particular care should be taken in the calculation of the weightings/coefficients ("a", "b", "c", ...), the total of which must not exceed unity)

"Ln", "En", "Mn", ... are the current cost indices or reference prices (stated in the Schedule of cost indexation) for period "n", expressed in the relevant currency of payment, each of which is applicable to the relevant tabulated cost element on the date 49 days prior to the last day of the period (to which the particular Payment Certificate relates); and

"Lo", "Eo", "Mo", ... are the base cost indices or reference prices, expressed in the relevant currency of payment, each of which is applicable to the relevant tabulated cost element on the Base Date.

The weightings (coefficients) for each of the factors of cost stated in the following table(s) of adjustment data shall only be adjusted if they have been rendered unreasonable, unbalanced or inapplicable, as a result of Variations.

Table(s) of adjustment data for payments each month/[*YEAR*] (delete as appropriate) in .... (currency)

Coefficient; scope of index	Country of origin; currency of index	Source of index; Title/definition	Value on stated date(s)* Value......Date
a= 0.10 Fixed	—	—	—
b= ____Labour	—	—	—
c=	—	—	—
d=	—	—	—
e=			

* These values and dates confirm the definition of each index, but do not define Base Date indices

成本指数表：费用变化调整范例公式

$$Pn = a + b\frac{Ln}{Lo} + c\frac{En}{Eo} + d\frac{Mn}{Mo} + ...$$

其中：

"Pn"是应用于估算的合同价值的调整乘数，以在"n"期为一个月内执行工作的相关货币表示；

"a"是固定系数，代表合同付款中不可调整的部分；

"b""c""d"……是代表与实施**工程**有关的每个成本要素的估计比例的系数，这种表列的成本要素可表示劳动力、设备和材料等资源；

[在计算权重/系数（"a""b""c"）时应特别注意，其总和不得超过 1]

"Ln""En""Mn"……是"n"期内（成本指数**数据表**中规定的）当前成本指数或参考价格，以相关的**支付货币**表示，每个均适用于（与特别**付款证书**相关的）期限的最后一天前第 49 天表列的相关费用表；（以及）

"Lo""Eo""Mo"……是基本成本指数或参考价格，以相关支付货币表示，每个均适用于**基准日期**的相关表列成本要素。

下列调整数据表中列出的每个成本要素的权重（系数），仅应在因**变动**而导致不合理、不平衡或不适用的情况下进行调整。

每月/[年]（酌情删除）的支付调整数据表，以 ………（货币）表示

系数； 指数范围	来源国； 指数货币	指数来源； 标题/定义	规定日期的价值* 价值 …… 日期
————————	————————	————————	————————
a= 0.10 固定	————————	————————	————————
————————	————————	————————	————————
b= ____ 劳动力	————————	————————	————————
c=	————————	————————	————————
d=	————————	————————	————————
e=	————————	————————	————————

* 这些值和日期确认了每个指数的定义，但未规定**基准日期**指数。

Clause 14 Contract Price and Payment

The procedures and timing for making payments under this Clause 14 should be checked to ensure that they are acceptable to both the Employer and any financing institution the Employer may be using to fund the project.

Sub-Clause 14.1 The Contract Price

When writing the Special Provisions, consideration should be given to the amount and timing of payment(s) to the Contractor. A positive cash-flow is clearly of benefit to the Contractor, and tenderers will take account of the interim payment procedures when preparing their tenders.

Cost-plus contracts, under which the actual Cost is determined and Cost Plus Profit is paid to the Contractor, are unusual and only used when (for reasons of urgency or otherwise) the Employer is willing to accept the risks involved. If the Contractor is to be paid Cost Plus Profit, Clause 12 should be replaced by provisions describing the method of determining the Cost and the Contract Price. As a result, the provisions in the General Conditions which entitle the Contractor to payment of additional Cost will generally be of no effect.

Additional Sub-Clauses may be required to cover any exceptions to the provisions set out in Sub-Clause 14.1, and any other matters relating to payment.

Lump sum contracts may be suitable if the tender documents include details which are sufficiently complete for construction and for significant Variations to be unlikely. From the information supplied in the tender documents, the Contractor can prepare any other details necessary, and construct the Works, without having to refer back to the Engineer for clarification or further information. However, as noted in the Introduction above, if significant design input by the Contractor is required, it is recommended that the Employer consider using FIDIC's Conditions of Contract for Plant and Design-Build, Second Edition 2017 (or, alternatively and if suitable for the circumstances of the project, FIDIC's Conditions of Contract for EPC/Turnkey Projects, Second Edition 2017).

> For a lump sum contract, the tender documents should include a schedule of payments (see Sub-Clause 14.4 [*Schedule of Payments*]).

EXAMPLE PROVISIONS FOR A LUMP SUM CONTRACT

> Delete the second paragraph of Sub-Clause 8.5 [*Extension of Time for Completion*].
>
> Delete Clause 12 [*Measurement and Valuation*].
>
> Under Sub-Clause 13.3.1 [*Variation by Instruction*]:
>
> - in sub-paragraph (c) delete the words "by valuing the Variation in accordance with Clause 12 [*Measurement and Valuation*], with supporting particulars (which shall include identification of any estimated quantities and, if the Contractor incurs or will incur Cost as a result of any necessary modification to the Time for Completion, shall show the additional payment (if any) to which the Contractor considers that the Contractor is entitled)" and replace with:
>
> "with supporting particulars. Whenever the omission of any work forms part (or all) of a Variation, and if:
>
> • the Contractor has incurred or will incur cost which, if the work had not been omitted, would have been deemed to be covered by a sum forming part of the Accepted Contract Amount; and

第 14 条　　　　　　合同价格和付款

应检查**第 14 条**规定的付款程序和时间安排，以确保**雇主**和**雇主**可能用于资助项目的任何融资机构都可以接受。

第 14.1 款　　　　合同价格

在编写**特别规定**时，应考虑向**承包商**付款的金额和时间。积极的现金流显然对**承包商**有利。投标人在编写投标书时将考虑期中付款程序。

确定实际成本并向**承包商**支付**成本加利润**的成本加合同并不常见，仅在**雇主**（出于紧急或其他原因）愿意接受所涉及的风险时才使用。如果要向**承包商**收取**成本加利润**，则**第 12 条**应替换为描述确定**成本**和**合同价格**方法的规定。因此，**通用条件**中赋予**承包商**支付额外**费用**权利的规定通常无效。

可能需要附加**条款**涵盖**第 14.1 款**中规定的任何例外情况，以及与付款有关的任何其他事项。

如果招标文件中包含的详细信息足以完成施工并且不太可能出现重大**变更**，那么总价合同可能是合适的。根据招标文件中提供的信息，**承包商**可以准备任何其他必要的详细资料，并进行**工程施工**，而不必向**工程师**寻求澄清或进一步信息。但是，如上文导言所述，如果需要**承包商**的大量设计投入，建议**雇主**考虑使用菲迪克（FIDIC）2017 年第 2 版《**生产设备和设计 - 施工合同条件**》[或，如适用于项目的具体情况，菲迪克（FIDIC）2017 年第 2 版《**设计采购施工 EPC/ 交钥匙工程合同条件**》]。

对总价合同而言，招标文件应包括付款计划表（见**第 14.4 款**[*付款计划表*]）。

范例 总价合同规定

删除**第 8.5 款**[*竣工时间的延长*]的第二段。

删除**第 12 条**[*测量和估价*]。

根据**第 13.3.1 项**[*指示变更*]:

- 删除第（c）段中"通过根据**第 12 条**[*测量和估价*]的规定对**变更**进行估价，并附上详细证明资料[应包括确定预估的数量，以及因对**竣工时间**进行必要修改而导致的**承包商费用**增加或即将增加的情况，应表明**承包商**认为其有权获得的额外付款（如果有）]"并代之以：

"以详细证明资料。在任何工作的删减构成**变更**的一部分（或全部）时，如果：

- **承包商**已产生或即将产生费用，如果工作没有删减，该费用应被认为已包括在构成部分**中标合同金额**的数额中；（以及）

- the omission of the work has resulted or will result in this sum not forming part of the Contract Price

this cost may be included in the Contractor's proposal (and, if so, shall be clearly identified). If the Parties have agreed to the omission of any work which is to be carried out by others, the Contractor's proposal may also include the amount of any loss of profit and other losses and damages suffered (or to be suffered) by the Contractor as a result of the omission"; and

- in sub-paragraph (ii): delete the words "(including valuation of the Variation in accordance with Clause 12 [*Measurement and Valuation*] using measured quantities of the varied work)" and replace with:

"and the Schedule of Payments (if any)".

Delete sub-paragraph (a) of Sub-Clause 14.1 and replace with:

"(a) the Contract Price shall be the lump sum Accepted Contract Amount and be subject to adjustments in accordance with the Contract;".

Delete sub-paragraph (c) of Sub-Clause 14.1 and replace with:

"(c) any quantities which may be set out in a Schedule are estimated quantities and are not to be taken as the actual and correct quantities of the Works which the Contractor is required to execute; and".

If the Contractor is not required to pay import duties on Goods imported by the Contractor into the Country, an additional Sub-Clause should be added and sub-paragraph (b) of Sub-Clause 14.1 [*The Contract Price*] of the General Conditions should be amended accordingly:

EXAMPLE SUB-CLAUSE FOR EXEMPTION FROM DUTIES

"All Goods imported by the Contractor into the Country shall be exempt from customs and other import duties, if the Employer's prior written approval is obtained for import. The Employer shall endorse the necessary exemption documents prepared by the Contractor for presentation in order to clear the Goods through Customs, and shall also provide the following exemption documents:

(*describe the necessary documents, which the Contractor will be unable to prepare*)

If exemption is not then granted, the customs duties payable and paid shall be reimbursed by the Employer.

All imported Goods, which are not incorporated in or expended in connection with the Works, shall be exported on completion of the Contract.

If not exported, the Goods will be assessed for duties as applicable to the Goods involved in accordance with the Laws of the Country.

However, exemption may not be available for:

- 工作的删减已导致或将导致该款项不构成**合同价格**的一部分。

该费用可包含在**承包商**的报价中（如果是这样，则应明确说明）。如果双方已同意工作的删减由其他**方**实施，则**承包商**的报价也可包括因删减而使**承包商**遭受的（或将要遭受的）任何利润损失和其他损失的金额"。以及

- 在第（ii）段中：删除"（包括使用测量的工作变更数量按照**第 12 条[测量和估价]**对**变更**进行的估价）"，并代之以：

"和付款计划表（如果有）"。

删除**第 14.1 款**（a）段，并代之以：

"（a）**合同价格**应为总价**中标合同金额**，并可根据合同进行调整；"

删除**第 14.1 款**（c）段，并代之以：

"（c）**资料表**中可能列出的任何数量均为估计数量，不应视为要求**承包商**实施**工程**的实际和正确数量；（以及）"。

如果不要求**承包商**对**承包商**进口到**工程所在国**的**货物**支付进口关税，则应增加一条附加**条款**，并且**通用条件**的第 14.1 款[**合同价格**]（b）段应相应修改为：

范例条款 关税豁免

"如果事先获得**雇主**的书面批准，**承包商**进口到**工程所在国**的所有**货物**均应免除关税及其他进口税。**雇主**应在**承包商**编写的必要的免税文件上背书，以便**货物**清关时提供，还应提供以下免税文件：

（说明**承包商**无法准备的必要的文件）

如果关税未予豁免，则**雇主**应支付税款。

所有未纳入**工程**或与**工程**有关的进口**货物**，均应在合同完成后出口。

如果**货物**未出口，则将根据**工程所在国**法律对**货物**进行相应关税评估。

但是，豁免可能不适用于：

(a) Goods which are similar to those locally produced, unless they are not available in sufficient quantities or are of a different standard to that which is necessary for the Works; and

(b) any element of duty or tax inherent in the price of goods or services procured in the Country, which shall be deemed to be included in the Accepted Contract Amount.

Port dues, quay dues and, except as set out above, any element of tax or duty inherent in the price of goods or services shall be deemed to be included in the Accepted Contract Amount."

If expatriate staff are exempted from paying local income tax, a suitable Sub-Clause should be added and sub-paragraph (b) of Sub-Clause 14.1 [*The Contract Price*] of the General Conditions should be amended accordingly. However, advice should be sought from a qualified tax expert before drafting any such additional Sub-Clause.

EXAMPLE SUB-CLAUSE FOR EXEMPTION FROM TAXES

"Expatriate (foreign) personnel shall not be liable for income tax levied in the Country on earnings paid in any foreign currency, or for income tax levied on subsistence, rentals and similar services directly furnished by the Contractor to Contractor's Personnel, or for allowances in lieu. If any Contractor's Personnel have part of their earnings paid in the Country in a foreign currency, they may export (after the conclusion of their term of service on the Works) any balance remaining of their earnings paid in foreign currencies.

The Employer shall seek exemption for the purposes of this Sub-Clause. If it is not granted, the relevant taxes paid shall be reimbursed by the Employer."

Sub-Clause 14.2   Advance Payment

When writing the Particular Conditions, consideration should be given to the benefits of the Employer making an advance payment to the Contractor. Unless this Sub-Clause is not to apply, the advance payment and the currencies in which it is to be paid must be specified in the Contract Data. The rate of deduction for the repayments should be checked to ensure that repayment is achieved before the Contractor's completion of the Works. The typical figures in sub-paragraphs (a) and (b) of Sub-Clause 14.2.3 [*Repayment of Advance Payment*] of the General Conditions are based on the assumption that the advance payment is less than 22% of the Accepted Contract Amount.

The acceptable form of Advance Payment Guarantee should be included in the tender documents, annexed to the Particular Conditions: an example form is included at the end of this publication (in the section "Sample Forms").

If the Employer wishes to provide the advance payment in instalments, the Contract Data and this Sub-Clause will need to be amended.

EXAMPLE   The Contract Data:

- delete the words "14.2 .... Amount of Advance Payment (percent of Accepted Contract Amount): ...." and replace with:

"14.2 ....Total Advance Payment ......% of the Accepted Contract Amount

(a) 与当地生产的类似**货物**，没有达到**工程**所需的数量或标准不同的情况除外；（以及）

(b) 在**工程所在国**采购的货物或服务的价格中固有的任何关税或税款要素，应视为已包含在**中标合同金额**中。

港口费、码头费以及除上述规定外的货物或服务价格中固有的任何税款或关税，应视为已包含在**中标合同金额**中。"

如果外籍员工免交当地所得税，则应增加适当条款，并应相应修改**通用条件**的**第 14.1 款**[*合同价格*]（b）段。但是，在编写任何此类附加条款前，应咨询合格的税务专家。

范例条款 税务免除

"外籍（外国）人员不应对在**工程所在国**以任何外币支付的收入缴纳所得税，也不应对**承包商**直接向**承包商人员**提供的生活、租金和类似服务以及补贴缴纳所得税。如果**承包商人员**在**工程所在国**的收入中有一部分是以外币支付的，则他们可以（在其服务期限结束后）输出以外币支付的任何余额。

就**本款**而言，**雇主**应寻求豁免。如果不予批准，则相关税款应由**雇主**报销。"

第 14.2 款   预付款

编写**专用条件**时，应考虑**雇主**向**承包商**支付预付款的好处。除非该**款**不适用，否则必须在**合同数据**中规定预付款和支付货币。应检查还款的折算率，以确保**承包商**完成**工程**前付还。**通用条件第 14.2.3 项**[*预付款付还*]第（a）和（b）段中的典型数字是基于预付款低于**中标合同金额** 22% 的假设。

可接受的**预付款担保**格式，应包含在**专用条件**所附的招标文件中：本文件末尾包含范例格式（在"格式"部分）。⊖

如果**雇主**希望分期提供预付款，则需要修改**合同数据**和此**款**。

范例   合同数据：

- 删除"第 14.2 款 ………… 预付款金额（**中标合同金额**的百分比）：…………"并代之以：

    "第 14.2 款…………预付款总额_____**中标合同金额**的_____%

⊖ 此处中文按照勘误表改正后的英文翻译。——译者注

14.2 ....Number and timing of instalments: _____".

- delete the words "14.2 .... Currencies of payment if different to the currencies quoted in the Contract ...." and replace with:

    "14.2 ....Currencies and proportions ..... ____% in _____ ____% in _____".

Delete the second sentence of the second paragraph of Sub-Clause 14.2 and replace with:

"The total amount of the advance payment, the number and timing of instalments, and the applicable currencies and proportions shall be as stated in the Contract Data.".

Sub-Clause 14.2.1 [*Advance Payment Guarantee*]:

- delete the words "equal to the advance payment" from the first sentence and replace with: "equal to the total amount of the advance payment".

- in the second paragraph before "the advance payment has been repaid" add the words: "the total amount of".

- in the third paragraph before "the advance payment has not been repaid" add the words: "the total amount of".

- in sub-paragraph (a) before "the advance payment has been repaid" add the words: "the total amount of".

- in the last sentence, before "the advance payment" add the words: "the first instalment of".

Sub-Clause 14.2.2 [*Advance Payment Certificate*]:

- in the first sentence before "the advance payment" add the words: "the first instalment of".

- delete sub-paragraph (b) replace with: "(b) the Engineer has received a copy of the Contractor's application for the first instalment of the advance payment".

- at the end of this Sub-Clause add the following wording: "Thereafter, the Engineer shall issue an Advance Payment Certificate for each subsequent instalment of the advance payment, which the Contractor is entitled to under the Contract, within 14 days after the Engineer has received the Contractor's application (in the form of a Statement) for that instalment of advance payment."

Sub-Clause 14.2.3 [*Repayment of Advance Payment*]:

- in the last sentence before "the advance payment has not been repaid" add the words: "the total amount of".

- 第 14.2 款…………分期付款的**次数**和时间：_____"。

- 删除"**第 14.2 款**…………支付**货币**，如果不同于**合同**中规定的货币…………"并代之以：

  "**第 14.2 款**…………**货币**和比例………… ____%, ____% _____"。

删除**第 14.2 款**第二段的第二句，并代之以：

"预付款总额、分期付款的次数和时间以及适用货币和比例，应在**合同数据**中予以规定。"

第 14.2.1 项 [*预付款保函*]：

- 从第一句中删除"等于预付款"，并代之以"等于预付款总额"。

- 在"已偿还预付款"的第二段中，增加"……的总额"。

- 在"尚未还清预付款"的第三段中，增加"……的总额"。

- 在"已付还预付款"前的（a）段中，增加："……的总额"。

- 在"预付款"前的最后一句中，增加"……的第一期付款"。

第 14.2.2 项 [*预付款证书*]：

- 在"预付款"前的第一句中，增加"……的第一期付款"。

- 删除（b）段，并代之以"（b）**工程师**已收到**承包商**申请支付第一笔预付款的副本"。

- 在本款的末尾增加以下措辞："而后，**工程师**应在收到**承包商**关于支付第一期预付款的申请（以**报表**的形式）14 天内，向**承包商**签发其合同规定的每期预付款的**预付款证书**。"

第 14.2.3 项 [*预付款付还*]：

在"尚未付还的预付款"的最后一句中，增加"……的总额"。

Sub-Clause 14.4   **Schedule of Payments**

The General Conditions contains provisions for interim payments to the Contractor, which may be based on a Schedule of Payments. If another basis is to be used for determining interim valuations, details should be added in the Special Provisions.

If payments are to be specified in a Schedule of Payments, the "minimum amount of interim certificates" may be omitted from the Contract Data, and consideration may be given to the Schedule of Payments being in one of the following forms:

a)   an amount, or percentage of the estimated final Contract Price, for each month (or other period) during the Time for Completion (but this can prove unreasonable if the Contractor's progress differs significantly from the expectation on which the Schedule was based); or

b)   amounts based on the actual progress achieved by the Contractor in executing the Works, which necessitates careful definition of the payment milestones (but disagreements may arise when the work required for a payment milestone is nearly achieved but the balance of the work, albeit minor, cannot be completed until some months later).

The figures inserted by a tenderer in the Schedule of Payments may be compared with his/her tender programme (if any), in order to assess whether they are reasonably consistent with each other.

Sub-Clause 14.7   **Payment**

Periods for payment should be long enough for the Employer to meet, but not so long as to prejudice the Contractor's positive cash-flow.

If the country or countries of payment need to be specified, details may be included in a Schedule.

Sub-Clause 14.8   **Delayed Payment**

As an alternative to the second paragraph of this Sub-Clause, consideration may be given to the payment of the Contractor's actual financing Costs, taking account of local financing arrangements.

Sub-Clause 14.9   **Release of Retention Money**

If part of the Retention Money is to be released and substituted by an appropriate guarantee, an additional Sub-Clause may be added. The acceptable form(s) of guarantee should be included in the tender documents, annexed to the Particular Conditions: an example form is included at the end of this publication (in the section "Sample Forms"). Also, a limit of Retention Money should be stated in the Contract Data.

EXAMPLE SUB-CLAUSE FOR RELEASE OF RETENTION

> When the Retention Money has reached three-fifths (60%) of the limit of Retention Money stated in the Contract Data, after the Employer has received the guarantee referred to below the Engineer shall certify and the Employer shall make payment of half (50%) of the limit of Retention Money to the Contractor.
>
> The Contractor shall obtain (at the Contractor's cost) a guarantee in amounts and currencies equal to half (50%) of the limit of Retention Money stated in the Contract Data, and shall submit it to the Employer with a copy to the Engineer. This guarantee shall be issued by an entity and from within a country (or other jurisdiction)

第 14.4 款　　　　付款计划表

通用条件包含可能基于**付款计划表**向**承包商**支付期中付款的规定。如使用其他依据确定期中估价，则应在**特别规定**中增加详细信息。

如在**付款计划表**中规定付款期数，**合同数据**可省略"最低期中付款证书"，并可考虑采用以下形式的**付款计划表**：

a)　　**竣工时间**期间每个月估算的最终**合同价格**的金额或比例（但是，如果**承包商**的进度与**资料表**预期的相差很大，这可能是不合理的）；（或）

b)　　根据**承包商**实施**工程**中所取得的实际进度的金额，需要详细规定付款**里程碑**（但是，当付款**里程碑**所需的工作近乎完成，但剩余工作，尽管很小，在几个月后才能完成的情况下，可能会产生分歧）。

投标人在**付款计划表**中插入的数字可与其投标方案（如果有）进行比较，以评估其相互之间是否保持合理一致。

第 14.7 款　　　　付款

付款期限应足够长，以满足**雇主**的要求，但又不损害**承包商**的正现金流。

如需规定付款国家，则在**资料表**中应详细说明。

第 14.8 款　　　　延误的付款

作为本款第二段的替代项，可在当地资金安排的基础上，考虑支付**承包商**的实际融资**成本**。

第 14.9 款　　　　保留金的发放

如需发放部分**保留金**并代之以适当的担保，则可增加一项附加**条款**。**专用条件**所附的招标文件应包括可接受的担保格式：本文件末尾包括范例格式（在"**格式**"部分）。另外，应在**合同数据**中规定**保留金**的限额。[⊖]

范例条款　　　　保留金的发放

> 当**保留金**达到**合同数据**规定的限额五分之三（60%）时，**工程师**应在**雇主**收到下述保函后，进行确认，并且**雇主**应向**承包商**支付**保留金**限额的一半（50%）。
>
> **承包商**应获得（由**承包商**承担费用）**合同数据**中规定的**保留金**限额的一半（50%）的金额和货币的保函，并提交**雇主**、抄送**工程师**。该担保应由**雇主**同意的国家（或其他司法管辖区）的实体出具，并应以招标文件中包含的范例

⊖ 此处中文按照勘误表改正后的英文翻译。——译者注

to which the Employer gives consent, and shall be based on the sample form included in the tender documents or on another form agreed by the Employer (but such consent and/or agreement shall not relieve the Contractor from any obligation under this Sub-Clause).

The Contractor shall ensure that the guarantee is valid and enforceable until the Contractor has executed the Works, as specified for the Performance Security in Sub-Clause 4.2.1. If the terms of the guarantee specify an expiry date, and the Contractor has not so executed the Works by the date 28 days before the expiry date, the Contractor shall extend the validity of the guarantee.

The release of Retention Money under this Sub-Clause shall be in lieu of the release of the second half of the Retention Money under the second paragraph of Sub-Clause 14.9 [*Release of Retention Money*].

Sub-Clause 14.15   **Currencies of Payment**

If all payments are to be made in Local Currency, this currency must be named in the Letter of Tender, and this Sub-Clause may be replaced:

EXAMPLE SUB-CLAUSE FOR A SINGLE CURRENCY CONTRACT

> All payments made in accordance with the Contract shall be in Local Currency. The Local Currency payments shall be fully convertible, except those for local costs. The percentage attributed to local costs shall be as stated in the Contract Data.

Financing Arrangements

For major contracts in some markets, there may be a need to secure finance from entities such as aid agencies, development banks, export credit agencies, or other international financing institutions. If financing is to be procured from any of these sources, the Special Provisions may need to incorporate the financing institution's special requirements. The exact wording will depend on the relevant institution, so reference will need to be made to the institution to ascertain its requirements, and to seek approval of the draft tender documents.

These requirements may include tendering procedures which need to be adopted in order to render the eventual contract eligible for financing, and/or additional Sub-Clauses which may need to be incorporated into the Special Provisions. The following examples indicate some of the topics which the institution's requirements may cover:

- prohibition from discrimination against the shipping companies of any one country;

- ensuring that the Contract is subject to a widely-accepted neutral law;

- provision for arbitration under recognised international rules and at a neutral location;

- giving the Contractor the right to suspend/terminate in the event of default under the financing arrangements;

- restricting the right to reject Plant;

- specifying the payments due in the event of termination;

- specifying that the Contract does not become effective until certain conditions precedent have been satisfied, including pre-disbursement conditions for the financing arrangements; and

格式或雇主同意的其他格式为基础（但这样的同意和/或商定不得免除本款规定的**承包商**的任何义务）。

承包商应保证保函是有效且可执行的，直到**承包商**实施了第 4.2.1 项中履约担保规定的工程。如果保函条款中规定了有效期，而**承包商**在到期前 28 天内仍未实施工程，则**承包商**应延长保函的有效期。

本款规定的**保留金**发放应代替第 14.9 款［*保留金的发放*］第二段规定的另外一半**保留金**的发放。

第 14.15 款　　支付的货币

如果所有付款均以**当地货币**支付，则必须在**招标书**中注明该货币，并且该**款**可替换为：

范例条款 单一货币合同

根据合同规定支付的所有款项均应以**当地货币**支付。**当地货币**付款应完全可兑换，当地成本的支付除外。**合同数据**中应规定当地成本的支付比例。

资金安排

对于某些国家的主合同，可能需要从援助机构、开发银行、出口信贷机构或其他国际融资机构等实体获得融资。如从这些渠道获得融资，则**特别规定**可能需包含融资机构的特殊要求。确切措辞将取决于相关机构，因此，需咨询该机构以明确其要求，并寻求对招标文件草案的批准。

这些要求可能包括需要采用的招标程序，使最终合同有资格获得融资，和/或可能需要在**特别规定**中纳入附加**条款**。以下范例列出了机构要求可能包括的一些议题：

- 禁止歧视任何一个国家的运输公司；

- 确保**合同**遵守公认的中立法律；

- 根据公认的国际规则在中立地点进行仲裁的规定；

- 授予**承包商**违反资金安排时暂停/终止的权利；

- 对拒收**生产设备**的权利予以限制；

- 终止时明确应付款项；

- 规定在满足某些先决条件（包括资金安排的预付款条件）之前，合同不能生效；（以及）

- obliging the Employer to make payments from his own resources if, for any reason, the funds under the financing arrangements are insufficient to meet the payments due to the Contractor, whether due to a default under the financing arrangements or otherwise.

The financing institution may wish the Contract to include references to the financing arrangements, especially if funding from more than one source is to be arranged to finance different elements of supply. It is not unusual for the Special Provisions to include particular provisions identifying different categories of Works and specifying the documents to be presented to the relevant financing institution to obtain payment. If the financing institution's requirements are not met, it may be difficult (or even impossible) to secure suitable financing for the project, and/or the institution may decline to provide finance for part or all of the Contract.

Where the financing is not tied to the export of goods and services from any particular country but is simply provided by commercial banks lending to the Employer, those banks may wish to satisfy themselves regarding the extent of the Contractor's rights under the Contract.

Alternatively, the Contractor may be prepared to initiate financing arrangements for the Contract and retain responsibility for them, although the Contractor would probably be unable or unwilling to provide finance from the Contractor's own resources. The Contractor's financing bank's requirements are then likely to affect the Contractor during contract negotiations. For example, the financing bank may require the Employer to make interim payments, although a large proportion of the Contract Price might be withheld until the Works are complete. Since the Contractor would then have to arrange financing to cover the shortfall between the payments and the Contractor's outgoings, the Contractor (and the financing bank) would probably require some form of security from the Employer, guaranteeing payment when due.

It may be appropriate for the Employer, when preparing the tender documents, to anticipate the latter requirement by undertaking to provide a guarantee for the element of payment which the Contractor is to receive when the Works are complete. The acceptable form(s) of guarantee should be included in the tender documents, annexed to the Particular Conditions: an example form is included at the end of this publication (in the section "Sample Forms"). The following Sub-Clause may be added.

EXAMPLE SUB-CLAUSE FOR CONTRACTOR FINANCE

"The Employer shall obtain (at the Employer's cost) a payment guarantee in the amount and currencies, and provided by an entity, as stated in the Contract Data. The Employer shall deliver the guarantee to the Contractor within 28 days after both Parties have entered into the Contract Agreement. The guarantee shall be in the form annexed to these Special Provisions, or in another form acceptable to the Contractor. Unless and until the Contractor receives the guarantee, the Engineer shall not give the notice under Sub-Clause 8.1 [*Commencement of Works*].

The guarantee shall be returned to the Employer at the earliest of the following dates:

(a) when the Contractor has been paid the Accepted Contract Amount;

(b) when obligations under the guarantee expire or have been discharged; or

(c) when the Employer has performed all obligations under the Contract."

- 无论出于何种原因，如果资金安排中的资金不足以支付应付**承包商**的款项，无论是由于违反资金安排还是其他原因，**雇主**有义务用自己的资源付款。

融资机构可能希望在**合同**中提及资金安排，尤其在安排多渠道资金资助不同供应要素的情况下。**特别规定**经常包括专用条件，明确不同类型的**工程**，规定为获取付款而提交给相关融资机构的文件。如果未满足融资机构的要求，则可能难以（甚至不可能）获得项目合适的融资，和／或融资机构可能拒绝为部分或全部**合同**提供融资。

如果融资不与任何特定国家的货物和服务的出口挂钩，而只是由商业银行向**雇主**提供贷款出口，这些银行可能希望满足自己关于**合同**规定的**承包商**权利范围。

另外，**承包商**可能会起草合同的资金安排并对其承担责任，尽管**承包商**可能无法或不愿使用**承包商**自己的资源提供资金。**承包商**的融资银行的要求可能会在合同谈判期间影响**承包商**。例如，尽管在**工程**竣工之前可能会扣留大部分**合同价格**，但融资银行可能会要求**雇主**支付期中付款。由于要求**承包商**必须安排资金弥补付款与**承包商**支出之间的差额，**承包商**（和融资银行）可能需要**雇主**提供某种形式的担保，以保证到期时付款。

编写招标文件时，**雇主**可以通过承诺为**承包商**在**工程**完工时收到的付款要素来预期后一项要求。可接受的保函格式应包含在**专用条件**所附的招标文件中：本文件末尾包含范例格式（在"**格式**"部分）。可增加以下条款。㊀

范例条款 承包商融资

"**雇主**应（由**雇主**承担费用）按照**合同数据**规定，获得由实体提供的金额和货币的付款保函。**雇主**应在双方签订合同协议书之日起 28 天内将保函提交给**承包商**。保函应采用该**特别规定**所附的格式，或采用**承包商**可以接受的另一种格式。除非并直到**承包商**收到保函，否则**工程师**不得按照**第 8.1 款**[**工程的开工**]的规定发出通知。

保函应在以下最早日期退还给**雇主**：

（a） 已向**承包商**支付**中标合同金额**；

（b） 保函规定的义务到期或已解除；（或）

（c） **雇主**已履行了**合同**规定的所有义务。"

㊀ 此处中文按照勘误表改正后的英文翻译。——译者注

Clause 15 — Termination by Employer

Sub-Clause 15.2 — Termination for Contractor's Default

Before inviting tenders, the Employer should verify that the wording of this Sub-Clause, and each anticipated ground for termination, is consistent with the law governing the Contract.

Sub-Clause 15.2.1 — Notice

Sub-paragraph (h) in this Sub-Clause is intended to include situations, where the Contractor or any of the Contractor's employees, agents, Subcontractors or Contractor's Personnel gives or offers to give (directly or indirectly) to any person any bribe, gift, gratuity, commission or other thing of value, as an inducement or reward for showing or forbearing to show favour or disfavour to any person in relation to the Contract. However, this is not intended to include lawful inducements and rewards by the Contractor to the Contractor's Personnel.

Sub-Clause 15.2.3 — After termination

If the Employer has made available any Employer-Supplied Materials and/or Employer's Equipment in accordance with Sub-Clause 2.6, consideration should be given to an additional sub-paragraph under this Sub-Clause:

EXAMPLE — Insert the following new sub-paragraph at the end of sub-paragraph (b) of Sub-Clause 15.2.3:

"(iv) all Employer-Supplied Materials and/or Employer's Equipment made available to the Contractor in accordance with Sub-Clause 2.6 [*Employer-Supplied Materials and Employer's Equipment*], and".

Sub-Clause 15.5 — Termination for Employer's Convenience

In many jurisdictions under the applicable law it may not be permissible for the Employer to terminate the Contract for convenience (termination of the Contract only being permitted in the event of default on the part of the Contractor and, thereafter, arranging an equitable, non-discriminatory and transparent procurement process to select a replacement contractor).

Therefore, before inviting tenders the Employer should verify that the wording of this Sub-Clause is consistent with the law governing the Contract.

Clause 16 — Suspension and Termination by Contractor

Sub-Clause 16.2 — Termination by Contractor

Before inviting tenders, the Employer should verify that the wording of this Sub-Clause is consistent with the law governing the Contract.

Sub-Clause 16.2.1 — Notice

Sub-paragraph (j) in this Sub-Clause is intended to include situations, where the Employer or any of the Employer's employees, agents or Employer's Personnel gives or offers to give (directly or indirectly) to any person any bribe, gift, gratuity, commission or other thing of value, as an inducement or reward for showing or forbearing to show favour or disfavour to any person in relation to the Contract. However, this is not intended to include lawful inducements and rewards by the Employer to the Employer's Personnel.

Sub-Clause 16.3 — Contractor's Obligations After termination

If the Employer has made available any Employer-Supplied Materials and/or Employer's Equipment in accordance with Sub-Clause 2.6, consideration should be given to amending this Sub-Clause:

第 15 条 由雇主终止

第 15.2 款 因承包商违约的终止

邀标前，**雇主**应保证，**本款**的措辞以及终止的预期理由均符合管辖**合同**的法律。

第 15.2.1 项 通知

本款（h）段旨在包括以下情况：**承包商**或**承包商**的任何雇员、代理人、**分包商**或**承包商人员**向任何人（直接或间接）行贿、送礼、送酬金、佣金或其他有价物，以此作为激励或奖励或不向与合同有关的任何人表示赞成或不赞成。但是，这并不包括**承包商**对**承包商人员**的合法激励和奖励。

第 15.2.3 项 终止后

如果**雇主**根据**第 2.6 款**规定提供了任何**雇主提供的材料**和／或**雇主设备**，则应考虑在**本款**中增加一段：

范例 在**第 15.2.3 项**（b）段之后，插入以下这段：

"（iv）按照**第 2.6 款**［*雇主提供的材料和雇主设备*］的规定向**承包商**提供所有雇主提供的材料和／或雇主设备，以及"。

第 15.5 款 为雇主便利的终止

在很多适用法律的司法管辖区，不允许**雇主**为了自身便利而终止合同（只有在**承包商**违约的情况下才允许终止**合同**，此后，安排公平的、非歧视性的和透明的采购流程以选择替代承包商）。

因此，在招标之前，**雇主**应核实**本款**的措辞是否符合管辖**合同**的法律。

第 16 条 由承包商暂停和终止

第 16.2 款 由承包商终止

邀标前，**雇主**应保证，**本款**的措辞符合管辖**合同**的法律。

第 16.2.1 项 通知

本款（j）段旨在包括以下情况：**雇主**或**雇主**的任何雇员、代理人或**雇主人员**向任何人（直接或间接）行贿、送礼、送酬金、佣金或其他有价物，以此作为激励或奖励或不向与**合同**有关的任何人表示赞成或不赞成。但是，这并不包括**雇主**对**雇主人员**的合法激励和奖励。

第 16.3 款 终止后承包商的义务

如果**雇主**按照**第 2.6 款**提供了任何**雇主提供的材料**和／或**雇主设备**，则应考虑修改**此款**：

EXAMPLE   Delete sub-paragraph (c) in Sub-Clause 16.3 and replace with:

"(c) deliver to the Engineer all Employer-Supplied Materials and/or Employer's Equipment made available to the Contractor in accordance with Sub-Clause 2.6 [*Employer-Supplied Materials and Employer's Equipment*]; and

(d) remove all other Goods from the Site, except as necessary for safety, and leave the Site."

Clause 17   Care of the Works and Indemnities

Sub-Clause 17.4   Indemnities by Contractor

In respect of the Contractor's obligation to indemnify under the second paragraph of this Sub-Clause, it should be noted that this applies only to the extent, if any, that the Contractor is required to carry out any design of the Permanent Works under the Contract. If this is the case:

a) as this liability is not excluded under the first paragraph of Sub-Clause 1.15 [*Limitation of Liability*], the Contractor has no liability for any indirect or consequential loss or damage suffered by the Employer as a result of any negligence by the Contractor in designing the Works to be fit for purpose;

b) as this liability is not excluded under the second paragraph of Sub-Clause 1.15 [*Limitation of Liability*], it falls within the Contractor's limitation of liability under Sub-Clause 1.15; and

c) this liability may be covered by the insurance to be taken out by the Contractor under Sub-Clause 19.2.3 [*Liability for breach of professional duty*], in which case a statement to this effect should be included in the Contract Data.

Additional Sub-Clause

If the Contractor is to be allowed to use and/or occupy any of the Employer's facilities and/or accommodation temporarily during the Contract, it is recommended that an additional sub-clause be added to Clause 17 to cover the responsibility for care of such facilities and/or accommodation:

EXAMPLE SUB-CLAUSE FOR CONTRACTOR'S USE OF EMPLOYER'S FACILITIES/ACCOMMODATION

"The Contractor shall take full responsibility for the care of the items of the Employer's facilities and/or accommodation listed below, from the date of use and/or occupation by the Contractor until the date on which such use and/or occupation is re-vested in the Employer.

[*List of items and details*]

If any loss or damage happens to any of the above items during a time when the Contractor is responsible for its care, arising from any cause other than a cause for which the Employer is responsible or liable, the Contractor shall promptly rectify the loss or damage at the Contractor's risk and cost."

范例	删除第 16.3 款（c）段，并代之以：

"（c）将所有**雇主**提供的**材料**和 / 或**雇主设备**交付给**工程师**，并按照**第 2.6 款**[*雇主提供的材料和雇主设备*]的规定提供给**承包商**；（以及）

（d）从**现场**移除除安全需要以外的所有其他**货物**，并离开**现场**。"

第 17 条　　　　　　　　工程照管和保障

第 17.4 款　　　　　　由承包商保障

关于**本款**第二段规定的**承包商**的保障义务，应注意的是，这仅在要求**承包商**根据合同进行永久工程的任何设计的范围内适用。如果是这种情况：

a）　由于**第 1.15 款**[*责任限度*]第一段并未排除此项责任，**承包商**对因设计适合其用途的工程时的任何疏忽，而导致**雇主**遭受的任何间接或衍生性损失或损害不承担任何责任；

b）　由于未在**第 1.15 款**[*责任限度*]的第二段中排除，该责任属于**第 1.15 款**规定的**承包商**的责任范围；（以及）

c）　该责任可由**承包商**按照**第 19.2.3 项**[*违反职业职责的责任*]的规定所购买的保险承保，在这种情况下，应在**合同数据**中包含对此的说明。

附加条款

如果在合同期间允许**承包商**临时使用和 / 或占用**雇主**的任何设施和 / 或住所，建议在**第 17 条**中增加一项附加条款，以包括照管此类设施和 / 或住所的责任：

范例条款 承包商使用雇主设施 / 住所

"从**承包商**使用和 / 或占有之日起至**雇主**收回使用和 / 或占有权为止，**承包商**应对下列**雇主设施**和 / 或住所的物品负全责。

[*项目清单和详细信息*]

承包商负责照管期间，如因**雇主**负责以外的原因而使上述任何物品发生任何损失或损害，**承包商**应立即纠正损失或损害，风险和费用由**承包商**承担。"

Clause 18 Exceptional Events

Sub-Clause 18.1 Exceptional Events

In respect of sub-paragraph (f) of this Sub-Clause, it should be noted that any event of "exceptionally adverse climatic conditions" is excluded from the definition of what constitutes an Exceptional Event. While this means that there is no right for either Party to suspend the Works in the case of an event of "exceptionally adverse climatic conditions", if this type of event has the effect of delaying completion and taking-over of the Works or Section the Contractor shall be entitled to EOT under sub-paragraph (c) of Sub-Clause 8.5 [*Extension of Time*].

Clause 19 Insurance

If the Employer wishes to change the insurance provisions of this Clause – for example by providing some of the insurance cover under the Employer's own policy(ies) – it will be necessary to review and revise the relevant Sub-Clause(s) under this Clause 19.

If this is the case, it is strongly recommended that:

a) the tender documents include details of such insurances as an annex to the Special Provisions so that tenderers can estimate what other insurances they may wish to have for their own protection. The details should include the conditions of insurance, limits, exceptions and deductibles; preferably in the form of a copy of each insurance policy; and

b) the Employer is advised by a professional with extensive experience in construction insurance and liability in the preparation of the wording of the revised sub-clauses. If the insurance provisions are changed without due care and attention, there is a risk that the Employer will inadvertently carry liabilities for which the Employer is neither prepared nor covered by insurance.

Clause 20 Employer's and Contractor's Claims

Sub-Clause 20.1 Claims

In respect of any Claim under sub-paragraph (c) of this Sub-Clause, it should be noted that "another entitlement or relief …or the execution of the Works" may include such matters as:

- interpretation of a provision of the Contract,

- rectification of an ambiguity or discrepancy found in the Contract documents,

- a declaration in favour of the claiming Party,

- access to the Site or to places where the Works are being (or to be) carried out, and/or

- any other matter of entitlement under the Conditions of Contract or in connection with, or arising out of, the Contract that does not involve payment by one Party to the other Party and/or EOT and/or extension of the DNP.

Consideration may be given to replacing the words "within a reasonable time" in the last paragraph of this Sub-Clause by a specified time period.

Clause 21 Disputes and Arbitration

Sub-Clause 21.1 Constitution of the DAAB

It is generally accepted that construction projects depend for their success on the avoidance of Disputes between the Employer and the Contractor and, if Disputes do arise, the timely resolution of such Disputes.

第 18 条　　　　例外事件

第 18.1 款　　　例外事件

关于本款（f）段，应注意，"特别不利的气候条件"的任何事件均不包括在**例外事件**的定义内。虽然这意味着在发生"特别不利的气候条件"事件时，任一方均无权暂停工程，但如果此类事件造成了**工程**或**分项工程**竣工和接收的延误，**承包商**应有权按照**第 8.5 款**［**竣工时间的延长**］（c）段的规定获得**竣工时间**的延长。

第 19 条　　　　保险

如果**雇主**希望修改本款的保险条款（例如，通过提供**雇主**自身保单中的某些保险），则有必要审核和修订**第 19 条**的相关**条款**。

在此情况下，强烈建议：

a)　　招标文件包括此类保险的详细信息，并作为**特别规定**的附件，以便投标人可以评估为保护自身而需购买的其他保险。详细信息应包括保险条件、限额、例外情况和免赔额；最好以每份保单副本的形式；（以及）

b)　　编制修订条款时，建议**雇主**咨询在施工保险和责任方面具有丰富经验的专业人员的意见。如果在没有适当注意和关注的情况下修改保险条款，**雇主**有可能会无意间承担未提供也未投保的责任。

第 20 条　　　　雇主和承包商的索赔

第 20.1 款　　　索赔

关于**本款**（c）段规定的任何**索赔**，应当注意，"另一种权利或救济……或**工程**的实施"可包括以下事项：

-　　对**合同条款**的解释；

-　　纠正**合同**文件中的歧义或差异；

-　　有利于索赔方的声明；

-　　**现场**进入权或进入正在实施（或将要实施）**工程**的地方；（和／或）

-　　**合同条件**规定的，或与**合同**有关或由**合同**引起的，不涉及一方向另一方付款的，和／或**竣工时间的延长**和／或**缺陷通知期限**延长的任何其他权利事项。

可以考虑将本款最后一段的"在合理时间内"代替为特定时间段。

第 21 条　　　　争端和仲裁

第 21.1 款　　　争端避免／裁决委员会的组成

普遍认为，施工项目的成功与否取决于避免**雇主**与**承包商**之间的**争端**，如果确实发生**争端**，需及时解决。

Therefore, the Contract should include the provisions under Clause 21 which, while not discouraging the Parties from reaching their own agreement on Disputes as the Works proceed, allow them to bring contentious matters to an independent and impartial dispute avoidance/adjudication board ("DAAB") for resolution.

The provisions of this Sub-Clause are intended to provide for the appointment of the DAAB, and FIDIC strongly recommends that the DAAB be appointed, as a 'standing DAAB'

– that is, a DAAB that is appointed at the start of the Contract who visits the Site on a regular basis and remains in place for the duration of the Contract to assist the Parties:

a) in the avoidance of Disputes,

and

b) in the 'real-time' resolution of Disputes if and when they arise

to achieve a successful project.

It is for this reason that, under the first paragraph of this Sub-Clause, the Parties are under a joint obligation to appoint the member(s) of the DAAB within 28 days after the Contractor receives the Letter of Acceptance if no other time is stated in the Contract Data. That said, it is preferable that the member(s) of the DAAB are appointed before the Letter of Acceptance is issued.

At an early stage in the Employer's planning of the project, consideration should be given as to whether a sole-member DAAB or a three-member DAAB is preferable for a particular project, taking account of its size, duration and the fields of expertise which will be involved.

This Sub-Clause provides for two alternative arrangements for the DAAB:

- a sole-member DAAB of one natural person, who has entered into a tripartite agreement with both Parties;

or

- a three-member DAAB of three natural persons, each of whom has entered into a tripartite agreement with both Parties.

The tripartite agreement above is referred to as the *DAAB Agreement* under the Conditions of Contract. It is recommended that the form of this agreement be one of the two alternative example forms included at the end of this publication (in the section "Sample Forms"), as appropriate to the arrangement adopted.

It should be noted that both forms of the DAAB Agreement incorporate (by reference) the *General Conditions of Dispute Avoidance/Adjudication Agreement* with its Annex *DAAB Procedural Rules*, which are included as the Appendix to the General Conditions in this publication.

Under either of these alternative forms of DAAB Agreement, each natural person of the DAAB is referred to as a *DAAB Member*.

A very important factor in the success of the dispute avoidance/adjudication procedure is the Parties' confidence in the agreed individual(s) who will serve on the DAAB. Therefore, it is essential that candidates for this position are not imposed by either Party on the other Party. The appointment of the DAAB is facilitated by the provision in the Contract Data for each Party to name potential DAAB Members. It is important that the Employer and the Contractor each avail himself/herself, of the opportunity at the tender stage of the Contract to name potential DAAB Members in the Contract Data.

因此，**合同**应包括**第 21 条**中的规定，该规定在不妨碍双方在**工程**进行过程中就**争端**达成协议的同时，允许其将争端事项提交给独立和公正的**争端避免/裁决委员会**进行解决。

本款的规定旨在**争端避免/裁决委员会**的任命，菲迪克（FIDIC）强烈建议任命**争端避免/裁决委员会**，作为"常设**争端避免/裁决委员会**"

——即在**合同**开始时就任命**争端避免/裁决委员会**，**争端避免/裁决委员会**定期考察**现场**，并在合同期间一直存在，以协助双方：

a) 避免**争端**；

（以及）

b) 如果并在出现**争端**时，予以"实时"解决。

以保证项目成功。

因此，按照本款第一段规定，如果**合同数据**中未规定其他时间，双方有共同义务在**承包商**收到**中标函**后 28 天内任命**争端避免/裁决委员会**成员。也就是说，最好在签发**中标函**前任命**争端避免/裁决委员会**成员。

在**雇主**进行项目规划的早期阶段，根据特定项目的规模、持续时间和涉及的专业领域，应考虑唯一成员的**争端避免/裁决委员会**还是三人成员的**争端避免/裁决委员会**更合适。

本款提供了两种**争端避免/裁决委员会**的替代安排：

- 由与双方达成三方协议的一个自然人组成的唯一成员**争端避免/裁决委员会**；

（或）

- 由三个自然人组成的三人成员**争端避免/裁决委员会**，每个人均与双方达成了三方协议。

按照**合同条件**，上述三方协议称为**争端避免/裁决委员会协议书**。建议本协议书的格式为本文件末尾（见"格式"一节）中的两种备选范例格式之一，视所采用的安排而定。㊀

应当指出的是，**争端避免/裁决委员会协议书**的两种格式将（通过引用）**通用条件争端避免/裁决委员会协议书**与**附件争端避免/裁决委员会程序规则**合并在一起，作为**附录**包含在本文件的**通用条件**中。

根据**争端避免/裁决委员会协议书**的两种备选范例格式任意一种，**争端避免/裁决委员会**的每个自然人均称为**争端避免/裁决委员会成员**。

争端避免/裁决程序成功的一个重要因素，是双方对商定的、将在**争端避免/裁决委员会**任职的个人的信心。因此，至关重要的是，任一方都不得将这一职位的候选人强加给另一方。**争端避免/裁决委员会**的任命是由**合同数据**中每方提名的潜在的**争端避免/裁决委员会成员**支持的。**雇主**和**承包商**应在**合同**的招标阶段利用各自机会在**合同数据**中提名潜在的**争端避免/裁决委员会成员**，这一点很重要。

㊀ 此处中文按照勘误表改正后的英文翻译。——译者注

The Contract Data includes two lists for potential DAAB Members to be named: one for the Employer to list three names, the other for the Contractor to list three names. This ensures that both Parties have equal opportunity to put forward (the same number of) names for potential DAAB Members, and so avoid any question that DAAB Member(s) may be imposed on one Party by the other Party.

This provides a total of six potential DAAB Members from which the sole member or three members (as the case may be) can be selected by the Parties. If it is considered necessary to have a wider selection of DAAB Member(s) to choose from, then provision may be made for longer lists in the Contract Data for both the Contractor and the Employer to name (the same number of) additional DAAB Members.

If the Parties cannot agree on any DAAB Member, Sub-Clause 21.2 [*Failure to Appoint DAAB Member(s)*] applies and the selection and appointment of the DAAB Member(s) should be made by a wholly impartial entity with an understanding of the nature and purpose of a DAAB. The President of FIDIC is prepared to perform this role if this authority has been delegated in accordance with the example wording in the Contract Data. FIDIC maintains a list of approved and experienced adjudicators for this specific purpose: *The FIDIC President's List of Approved Dispute Adjudicators* (http://fidic.org/president-list[†]).

If no potential DAAB Members' names are given in the Contract Data, consideration should be given to stating a time period in the Contract Data that is greater than the default period of 28 days stated in the first paragraph of this Sub-Clause.

The period of 224 days stated in sub-paragraph (i) of this Sub-Clause has been arrived at by taking account of certain time allowances, as follows:

> date of termination + 28 days to give a Notice of Claim under Sub-Clause 20.2.1 [*Notice of Claim*]
>
> + 84 days to submit detailed particulars for the Claim under Sub-Clause 20.2.4 [*Fully detailed Claim*]
>
> + 84 days for the Engineer's agreement/determination of the Claim under Sub-Clause 3.7 [*Agreement or Determination*]
>
> + 28 days for a NOD under Sub-Clause 3.7 [*Agreement or Determination*]
>
> = 224 days after the date of termination.

As noted above, FIDIC strongly recommends that the DAAB be appointed at the start of the Contract and remain in place for the duration of the Contract. However, as an alternative to the 'standing DAAB' envisaged under this Sub-Clause, the Parties may prefer the dispute board to be appointed on an 'ad-hoc' basis. In such case, the dispute board would be appointed when a Dispute arises, its appointment would be limited to resolution of the Dispute, it would have no role to play in the avoidance of Disputes between the Parties, and its appointment would cease when it had given its decision on that Dispute. Should a new Dispute arise, a new ad-hoc DAAB would be appointed. Given that such a dispute board would have no role in dispute avoidance, it is more correctly referred to as a "Dispute Adjudication Board" or "DAB". If the Parties wish to provide for an 'ad-hoc DAB', rather than the recommended 'standing DAAB', then the following amendments will be needed:

EXAMPLE	Delete the definitions under Sub-Clause 1.1.22 and 1.1.23 and replace with:
	" "Dispute Adjudication Board" or "DAB" means the person or three persons (as the case may be) so named in the Contract, or appointed under Clause 21."

合同数据包括两个要提名的潜在**争端避免／裁决委员会成员**表格：一个由**雇主**列出三个姓名，另一个由**承包商**列出三个姓名。这确保双方都有平等机会提名（相同数量的）潜在的**争端避免／裁决委员会成员**，从而可以避免一方将**争端避免／裁决委员会成员**强加给另一方的任何问题。

这样一共提供了 6 名潜在的**争端避免／裁决委员会成员**，双方可从中选择唯一成员或三名成员（视情况而定）。如果认为有必要选择更多的**争端避免／裁决委员会成员**，则可在**合同数据**中规定更长的名单，以便**承包商**和**雇主**提名（相同数量的）额外**争端避免／裁决委员会成员**。

如果双方不能就任何**争端避免／裁决委员会成员**商定一致，则**第 21.2 款**［*未能任命争端避免／裁决委员会成员*］的规定适用，并且**争端避免／裁决委员会成员**的选择和任命应由了解**争端避免／裁决委员会**性质和目的、完全公正的实体进行。如果已根据**合同数据**中的范例措辞进行了授权，菲迪克（FIDIC）主席可履行这一职责。菲迪克（FIDIC）为此特别目的保留有一份经批准的且经验丰富的裁决员名单：***菲迪克（FIDIC）主席批准的争端裁决员名单***（http://fidic.org/president-list[†]）。

如果**合同数据**中未提供潜在的**争端避免／裁决委员会成员**的姓名，则应考虑在**合同数据**中规定一个时间段，该时间段大于**本款**第一段规定的 28 天的默认时间期限。

本款（i）段所述 224 天的期限是通过考虑某些时间宽限后得出的，具体如下：

终止日期 + 28 天根据**第 20.2.1 项**［*索赔通知*］规定发出**通知** +

84 天根据**第 20.2.4 项**［*充分详细的索赔*］规定提交**索赔**的详细说明材料 +

84 天**工程师**根据**第 3.7 款**［*商定或确定*］规定对**索赔**做出商定／确定 +

28 天根据**第 3.7 款**［*商定或确定*］规定发出**不满意通知**

= 终止日期后 224 天。

如上所述，菲迪克（FIDIC）强烈建议在合同开始时任命**争端避免／裁决委员会**，并在合同有效期内保持持续存在。但是，作为本款预期的"常设**争端避免／裁决委员会**"的替代方案，双方可能更愿意以"临时"方式任命争端委员会。在这种情况下，争端委员会将在发生**争端**时任命，其任命仅限于**争端**的解决，在避免双方发生**争端**方面没有任何作用，而在做出**争端**决定时其任命即终止。如果出现新的**争端**，将任命新的临时**争端避免／裁决委员会**。鉴于此类争端委员会在避免**争端**中没有作用，称为"**争端裁决委员会**"或"DAB"更为恰当。如果双方希望提供"临时**争端裁决委员会**"而非建议的"常设**争端避免／裁决委员会**"，则需要进行以下修改：

范例	删除**第 1.1.22 项**和**第 1.1.23 项**中的定义，并代之以：
	"'**争端裁决委员会**'或'DAB'系指合同中提名或根据**第 21 条**任命的一人或三人（视情况而定）。"

and replace all references in the Conditions of Contract to "DAAB" with "DAB".

Sub-Clause 21.1, plus the Appendix *General Conditions of Dispute Avoidance/Adjudication Agreement* with its Annex DAAB Procedural Rules, should be amended to comply with the wording contained in the corresponding provisions of FIDIC's Conditions of Contract for Plant and Design-Build, First Edition 1999.

Delete Sub-Clause 21.3 [*Avoidance of Disputes*];

in the first paragraph of Sub-Clause 21.4 [*Obtaining DAAB's Decision*]:

- after the words "If a Dispute arises between the Parties then, " add:
"after a DAB has been appointed";
- delete the words "(whether or not any informal discussions have been held under Sub-Clause 21.3 [*Avoidance of Disputes*])"; and

delete sub-paragraph (a) of Sub-Clause 21.4.1 [*Reference of a Dispute to the DAAB*].

To facilitate the appointment of the member(s) of an 'ad-hoc DAB', it is important that the Employer and the Contractor each avail themselves of the opportunity at the tender stage of the Contract to name three potential members in the Contract Data.

Sub-Clause 21.5   Amicable Settlement

In circumstances where the DAAB has given its decision but one or both Parties is/are dissatisfied with the decision, the provisions of this Sub-Clause are intended to encourage the Parties to settle a Dispute amicably, without the need for arbitration.

Rather than considering the 28 day period stated in this Sub-Clause as a 'cooling-off period', FIDIC recommends that the Parties avail themselves of this opportunity to actively engage with each other with a view to settling their Dispute.

Such active engagement may be by, for example:

- direct negotiation by senior executives from each of the Parties;

- mediation (please see below);

- expert determination (using, for example, the Expert Rules published by the International Chamber of Commerce (the "ICC", which is based at 33-43 Avenue du Président Wilson, 75116 Paris, France) https://iccwbo.org/publication/icc-expert-rules-english-version/[†]);

- or other form of alternative dispute resolution that is not as formal, time-consuming and costly as arbitration.

In this regard, it is recommended that consideration be given by the Parties to agree to a longer time period than the period of 28 days stated in this Sub-Clause in an effort to arrive at an amicable settlement procedure chosen by the Parties.

Amicable settlement procedures typically depend for their success on the consensual involvement of both Parties, on confidentiality and on both Parties' acceptance of the particular procedure. Therefore, while it is recommended that both Parties engage actively to settle the Dispute amicably, neither Party should seek to impose the procedure on the other Party.

并用"DAB"替换合同条件中对"争端避免/裁决委员会"的所有引用。

第 21.1 款，加上附录及其附件*争端避免/裁决委员会程序规则*，*争端避免/裁决委员会协议书附录的一般条件*，应加以修改，以符合菲迪克（**FIDIC**）1999 年第 1 版《生产设备和设计 - 施工合同条件》中相应条款的规定。

删除*第 21.3 款*［*争端避免*］；

在*第 21.4 款*［*取得争端避免/裁决委员会的决定*］的第一段中：

- 在"如果双方之间发生**争端**，"之后增加：

  "任命**争端裁决委员会**后"；
- 删除"（无论是否根据*第 21.3 款*［*争端避免*］进行了任何非正式讨论）"；（以及）

删除*第 21.4.1 项*［*争端提交给争端避免/裁决委员会*］（a）段。

为便于"临时**争端裁决委员会**"成员的任命，**雇主**和**承包商**应在合同招标阶段利用各自的机会在**合同数据**中提名三个潜在成员。

第 21.5 款   友好解决

在**争端避免/裁决委员会**做出决定但一方或双方对该决定不满意的情况下，本**款**的规定旨在鼓励双方友好解决**争端**，而无须仲裁。

菲迪克（**FIDIC**）建议双方不要认为本**款**规定的 28 天是"冷静期"，而应利用这一机会相互积极接触，友好解决**争端**。

这种积极接触可以是，例如：

- 双方高级主管直接谈判；

- 调解（请参见下文）；

- 专家确定（例如，使用**国际商会**，其总部位于 33-43 Avenue du Président Wilson, 75116 Paris, France，发布的**专家规则**，https://iccwbo.org/publication/ icc-expert-rules-english-version /[†]）；

- 或其他替代形式解决**争端**，不像仲裁那样正式、耗时且成本高昂。

建议双方就此考虑商定比本**款**规定的 28 天更长的期限，以期达成双方选择的友好解决程序。

友好解决程序的成功通常取决于双方的自愿参与、保密性以及双方对特定程序的接受程度。因此，尽管建议双方通过积极接触友好解决**争端**，但任一方均不应寻求将程序强加于另一方。

If the Parties wish to adopt a mediation procedure in their attempt to settle the Dispute amicably, then consideration may be given to the Mediation Rules, 2017 published by the International Chamber of Commerce (the "ICC", which is based at 33-43 Avenue du Président Wilson, 75116 Paris, France) https://iccwbo.org/dispute-resolution-services/mediation/mediation-rules/ [†]

or, alternatively, to the following mediation rules:

EXAMPLE   MEDIATION RULES

Appointment of Mediator

If the Parties are unable to agree on the choice of an independent and impartial mediator, or if the chosen mediator is unable or unwilling to act, then either Party may immediately apply to the appointing entity or official named in the Contract Data to appoint a mediator.

Once the mediator has been appointed, the Dispute shall immediately be referred to the mediator by the Parties or by either Party.

Obligations of the Parties

In addition to the Parties' obligations as set out in the provisions that follow below, during the mediation process the Parties shall engage with the mediator and with each other in co-operation, in a timely manner and in good faith.

Agreement of the Mediation Timetable

The mediator shall, within 7 days of his/her appointment, or other period as may be proposed by the mediator and agreed by both Parties, consult with the Parties to agree a timetable for the exchange of any relevant information, the date/time/venue of the mediation meeting, and the procedure to be adopted for the negotiations. The agreement of the timetable shall have regard to the period stated under Sub-Clause 21.5 [*Amicable Settlement*] of the Conditions of Contract or as may be amended by the agreement of the Parties.

Confidentiality

The Parties' negotiations facilitated by the mediator shall be conducted on a strictly private and confidential basis. Neither Party shall disclose to any third party any detail of the mediation process, including but not limited to: the fact of the mediation, the identity of the mediator, any matter discussed during the mediation process, any information or document exchanged with the mediator, any negotiations during the mediation meeting and/or the outcome of the mediation.

The mediation, and/or any negotiations taking place during the mediation meeting, shall not be referred to by either Party in any concurrent or subsequent proceedings, unless such negotiations conclude with a written legally binding agreement.

Written Agreement to be Binding

If the Parties accept the mediator's recommendations, or otherwise reach agreement on the settlement of the Dispute, such agreement shall be recorded in writing and, once signed by the designated representative(s) of both Parties, shall be binding on the Parties.

如果双方希望采用调解程序友好解决**争端**，则可以考虑使用**国际商会**（"ICC"，其总部位于 33-43 Avenue du Président Wilson, 73116 Paris, France）2017 年发布的《**调解规则**》https://iccwbo.org/dispute-resolution-services/mediation/mediation-rules/[†]。

或，遵循以下调解规则：

范例	调解规则
	调解员的任命
	如果双方无法就选择独立公正的调解员达成商定，或所选调解员不能或不愿行事，则任一方都可以立即向**合同数据**中指定的任命实体或官员申请任命调解员。
	调解员一旦任命，双方或任一方应立即将**争端**移交给调解员。
	双方的义务
	除以下条款中规定的双方义务外，双方在调解过程中应及时、真诚地与调解员和彼此合作。
	调解时间表的商定
	调解员应在其任命的 7 天内，或在调解员提议的并经双方商定的其他期限内，与双方协商，商定交换任何有关信息的时间表、调解会议的日期／时间／地点以及采用的谈判程序等。商定的时间表应考虑**合同条件第 21.5 款［友好解决］**规定的期限，或经双方商定而修改的期限。
	保密
	由调解员推动的**双方**谈判，应严格在私密和保密的基础上进行。任一方均不得向任何第三方透露调解程序的任何细节，包括但不限于：调解的事实、调解员的身份、在调解过程中讨论的任何事项、与调解员交换的任何信息或文件、调解会议期间的任何谈判和／或调解的结果等。
	双方不得在任何同时进行或随后的程序中提及调解和／或调解会议期间发生的任何谈判，该谈判以书面具有法律约束力的形式结束的情况除外。
	具有约束力的书面商定
	如果双方接受调解员的建议，或以其他方式就解决**争端**达成一致，则该协议应以书面形式记录，并由双方指定的代表签字后，对双方均具有约束力。

Mediator's Opinion

If no agreement is reached by the Parties after negotiations have been facilitated by the mediator, either Party may invite the mediator to provide to both Parties a non-binding opinion in writing. Such opinion shall not be used in evidence in any concurrent or subsequent proceedings.

Costs of Mediation

The Parties will bear their own costs of participating in the mediation, including but not limited to the costs of preparing and submitting evidence to the mediator and attending the mediation meeting.

Each Party shall be responsible for paying one-half of the remuneration of the mediator (and, if the mediator has been appointed by the appointing entity or official named in the Contract Data, the remuneration of such appointing entity or official).

However, if the mediator finds that a Party has initiated the mediation, or has engaged in the mediation, in a frivolous or vexatious manner, then the mediator shall have the power to order that Party to pay the reasonable costs of the other Party for preparing for and attending the mediation meeting. If these costs cannot be agreed, they will be assessed by the mediator, whose assessment shall be binding on the Parties.

Sub-Clause 21.6   Arbitration

The Contract should include provisions for the resolution by international arbitration of any Dispute which is not settled amicably by the Parties. In international contracts, international commercial arbitration has numerous advantages over litigation in national courts and is likely to be more acceptable to the Parties.

Careful consideration should be given to ensuring that the international arbitration rules that are chosen are compatible with the provisions of Clause 21 and with the other elements to be set out in the Contract Data. The Arbitration Rules published by the International Chamber of Commerce (the "ICC", which is based at 33-43 Avenue du Président Wilson, 75116 Paris, France) https://iccwbo.org/dispute-resolution-services/arbitration/rules-of-arbitration/[†] are frequently incorporated by reference in international contracts.

It is important that the Parties agree on the number of arbitrators and the language of arbitration. In the absence of specific stipulations as to the number of arbitrators and the place of arbitration in the Contract, the International Court of Arbitration of the ICC will decide these matters.

If the arbitration rules published by the United Nations Commission on International Trade Law ("UNCITRAL" which is based at the Vienna International Centre, A-1400 Vienna, Austria) http://www.uncitral.org/pdf/english/texts/arbitration/arb-rules-revised/arb-rules-revised-2010-e.pdf[†] or other non-ICC arbitration rules are preferred, it may be necessary to designate, in the Contract Data, an institution to appoint the arbitrators or to administer the arbitration, unless the institution is named (and their role specified) in the arbitration rules. It may also be necessary to ensure, before so designating an institution in the Contract Data, that the institution is prepared to appoint or administer.

For major projects tendered internationally, it is desirable that the place of arbitration be situated in a country other than that of the Employer or Contractor. This country should have a modern and liberal arbitration law and should have ratified a bilateral or multilateral convention (such as the 1958 New York Convention on the Recognition and Enforcement of Foreign Arbitral Awards), or both, that would facilitate the enforcement of an arbitral award in the states of the Parties.

调解员的意见

如果在调解员促成谈判之后双方未达成商定,则任一方均可邀请调解人以书面形式向双方提供无约束力的意见。此类意见不得在任何并行或后续程序中用做证据。

调解费用

双方将自行承担参加调解的费用,包括但不限于准备和向调解员提交证据以及参加调解会议的费用。

双方应负责各支付调解员酬金的一半(如果调解员是由**合同数据**中指定的任命实体或官员任命的,则也应承担该任命实体或官员的酬金)。

但是,如果调解员发现某**方**以轻率或无理的方式发起了调解或参与了调解,则调解员应有权命令该方支付另一方准备并参加调解会议的合理费用。如果无法商定这些费用,则由调解员估价,其估价对双方均具有约束力。

第 21.6 款　　　　**仲裁**

合同应包括通过国际仲裁解决双方未能友好解决的任何**争端**的条款。在国际合同中,与国家法院的诉讼相比,国际商事仲裁具有很多优势,并更可能被双**方**接受。

应仔细考虑确保选择的国际仲裁规则与**第 21 条**的规定以及**合同数据**中列出的其他要素相一致。**国际商会**(其总部位于 33-43 Avenue du Président Wilson, 75116 Paris, France)发布的**仲裁规则** https://iccwbo.org/dispute-resolution-services/arbitration/rules-of-arbitration/[†] 通常会被国际合同引用。

双**方**必须就仲裁员人数和仲裁语言达成一致,这一点很重要。在**合同**未具体规定仲裁员人数和仲裁地点的情况下,**国际商会**的**国际仲裁法院**将决定这些事项。

如果采用**联合国国际贸易法委员会**(总部位于奥地利维也纳 A-1400 维也纳国际中心)发布的仲裁规则 http://www.uncitral.org/pdf/english/texts/arbitration/arb-rules-revised/arb-rules-revised-2020-e.pdf[†],或其他非**国际商会**仲裁规则,可能有必要在**合同数据**中指定一家机构任命仲裁员或管理仲裁,仲裁规则已提名该机构(并规定其职能)的情况除外。在**合同数据**中如此指定一家机构前,可能还需确保该机构已就任命或管理准备就绪。

对于国际招标的重大项目,仲裁地最好位于**雇主**或**承包商**以外的国家。该国应有一部现代自由的仲裁法,并应已经批准了一项双边或多边公约(例如 1958 年《**承认和执行外国仲裁裁决的纽约公约**》),或者同时批准了这两项公约,这将有利于仲裁裁决的执行。

It may be considered desirable in some cases for other Parties to be joined into any arbitration between the Parties or for two or more pending arbitrations to be consolidated, thereby creating a multi-party arbitration. While the ICC Arbitration Rules address multi-party arbitration, such arbitration may in some cases be feasible only if the parties have included multi-party arbitration clauses in their contracts. Such clauses require skilful drafting, and usually need to be prepared on a case-by-case basis

It is not unusual that the arbitration of a complex dispute is concluded sometime after performance of the Contractor's obligations under the Contract has been completed, in which case it may not be fair and reasonable in the circumstances for any arbitral award which requires the payment of an amount by one Party to the other Party to be in the currencies of payment and in the proportions provided for under Sub-Clause 14.15 [*Currencies of Payment*]. For example, the Contractor may no longer have any need for local currency after completion of the Works; or the local currency may have declined in value in the interim or be a 'blocked currency' that, consequently, cannot be removed from the country concerned.

某些情况下，其他方可能更希望参加双方之间的任何仲裁，或者合并两个或多个待决的仲裁，从而形成多方仲裁。尽管《**国际商会仲裁规则**》涉及多方仲裁，但某些情况下，仅在当事方在其合同中包含多方仲裁条款时，这种仲裁才可行。此类条款的起草需要技巧，通常需要根据具体情况进行起草。

一般情况下，复杂争端的仲裁是在**承包商**履行其**合同**规定的义务后的某个时间完成的，在这种情况下，对于一方要求另一方按照**第 14.15 款**［*支付的货币*］规定的支付货币和比例，支付赔偿金的仲裁裁决可能是不公平合理的。例如，**工程**竣工后，**承包商**可能不再需要当地货币；或当地货币可能在此期间已贬值，或成为"封闭货币"，因此无法从相关国家 / 地区转出。

Advisory Notes to Users of FIDIC Contracts Where the Project is to Include Building Information Modelling Systems

Building Information Modelling (BIM), is a process which is changing many elements of the design profession, the construction industry and, possibly, even the operation and maintenance of a facility.

BIM is one of the digital data technologies used in all aspects of project planning, investigation, design, construction and operation. Digital data technologies include systems for: data acquisition; document management; design and process management; estimating, planning, and scheduling; contract management; performance management; and building information modelling.

BIM has varying degrees of complexity ranging from rather isolated use of computer aided design tools to full sharing of models and information by the entire project team. Currently, BIM is more often used and better understood in developed countries, many of which are encouraging or even mandating its use to improve quality, accuracy and delivery times for projects as well as to provide cost savings. BIM has the potential to dramatically improve productivity in the construction industry and reduce operational costs of facilities as well.

BIM is not a set of contract conditions; it is a mechanism to provide an environment where all parties have access to information relevant to their role in the design and construction of a project. Wherever possible, a combined (sometimes called federated or collaborative) model is developed for all parties to share, even if, as is often the case, various designers have used different computer aided design programs to develop their respective designs. Drawings and specifications are held in a common database accessible to everyone. These can be used for clash detection, coordination of designs, communication of changes, and construction sequencing.

Coordination of goals and effort is essential and is generally achieved by a BIM Protocol and a BIM Execution Plan, both key documents to access and understand work in this environment. A designer needs to understand and work to the Levels of Design (or Detail) (LOD) that will be spelled out in these documents to ensure that there is sufficient detail at each level to allow all designs to progress efficiently and avoid unnecessary changes.

BIM is founded on a team approach and successful projects utilising BIM encourage collaboration. FIDIC contracts are designed to be fair to all parties and are considered suitable for use with projects featuring the use of BIM - providing that the parties recognise the difference in approach and use the contract appropriately. This starts with proper planning and, unless an employer has appropriate expertise in house, they are well advised to retain an engineer who is appropriately qualified to assist them in the solicitation of interest, proposals, selection and negotiation of contracts with the selected project team. Legal advice is also necessary, of course, especially during the latter steps. The request for proposal (RFP) must clearly outline what the client's (employer's) expectations of the consultant are. The expectations should focus on the specific BIM goals and benefits desired. If properly developed, the RFP will help the proposers be responsive and the employer select the consultant (or in the case of a design build project, the project team) best qualified to deliver the desired BIM outcome. This process will in turn help all parties develop appropriate contract terms and conditions. Ideally the selection process would use FIDIC's Quality Based Selection (QBS) guideline.

It should be noted that the improved quality of information in projects utilising BIM can result in a significant reduction in variations. It is worth thinking about how the traditional roles of Contractor and Employer fit into this structure. In general, BIM is well suited for integrated project delivery, including Design Build and especially Design Build Operate projects where early proactive involvement of the design engineer, contractor and employer are essential. If advanced levels of BIM are anticipated for the project, the possibility of adding operation and maintenance elements of the constructed facility might be considered.

项目使用建筑信息模型系统的菲迪克（FIDIC）合同用户的建议说明

建筑信息模型（BIM）是一个改变设计行业、建筑行业，甚至可能改变设施运营和维护的许多要素的过程。

BIM 是项目规划、勘察、设计、施工和运营等各个方面使用的数字数据技术之一。数字数据技术包括用于以下方面的系统：数据采集；文档管理；设计和过程管理；估算、规划和进度安排；合同管理；绩效管理；建筑信息建模。

BIM 具有不同程度的复杂性，从相当孤立地使用计算机辅助设计工具到整个项目团队充分共享模型和信息。当前，BIM 在发达国家得到更广泛的使用和更好的理解，其中许多国家正在鼓励甚至强制要求使用 BIM，以提高项目质量、准确性和交付时间并节省成本。BIM 有潜力显著提高建筑业的生产力并降低设施的运营成本。

BIM 不是一套合同条件；它是一种机制，提供一种环境，使所有各方都可以获得与其在项目设计和施工中的角色相关的信息。在一切可能的情况下，都将开发一种组合的模型（有时称为联合模型或协作模型）供各方共享，即使在通常情况下，各种设计人员已使用不同的计算机辅助设计程序来开发各自的设计。图纸和规范要求保存在每个人都可以访问的通用数据库中。这些可用于碰撞检测、设计协调、变更沟通和施工排序。

目标和努力间的协调是必不可少的，通常通过 **BIM 协议**和 **BIM 执行计划**实现，这两个文件都是访问和理解该环境中工作的关键文件。设计人员需要理解并遵循这些文件中规定的**设计（或细节）级别（LOD）**，以确保每个级别都有足够的细节，使所有设计高效进行，并避免不必要的变更。

BIM 建立在团队合作的基础上，利用 BIM 的成功项目鼓励协作。菲迪克（FIDIC）合同旨在对各方公平，并被认为适合与使用 BIM 的项目一起使用——前提是各方应认识到方法上的差异，并恰当使用合同。首先要进行适当规划，除非雇主自身具有适当的专业知识，否则最好建议其聘请一名具有适当资格的工程师，以协助其与选定的项目团队开展兴趣书、建议书的征集、选聘及合同谈判等工作。当然，法律咨询也是必要的，尤其是在后续步骤中。征集建议书（RFP）必须明确概述客户（雇主）对咨询工程师的预期。预期应集中在特定的 BIM 目标以及期望的收益上。适当编写的征集建议书，将有助于建议书提交者做出响应，以及雇主选择最有资格提供期望的 BIM 成果的咨询工程师（对于设计—施工项目，则为项目团队）。此过程继而将帮助双方制定适当的合同条款和条件。理想情况下，选聘过程应使用《菲迪克（FIDIC）根据质量选择咨询工程师/咨询单位（QBS）指南》。

应注意的是，BIM 项目的信息质量的提升能够显著减少变更。值得思考**承包商和雇主**的传统角色如何适应这种结构。通常，BIM 非常适合集成项目交付，包括**设计—施工**，尤其是**设计—施工—运营**项目，在这些项目中，设计工程师、承包商和雇主必须尽早积极参与。如果该项目的 BIM 预计达到更高水平，则可考虑增加已建设施的运营和维护要素的可能性。

For advanced levels of BIM where fully open sharing of information is required, a qualified individual needs to be assigned the duty of managing the combined model. This should be separate from the project manager's role to ensure clear delineation of project responsibilities. This may require special outsourcing, if such skills are not available from the employer or engineer. Managing the BIM elements of a project may also involve risks that are beyond the normal coverage of professional indemnity (PI) insurance policies. Specialist advice should be sought from PI insurers if there is any doubt on coverage for this role. The competence and responsibilities of this position need to be understood and should include experience in data back-up and integrity, continuity planning and cyber security.

All parties involved in a project utilising BIM should take special care in checking their assigned scope and contract to ensure that they are aware of their BIM related responsibilities. The risks that FIDIC has identified in working in a BIM environment arise from these key features:

- misunderstanding of scope of services

- use of data for an inappropriate purpose and reliance on inappropriate data

- ineffective information, document or data management

- cyber security and responsibility for "holding" the models or data

- definition of deliverables, approval, and delivery

To manage these and other digital technology related risks, the consultant is encouraged to clearly define adequately their proposed services, addressing such issues as:

- the systems and versions or releases they propose and the management processes they will adopt

- access rights and limitations of the client, other consultants and contractors

- reliance other parties in the project may place on data in the digital environment

- limitations in terms of uptime and access and potential exclusion of liability for downtime

- potential exclusion of liability in the event of cyber attack

- potential exclusion or limitations on professional liability in respect of the actions of others

- access to the current and previous issues of the combined model and to the complete audit trail of changes to the model

At the completion of the project the model may need to be brought up to as-built status. This involves not only the drawing elements but also the embedded data. Experience shows that this is a significant effort, so responsibilities for completing this task should be clear and appropriate allowances provided.

The process for the delivery of contractual notices should be checked to determine if this will be through the common data environment or by more traditional means.

If sub-contractors are to be utilised, they should be bound by the BIM Protocol and Execution Plan.

The dispute resolution processes in the agreement should be appropriate, considering the collaborative nature of the BIM process. Professionals with engineering, construction and legal expertise should be consulted in this regard.

对于需要完全公开共享信息的高级 BIM，需要指派一名合格的人员来管理组合模型。这应与项目经理的角色分开，以确保明确划分项目责任。如果雇主或工程师不具备此类技能，则可能需要特别外包。管理项目的 BIM 要素还可能涉及职业责任（PI）保险保单以外的正常承保范围的风险。如果对此角色的承保范围有任何疑问，应咨询职业责任（PI）保险公司。需要了解该职位的能力和职责，并应包括数据备份和完整性、连续性规划以及网络安全方面的经验。

参与使用 BIM 项目的各方应特别注意检查其分配的范围和合同，以确保其了解与 BIM 相关的职责。**菲迪克（FIDIC）**已确定在 BIM 环境中工作的风险，来自以下主要特征：

- 对服务范围的误解；
- 将数据用于不适当的目的并依赖不适当的数据；
- 无效的信息、文件或数据管理；
- 网络安全和"持有"模型或数据的责任；
- 交付成果、同意和交付的定义。

为管理与这些及其他与数字技术相关的风险，建议咨询工程师明确定义其建议的服务，并解决以下问题：

- 咨询工程师建议的系统和版本或发行及其将采用的管理程序；

- 客户、其他咨询工程师和承包商的访问权和限制；
- 项目其他方对数字环境中的数据的可能依赖；
- 正常运行时间和访问限制以及可能的停机责任排除；
- 发生网络攻击时可能的责任排除；
- 他方行为方面职业责任的可能排除或限制；

- 访问当前和先前组合模型的版本以及对模型更改的完整审核。

项目完成时，可能需要将模型恢复为已建状态。这不仅涉及绘图元素，还涉及嵌入的数据。经验表明，这需付出巨大努力，因此应明确完成此任务的责任，并提供适当的津贴。

应检查合同通知的发送过程，以确定这是通过普通数据环境还是更传统的方式进行的。

如果使用分包商，则应受 **BIM 协议**和**执行计划**的约束。

考虑到 BIM 过程的协作性质，协议中的争端解决流程应恰当。在这方面，应咨询具有工程、施工和法律专业知识的专业人员。

Legal counsel should review the contract to ensure that it does not create an unintended joint venture which may be a risk in some jurisdictions.

If there is a fitness for purpose clause, make sure that it is clear who is responsible for ensuring compliance and how responsibility for rectification is spread between participants. Make sure that audit trails for modifications to the composite or federated model are controlled by a process which can generate an audit trail that is preserved during and after completion. The audit trail should be accessible by appropriate parties during and after project completion. Given that there will be many parties involved in the design and construction effort, ensure that there are appropriate limits of liability in place for each participant and for the project as a whole.

For construction or building projects involving BIM, in addition to considering the general principles as introduced above, the following (non-exhaustive) list of Sub-Clauses of the General Conditions of Contract for Construction [©FIDIC 2017 Second Edition] should be thoroughly reviewed when drafting the Particular Conditions:

Sub-Clause No.	Sub-Clause Title
1.1	Definitions
1.3	Notices and Other Communications
1.5	Priority of Documents
1.10	Employer's Use of Contractor's Documents
1.11	Contractor's Use of Employer's Documents
1.15	Limitation of Liability
2.3	Employer's Personnel and Other Contractors
2.5	Site Data and Items of Reference
3.2	Engineer's Duties and Authority
4.1	Contractor's General Obligations
4.4	Contractor's Documents
4.6	Co-operation
4.7	Setting Out
4.9	Quality Management and Compliance Verification Systems
4.20	Progress Reports
6.8	Contractor's Superintendence
6.10	Contractor's Records
8.3	Programme
9.1	Contractor's Obligations (Tests on Completion)
13.3	Variation Procedure
17.3	Intellectual and Industrial Property Rights
17.4	Indemnities by Contractor
17.5	Indemnities by Employer
17.6	Shared Indemnities
19.2.6	Other insurances required by Laws and by local practice

FIDIC intends to publish a "Technology Guideline" and a "Definition of Scope Guideline Specific to BIM" with the aim of providing further detailed support. These documents are under preparation and expected to be released shortly after the publication of the updates of the 1999 suite of FIDIC contract forms.

法律顾问应审核合同，以确保没有组建意外的合资公司，在某些法律辖区可能会有风险。

如果有适合目的的条款，一定明确由谁负责确保其合规性，以及如何给参与各方分配纠正措施。确保用于修改组合模型或联合模型的审计轨迹由一个过程控制，该过程可以生成完成期间和完成后保留的审计轨迹。在项目完成期间和之后，相关方应能获得审计轨迹。鉴于参与设计和施工的各方众多，确保每个参与者和整个项目都有适当的责任限制。

对于涉及 BIM 的施工或建筑项目，除了考虑上述基本原则外，起草**专用条件**时还应仔细阅读以下《**施工合同通用条件**》［©FIDIC 2017 年第 2 版］（非详尽的）**条款**清单：

条款序号	条款标题
1.1	定义
1.3	通知及其他通信交流
1.5	文件优先次序
1.10	雇主使用承包商文件
1.11	承包商使用雇主文件
1.15	责任限度
2.3	雇主人员和其他承包商
2.5	现场数据和参考事项
3.2	工程师的任务和权利
4.1	承包商的一般义务
4.4	承包商文件
4.6	合作
4.7	放线
4.9	质量管理和合规验证体系
4.20	进度报告
6.8	承包商的监督
6.10	承包商的记录
8.3	进度计划
9.1	承包商的义务（竣工试验）
13.3	变更程序
17.3	知识产权和工业产权
17.4	由承包商保障
17.5	由雇主保障
17.6	保障分担
19.2.6	法律和当地惯例要求的其他保险

菲迪克（FIDIC）计划出版《**技术指南**》和《**BIM 专用范围指南定义**》，目的是提供进一步详细资料。这些文件正在编写中，预计将在 1999 版 FIDIC 系列合同格式更新出版后不久发布。

Annexes   FORMS OF SECURITIES

Acceptable form(s) of security should be included in the tender documents: for Annex A and/or B, in the Instructions to Tenderers; and for Annexes C to G, annexed to the Particular Conditions. The following example forms, which (except for Annex A) incorporate Uniform Rules published by the International Chamber of Commerce (the "ICC", which is based at 33-43 Avenue du Président Wilson, 75116 Paris, France, www.iccwbo.org), need to be carefully reviewed against, and may have to be amended to comply with, applicable law. Although the ICC publishes guides to these Uniform Rules, legal advice should be taken before the securities are written. Note that the guaranteed amounts should be quoted in all the currencies, as specified in the Contract, in which the guarantor pays the beneficiary.

附件　　　　担保函格式

招标文件中应包括认可的担保函格式：**附件 A 和 / 或 B** 附于**投标人须知**；**附件 C 至 G** 附于**专用条件**。下列范例格式（**附件 A 除外**）体现了**国际商会**（其总部位于 33-43 Avenue du Président Wilson, 75116 Paris, France, www.iccwbo.org）公布的**统一规则**，应用时可能需要修改，以符合适用法律。虽然**国际商会**出版了对这些**统一规则**的指南，在起草担保函前还应听取法律咨询建议。还应注意，保证金额应按合同中规定的，担保人向受益人支付的所有币种分别列出。

Annex A   EXAMPLE FORM OF PARENT COMPANY GUARANTEE

Name of Contract/Contract No.: ....

Name and address of Employer: ....
.... (together with successors and assigns).

We have been informed that .... (hereinafter called the "Contractor") is submitting an offer for such Contract in response to your invitation, and that the conditions of your invitation require his/her offer to be supported by a parent company guarantee.

In consideration of you, the Employer, awarding the Contract to the Contractor, we (name of parent company) irrevocably and unconditionally guarantee to you, as a primary obligation, the due performance of all the Contractor's obligations and liabilities under the Contract, including the Contractor's compliance with all its terms and conditions according to their true intent and meaning.

If the Contractor fails to so perform his/her obligations and liabilities and comply with the Contract, we will indemnify the Employer against and from all damages, losses and expenses (including legal fees and expenses) which arise from any such failure for which the Contractor is liable to the Employer under the Contract.

This guarantee shall come into full force and effect when the Contract comes into full force and effect. If the Contract does not come into full force and effect within a year of the date of this guarantee, or if you demonstrate that you do not intend to enter into the Contract with the Contractor, this guarantee shall be void and ineffective. This guarantee shall continue in full force and effect until all the Contractor's obligations and liabilities under the Contract have been discharged, when this guarantee shall expire and shall be returned to us, and our liability hereunder shall be discharged absolutely.

This guarantee shall apply and be supplemental to the Contract as amended or varied by the Employer and the Contractor from time to time. We hereby authorise them to agree any such amendment or variation, the due performance of which and compliance with which by the Contractor are likewise guaranteed hereunder. Our obligations and liabilities under this guarantee shall not be discharged by any allowance of time or other indulgence whatsoever by the Employer to the Contractor, or by any variation or suspension of the works to be executed under the Contract, or by any amendments to the Contract or to the constitution of the Contractor or the Employer, or by any other matters, whether with or without our knowledge or consent.

This guarantee shall be governed by the law of the same country (or other jurisdiction) as that which governs the Contract and any dispute under this guarantee shall be finally settled under the Rules of Arbitration of the International Chamber of Commerce by one or more arbitrators appointed in accordance with such Rules. We confirm that the benefit of this guarantee may be assigned subject only to the provisions for assignment of the Contract.

SIGNED by: ....   SIGNED by[1]: ....
(signature)   *(signature)*

....
(name)   *(name)*

....
(position in the company)   *(position in the company)*

Date:

[1] *Whether one or more signatories for the parent company are required will depend on the parent company and/or applicable law.*

附件 A　　母公司保函范例格式

合同名称 / 合同编号：＿＿＿＿＿＿＿＿＿＿＿＿＿＿＿＿＿＿＿＿＿＿＿＿＿＿＿＿＿＿＿＿＿

雇主名称和地址：＿＿＿＿＿＿＿＿＿＿＿＿＿＿＿＿＿＿＿＿＿＿＿＿＿＿＿＿＿＿＿＿＿＿
＿＿＿＿＿＿＿＿＿＿＿＿＿＿＿＿＿＿＿＿＿＿＿＿＿＿＿＿（连同继承人和受让人）。

我方已获知＿＿＿＿＿＿＿＿＿（以下称"**承包商**"）正响应你方邀请对上述**合同**提交报价，你方邀请条件要求报价要有一份母公司保函支持。

考虑到你方，**雇主**，将向**承包商**授予**合同**，我方（母公司名称）不可撤销和无条件地，作为一项首要义务向你方保证，**承包商**根据**合同**规定的所有应履行的义务和责任，包括**承包商**按照其真实意图和含义遵守所有**合同**条款和条件。

如果**承包商**未能如上履行其义务和责任，未能遵守**合同**，我方将保障**雇主**免受因**承包商**根据合同应对**雇主**负责的任何该类违约造成的所有损害赔偿费、损失和开支（包括法律费用和开支）。

本保函将在**合同**全面实施和生效时，全面实施和生效。如果在本保函日期后一年内，**合同**没有全面实施和生效，或如果你方表明不想与**承包商**签订**合同**，本保函将作废和无效。本保函将持续全面实施和有效直到**承包商**根据**合同**规定的义务和责任全部解除为止，届时本保函将期满，应退还我方，我方在其下的责任应完全解除。

雇主和**承包商**有时对**合同**进行修改或变更时，本保函仍适用并作为**合同**的补充。我方在此授权他们商定任何此类修改或变更，对**承包商**应履行和应遵守的修改或变更在此同样予以保证。我方根据本保函应负的义务和责任，不应因**雇主**对**承包商**做出的任何时限允许或其他宽让，或根据合同要实施的工程的任何变更或暂停，或对**合同**、**承包商**或**雇主**的组成的任何修改，或任何其他事项而解除，不论这些事项是否经我方知晓或同意。

本保函应由管辖**合同**的同一国家（或其他司法管辖区）的法律管辖，关于本保函的任何争端，应根据**国际商会仲裁规则**，由按该**规则**任命的一位或几位仲裁员最终解决。我方确认，本保函的权益仅可按照**合同**转让的条款进行转让。

签字人：＿＿＿＿＿＿＿＿＿＿＿＿＿＿＿＿　　签字人[1]：＿＿＿＿＿＿＿＿＿＿＿＿＿＿＿＿
　　　　　　　（签字）　　　　　　　　　　　　　　　　　　（签字）

＿＿＿＿＿＿＿＿＿＿＿＿＿＿＿＿＿＿＿＿＿　　＿＿＿＿＿＿＿＿＿＿＿＿＿＿＿＿＿＿＿＿＿
　　　　　　　（名称）　　　　　　　　　　　　　　　　　　（名称）

＿＿＿＿＿＿＿＿＿＿＿＿＿＿＿＿＿＿＿＿＿　　＿＿＿＿＿＿＿＿＿＿＿＿＿＿＿＿＿＿＿＿＿
　　　　　（在公司的职位）　　　　　　　　　　　　　　（在公司的职位）

日期：

[1] *需要母公司的一人或多人签字将取决于母公司和 / 或适用法律。*

Annex B   EXAMPLE FORM OF TENDER SECURITY

Guarantee No.: _____ [*insert guarantee reference number*]

The Guarantor: _____
[*insert name and address of place of issue, unless indicated in the letterhead*]

Name of Contract/Contract No.:: _____ [*insert reference number or other information identifying the contract with regard to which the tender is submitted*]

The Beneficiary (the "Employer"): _____ [*insert name and address of the Beneficiary*]

We have been informed that _____ [*insert name and address of the Tenderer*] (hereinafter called the "Applicant") is submitting an offer for such Contract in response to your invitation, and that the conditions of your invitation (the "Conditions of Invitation", which are set out in a document entitled Instructions to Tenderers) require his/her offer to be supported by a tender security.

At the request of the Applicant, we _____ [*insert name of Guarantor*] hereby irrevocably undertake to pay you, the Beneficiary/Employer, any sum or sums not exceeding in total the amount of _____ [*insert in figures and words the maximum amount payable and the currency in which it is payable*] upon receipt by us of your demand in writing and your written statement (in the demand) that:

(a) the Applicant has, without your agreement, withdrawn his/her offer after the latest time specified for its submission and before the expiry of its period of validity, or

(b) the Applicant has refused to accept the correction of errors in his/her offer in accordance with such Conditions of Invitation, or

(c) you awarded the Contract to the Applicant and he/she has failed to comply with Sub-Clause 1.6 of the Conditions of Contract, or

(d) you awarded the Contract to the Applicant and he/she has failed to comply with Sub-Clause 4.2.1 of the Conditions of Contract.

Any demand for payment must contain your signature(s) which must be authenticated by your bankers or by a notary public. The authenticated demand and statement must be received by us at the following office [*insert address of office*] on or before _____ [*insert the date 35 days after the expiry of the validity of the Letter of Tender*], when this guarantee shall expire.

The party liable for the payment of any charges: _____
_____ [*insert the name of the party*].

This guarantee shall be governed by the laws of _____ [*insert the law governing the guarantee*], and shall be subject to the Uniform Rules for Demand Guarantees (URDG) 2010 Revision, ICC Publication No. 758.

SIGNED by: _____   SIGNED by[(1)]: _____
           (signature)   (signature)

_____   _____
           (name)   (name)

Date: _____

[(1)] *Whether one or more signatories for the bank are required will depend on the bank and/or applicable law.*

附件 B　　投标保函范例格式

保函编号：_____ ［插入保函标号］

担保人：_____
［插入名称和签署地地址，纸头显示地址的情况除外］

合同名称/合同编号：_____ ［插入编号或其他能够说明已投标的合同的信息］

受益人（"雇主"）：_____ ［插入受益人名称和地址］

我方已获知 _____ ［插入**投标人**名称和地址］（以下称"**申请人**"）正响应你方邀请对上述合同提交一份报价，你方邀请条件（在题为**投标人须知**的文件中规定的"**邀请条件**"）要求其报价要有一份投标保函支持。

应**申请人**请求，我方 _____ ［插入**担保人**名称］在此不可撤销地承诺，在我方收到你方的书面要求和关于（在要求中的）下列情况的书面说明后，向你方，**受益人／雇主**，支付总额不超过 _____ ［插入数字和文字，表明可付最高金额和货币］的任何一笔或几笔款额：

(a)　**申请人**未经你方同意，在规定的提交报价的最终时间后和其有效期限期满前，已撤回其报价；（或）

(b)　**申请人**已拒绝接受对其按照上述**邀请条件**所做报价中的错误的改正；（或）

(c)　你方将**合同**授予了**申请人**，但**申请人**未能遵守**合同条件**第 1.6 款；（或）

(d)　你方将**合同**授予了**申请人**，但**申请人**未能遵守**合同条件**第 4.2.1 项。

任何付款的要求，都必须有经你方银行或公证人确证的你方签字。经确证的要求和说明必须在 _____ ［插入**招标函**有效期期满后 35 天的日期］或其以前，由我方在本办公地点［插入办公地点］收到。

负责支付任何费用的一方：_____ ［插入该方名称］。

本保函应由 _____ 法律［插入管辖保函的法律］管辖，并应遵守**国际商会**以 2010 年修订出版 758 号文公布的即付保函统一规则（URDG）的规定。

签字人：_____　　　　签字人[1]：_____
　　　　　　（签字）　　　　　　　　　　　　　　（签字）

　　　_____　　　　　　　_____
　　　　　　（名称）　　　　　　　　　　　　　　（名称）

日期：_____

---

[1] 需要银行一人签字还是多人签字将取决于银行和／或使用法律。

Annex C **EXAMPLE FORM OF PERFORMANCE SECURITY - DEMAND GUARANTEE**

Guarantee No.: _____ [*insert guarantee reference number*]

The Guarantor: _____
[*insert name and address of place of issue, unless indicated in the letterhead*]

Name of Contract/Contract No.: _____ [*insert reference number or other information identifying the contract between the Applicant and the Beneficiary on which the guarantee is based*]

The Beneficiary (the "Employer"): _____
_____ [*insert name and address of the Beneficiary*]

We have been informed that _____ [*insert name and address of the Contractor*] (hereinafter called the "Applicant") is your Contractor under such Contract, which requires him/her to obtain a Performance Security.

At the request of the Applicant, we _____ [*insert name of Guarantor*] hereby irrevocably undertake to pay you, the Beneficiary/Employer, any sum or sums not exceeding in total the amount of _____ [*insert in figures and words the maximum amount payable and the currency in which it is payable*] (the "Guaranteed Amount") upon receipt by us of your demand in writing and your written statement indicating in what respect the Applicant is in breach of its obligations under the Contract.

[*Following receipt by us of an authenticated copy of the Taking-Over Certificate for the whole of the Works under clause 10 of the Conditions of Contract, the Guaranteed Amount shall be reduced by _____ % and we shall promptly notify you that we have received such certificate and have reduced the Guaranteed Amount accordingly.*] [1]

Following receipt by us of an authenticated copy of a statement issued by _____
_____
[*insert name and address of the Engineer under the Contract*] that, pursuant to Sub-Clause 4.2.1 of the Conditions of Contract, variations or adjustments under Clause 13 of the Conditions of Contract have resulted in an accumulative increase or decrease of the Contract Price by more than twenty percent (20%) of the Accepted Contract Amount, and that therefore the Guaranteed Amount should be adjusted by the percentage specified in the statement equal to the accumulative increase or decrease, respectively, we shall promptly inform you that we have received such statement and have adjusted the Guaranteed Amount accordingly. In the case of a request for a decrease of the amount of the Performance Security, the above statement shall be accompanied by your written consent to such decrease.

Any demand for payment must contain your signature(s) which must be authenticated by your bankers or by a notary public. The authenticated demand and statement must be received by us at the following office [*insert address of office*] on or before _____ (*insert the date 70 days after the expected expiry of the Defects Notification Period for the Works*) (the "Expiry Date"), when this guarantee shall expire.

The party liable for the payment of any charges: [*insert the name of the party*].

This guarantee shall be governed by the laws of _____ [*insert the law governing the guarantee*], and shall be subject to the Uniform Rules for Demand Guarantees, (URDG) 2010 Revision, ICC Publication No. 758.

附件 C　　履约担保函——即付保函范例格式

保函编号：＿＿＿＿＿＿＿＿＿＿＿＿＿＿＿＿［插入保函标号］

担保人：＿＿
［插入姓名和签署地地址，纸头显示地址的情况除外］

合同名称 / 合同编号：＿＿＿＿＿＿＿＿＿＿［插入编号或其他能够说明保函所依据的**申请人**与**受益人**之间合同的信息］

受益人（"雇主"）：＿＿＿＿＿＿＿＿＿＿［插入**受益人**名称和地址］

我方已获知 ＿＿＿＿＿＿＿＿＿＿［插入**承包商**的名称和地址］（以下称"申请人"）是此类合同的**承包商**，合同要求其取得一份**履约担保函**。

应**申请人**请求，我方 ＿＿＿＿＿＿＿＿＿＿［插入**担保人**名称］在此不可撤销地承诺，在我方收到你方的书面要求以及**申请人**在哪些方面违反了**合同**规定义务的书面说明后，向你方，**受益人 / 雇主**，支付全部总额不超过 ＿＿＿＿＿＿＿＿＿＿［插入最高可付金额的数字和文字以及可付货币］（"保证金额"）。

[在我方收到经确证的根据**合同条件**第 10 条规定签发的整个**工程接收证书**的抄件后，此项**保证金额**应减少 ＿＿＿＿＿＿％，我方将立即通知你方，我方已收到该证书并已相应减少了**保证金额**。]⁽¹⁾

在我方收到由＿＿＿＿＿＿＿＿＿＿［插入**合同**规定的**工程师**的名称和地址］根据合同条件第 **4.2.1 项**签发的经确证的抄件，说明按照**合同条件**第 13 条规定所做的变更或调整导致合同价格累计增加或减少超过**中标合同金额**的百分之二十（20%），并且因此使**保证金额**应按说明中规定的分别与增额或减额相同的比例调整后，我方将立即通知你方，我们已收到该说明并已相应调整了**保证金额**。如果要求减少**履约担保函**金额，则上述说明应附上你方对减额的书面同意。

任何付款的要求都必须有经你方银行或公证人确证的你方的签字。经确证的要求和说明必须在 ＿＿＿＿＿＿＿＿＿＿（插入预定**工程缺陷通知期限**期满后 70 天的日期）（"期满日期"）或其以前，由我方在以下办公地点 ＿＿＿＿＿＿＿＿＿＿［插入办公地点］收到，届时本保函应期满。

支付任何费用的责任方：[插入该方的名称]。

本保函应受 ＿＿＿＿＿＿＿＿＿＿［插入管辖保函的法律]法律管辖，并应遵守 2010 年修订出版的国际商会 758 号文公布的即付保函统一规则的规定。

SIGNED by: _____   SIGNED by[2]: _____
                   (signature)   *(signature)*

_____   _____
           (name)   *(name)*

Date: _____

[1] *When drafting the tender documents, the writer should ascertain whether to include the optional text, shown in parentheses [ ].*

[2] *Whether one or more signatories for the bank are required will depend on the bank and/or applicable law.*

签字人：_____ 签字人 (2)：_____
　　　　　　（签字）　　　　　　　　　　　　　（签字）

　　　　　_____　　　　　_____
　　　　　　（名称）　　　　　　　　　　　　　（名称）

日期：_____

(1) 起草人在起草招标文件时，应确定是否要包括方括号［　］中的备选文字。

(2) 需要银行一人或多人签字将取决于银行和/或使用法律。

Annex D   EXAMPLE FORM OF PERFORMANCE SECURITY - SURETY BOND

Name of Contract/Contract No.: ....

Name and address of Beneficiary (the "Employer"): ....
....

We have been informed that .... [insert name of the Contractor] (hereinafter called the "Principal") is your contractor under such Contract, which requires him/her to obtain a Performance Security.

By this Bond, ...
.... [insert name and address of contractor]

(who is your Contractor under such Contract) as Principal and: ....
.... [insert name and address of Guarantor] as Guarantor are irrevocably held and firmly bound to the Beneficiary in the total amount of .... [insert in figures and words the maximum amount payable and the currency in which it is payable] (the "Bond Amount") for the due performance of all such Principal's obligations and liabilities under the above named Contract.

[Such Bond Amount shall be reduced by .... % upon the issue of the Taking-Over Certificate for the whole of the Works under Clause 10 of the Conditions of Contract.][1]

This Bond shall become effective on the Commencement Date defined in the Contract.

Upon Default by the Principal to perform any contractual obligation, or upon the occurrence of any of the events and circumstances listed in Sub-Clause 15.2.1 of the Conditions of Contract, the Guarantor shall satisfy and discharge the damages sustained by the Beneficiary due to such Default, event or circumstances.[2] However, the total liability of the Guarantor shall not exceed the Bond Amount.

The obligations and liabilities of the Guarantor shall not be discharged by any allowance of time or other indulgence whatsoever by the Beneficiary to the Principal, or by any variation or suspension of the Works to be executed under the Contract, or by any amendments to the Contract or to the constitution of the Principal or the Beneficiary, or by any other matters, whether with or without the knowledge or consent of the Guarantor.

Any claim under this Bond must be received by the Guarantor on or before ....
[insert the date six months after the expected expiry of the Defects Notification Period for the Works] (the "Expiry Date"), when this Bond shall expire and shall be returned to the Guarantor.

The benefit of this Bond may be assigned subject to the provisions for assignment of the Contract, and subject to the receipt by the Guarantor of evidence of full compliance with such provisions.

This Bond shall be governed by the law of .... [insert the law governing the bond] being the same country (or other jurisdiction) as that which governs the Contract. This Bond incorporates and shall be subject to the Uniform Rules for Contract Bonds, published as number 524 by the International Chamber of Commerce, and words used in this Bond shall bear the meanings set out in such Rules.

Whereas this Bond has been issued by the Principal and the Guarantor on .... [date]

附件 D　　履约担保函——担保保证范例格式

合同名称／合同编号：_____

受益人（"雇主"）名称和地址：_____

我方已获知_____［插入*承包商*的名称和地址］（以下称"**委托人**"）是此类**合同**的**承包商**，**合同**要求其取得一份**履约担保函**。

根据本**保证书**_____（插入承包商名称和地址）（此类合同规定的**承包商**）作为**委托人**与_____［插入**担保人**名称和地址］作为**担保人**，对该**委托人**根据上述合同应履行的全部义务和责任，以总金额_____［插入可付最高金额的数字和文字以及可付货币］（"**保证金额**"），向**委托人**不可撤销地保持和坚定地担保。

［上述**保证金额**在根据合同条件第 10 条签发整个工程接收证书后，应减少_____%］ (1)

本**保证书**自合同中规定的**开工日期**起生效。

在**委托人**履行任何合同义务中发生**违约**，或出现任何合同条件第 15.2.1 项所列举的事件或情况时，**担保人**应满足并偿清**受益人**因该项**违约**、事件或情况遭受的损害赔偿费 (2)，但**担保人**的全部责任不应超过**保证金额**。

担保人的义务和责任不因**受益人**对**委托人**做出的任何时限允许或其他宽让，或对根据合同应实施工程的任何变更或暂停，或对**合同**或对**委托人**或**受益人**的组成的任何修改，或任何其他事项而解除，不论是否经**担保人**知晓或同意。

根据本**保证书**提出的任何索赔必须由**担保人**在_____［插入预定*工程缺陷通知期限*期满后 6 个月的日期］（"**期满日期**"）或其以前收到，届时本**保证书**应期满，应退回**担保人**。

本**保证书**的权益可以依照**合同**转让的条款，以及**担保人**收到完全符合该项条款的证据，进行转让。

本**保证书**应由管辖合同的同一国家（或其他司法管辖区）的_____［插入管辖本保证书的法律］法律管辖。本**保证书**体现并应遵守**国际商会**以 524 号文公布的合同保证书统一规则的规定，以及本**保证书**使用的词语应具有该**规则**规定的含义。

本**保证书**于_____［日期］由**委托人**和**担保人**签署

Signatures for and on behalf of the Principal[3]:

_____   _____
          (signature)   *(signature)*

_____   _____
            (name)   *(name)*

Signatures for and on behalf of the Guarantor[4]:

_____   _____
          (signature)   *(signature)*

_____   _____
            (name)   *(name)*

[1] *When writing the tender documents, the writer should ascertain whether to include the optional text, shown in parentheses [ ].*

[2] *Insert: [and shall not be entitled to perform the Principal's obligations under the Contract.]*

Or: [or at the option of the Guarantor (to be exercised in writing within 42 days of receiving the claim specifying such Default) perform the Principal's obligations under the Contract.]

[3] *Whether one or more signatories for the Principal are required will depend on the Principal and/or applicable law.*

[4] *Whether one or more signatories for the Guarantor are required will depend on the Guarantor and/or applicable law.*

委托人⁽³⁾代表签字：

_____ （签字）　　_____ （签字）

_____ （名称）　　_____ （名称）

担保人⁽⁴⁾代表签字：

_____ （签字）　　_____ （签字）

_____ （名称）　　_____ （名称）

⁽¹⁾ 起草人起草招标文件时，应确定是否要包括方括号［　］内的备选文字。

⁽²⁾ 此处插入：[并不得履行**委托人**合同规定的义务。]

或：[或由**担保人**选择（要在收到提出该项**违约**索赔42天内以书面形式提出）履行**委托人**合同规定的义务。]

⁽³⁾ 需要一个或更多**委托人**签字将取决于**委托人**和/或适用法律。

⁽⁴⁾ 需要一个或更多**担保人**签字将取决于**担保人**和/或适用法律。

Annex E EXAMPLE FORM OF ADVANCE PAYMENT GUARANTEE

Guarantee No.: _____ [insert guarantee reference number]

The Guarantor: _____
_____ [insert name and address of place of issue, unless indicated in the letterhead]

Name of Contract/Contract No.: _____
_____ [insert reference number or other information identifying the contract between the Applicant and the Beneficiary on which the guarantee is based]

The Beneficiary (the "Employer"): _____
_____ [insert name and address of the Beneficiary]

We have been informed that _____
_____ [insert name and address of the Contractor] (hereinafter called the "Applicant") is your Contractor under such Contract and wishes to receive an advance payment, for which the Contract requires him/her to obtain a guarantee.

At the request of the Applicant, we _____ [insert name of Guarantor] hereby irrevocably undertake to pay you, the Beneficiary/Employer, any sum or sums not exceeding in total the amount of _____ [insert in figures and words the maximum amount payable and the currency in which it is payable] (the "Guaranteed Amount") upon receipt by us of your demand in writing and your written statement that:

(a) the Applicant has failed to repay the advance payment in accordance with the Conditions of Contract, and

(b) the amount of the advance payment which the Applicant has failed to repay.

This guarantee shall become effective upon receipt [of the first instalment] of the advance payment by the Applicant. The Guaranteed Amount shall be reduced by the amounts of the advance payment repaid to you, as evidenced by interim payment certificates issued under Sub-Clause 14.6 of the Conditions of Contract. Following receipt of a copy of each interim payment certificate, we shall promptly notify you of the revised Guaranteed Amount accordingly.

Any demand for payment must contain your signature(s) which must be authenticated by your bankers or by a notary public. The authenticated demand and statement must be received by us at the following office [insert address of office] on or before _____ [insert the date 70 days after the expected expiry of the Time for Completion] (the "Expiry Date"), when this guarantee shall expire.

The party liable for the payment of any charges: _____ [insert the name of the party].

This guarantee shall be governed by the laws of _____ [insert the law governing the guarantee], and shall be subject to the Uniform Rules for Demand Guarantees (URDG) 2010 Revision, ICC Publication No. 758.

Signed by: _____   Signed by[(1)]: _____
                    (signature)   (signature)

_____   _____
                    (name)   (name)

Date: _____

[(1)] Whether one or more signatories for the bank are required will depend on the bank and/or applicable law.

附件 E　　　预付款保函范例格式

保函编号： _____ [插入保函编号]

担保人： _____ [插入签发地的名称和地址，信头纸有显示的除外]

合同名称/合同编号： _____ [插入说明保函依据的**申请人**和**受益人**之间合同编号和其他信息。]

受益人（"雇主"）： _____ [插入**受益人**名称和地址]

我方已获知 _____ [插入**承包商**名称和地址]（以下称为"**申请人**"）是你方上述合同规定的**承包商**，希望得到一笔预付款，为此，**合同**要求其取得一份保函。

应**申请人**请求，我方 _____ [插入**担保人**名称] 在此不可撤销地承诺，在我方收到你方书面要求和关于以下情况的书面说明后，向你方，**受益人/雇主**，支付全部总额不超过 _____ [插入最高可付金额的数字和文字以及可付货币]（"**保证金额**"）的任何一笔或几笔款额：

（a）　**申请人**未能按照**合同条件**付还预付款；（以及）

（b）　**申请人**未能付还的预付款额。

本保函在**申请人**收到预付款的[首次分期付款]时开始生效。该**保证金额**应按你方根据**合同条件**第 **14.6** 款规定发出期中付款证书中证明已向你方付还的款额，逐步减少。我方每次收到期中付款的抄件后，将立即将相应修改的**保证金额**通知你方。

任何关于付款的要求都必须有经你方银行或公证人确证的你方签字。经确证的要求和说明必须在 _____ [插入预定**竣工时间**期满后 70 天的日期]（"**期满日期**"）或其以前，由我方在下述办公地点[插入办公地点]收到，届时本保函将期满。

负责支付任何费用的一方 _____ [插入该方名称]。

本保函应受 _____ [插入管辖担保人的法律名称] 法律管辖，并应遵守**国际商会** 2010 年修订出版的第 758 号公布的**即付保函统一规则**的规定。

签字人：_____　　　　签字人[1]：_____
　　　　　　　（签字）　　　　　　　　　　　　　　　　　（签字）

　　　　　　_____　　　　　　　　　　_____
　　　　　　　（名称）　　　　　　　　　　　　　　　　　（名称）

日期 _____

[1] 需要银行一人签字还是多人签字取决于银行和/或适用法律。

Annex F   EXAMPLE FORM OF RETENTION MONEY GUARANTEE

Guarantee No.: _____ [*insert guarantee reference number*]

The Guarantor: _____
_____ [*insert name and address of place of issue, unless indicated in the letterhead*]

Name of Contract/Contract No.: _____
_____ [*insert reference number or other information identifying the contract between the Applicant and the Beneficiary on which the guarantee is based*]

The Beneficiary (the "Employer"): _____
_____ [*insert name and address of the Beneficiary*]

We have been informed that _____
_____ [*insert name and address of the Contractor*] (hereinafter called the "Applicant") is your Contractor under such Contract and wishes to receive an early payment of [part of] the retention money, for which the Contract requires him/her to obtain a guarantee.

At the request of the Applicant, we _____ [*insert name of Guarantor*] hereby irrevocably undertake to pay you, the Beneficiary/Employer, any sum or sums not exceeding in total the amount of _____ [*insert in figures and words the maximum amount payable and the currency in which it is payable*] (the "Guaranteed Amount") upon receipt by us of your demand in writing and your written statement that the Applicant has failed to carry out his/her obligation(s) to rectify the following defect(s) for which he/she is responsible under the Contract [*state the nature of the defect(s)*].

At any time, our liability under this guarantee shall not exceed the total amount of retention money released to the Applicant by you, as evidenced by Interim Payment Certificates issued under Sub-Clause 14.6 of the Conditions of Contract with a copy being submitted to us.

Any demand for payment must contain your signature(s) which must be authenticated by your bankers or by a notary public. The authenticated demand and statement must be received by us at the following office [insert address of office] on or before _____
[*insert the date 70 days after the expected expiry of the Defects Notification Period for the Works*], (the "Expiry Date"), when this guarantee shall expire.

We have been informed that the Beneficiary may require the Applicant to extend this guarantee if the Performance Certificate under the Contract has not been issued by the date 28 days prior to such Expiry Date. We undertake to pay you the Guaranteed Amount upon receipt by us, within such period of 28 days, of your demand in writing and your written statement that the Performance Certificate has not been issued, for reasons attributable to the Applicant, and that this guarantee has not been extended.

The party liable for the payment of any charges: _____ [*insert the name of the party*].

This guarantee shall be governed by the laws of _____ [*insert the law governing the guarantee*] and shall be subject to the Uniform Rules for Demand Guarantees (URDG) 2010 Revision, ICC Publication No. 758.

附件 F　　　保留金保函范例格式

保函编号：_____［插入保函编号］

担保人：_____［插入签发地的名称和地址，信头纸有显示的除外］

合同名称/合同编号：_____［插入说明保函依据的**申请人**和**受益人**之间的合同编号和其他信息］

受益人（"雇主"）：_____［插入**受益人**名称和地址］

我方已获知_____［插入**承包商**名称和地址］（以下称为"**申请人**"）是你方上述合同规定的**承包商**，希望收到提前付给的［部分］保留金，为此，合同要求其取得一份保函。

应**申请人**请求，我方_____［插入**担保人**名称］在此不可撤销地承诺，在我方收到你方书面要求和关于**申请人**未能履行合同规定［说明缺陷性质］的其应负责的改正某些缺陷的义务的书面说明后，向你方，**受益人/雇主**，支付全部总额不超过_____［插入最高可付金额的数字和文字以及可付货币］（"**保证金额**"）的任何一笔或几笔款额。

本保函规定的我方的责任任何时候都不应超过、经你方按照**合同条件**第 14.6 款规定发出的**期中付款证书**并给我方一份抄件证明的、你方放还给**申请人**的保留金的总额。

任何关于付款的要求都必须有经你方银行或公证人确证的你方签字。经确证的要求和说明必须在_____［插入预定**工程缺陷通知期限**期满后 70 天的日期］（"**期满日期**"）或其以前，由我方在下述办公地点［插入办公地址］收到，届时本保函将期满。

我方已获知，如果到该**期满日期** 28 天前还没有签发合同规定的**履约证书**，**受益人**可以要求**申请人**延长本保函。我方承诺，根据我方在该 28 天期限内收到你方的书面要求和关于**履约证书**因**申请人**应负责的原因尚未签发、以及本保函尚未延长的书面说明，向你方支付该**保证金额**。

负责支付任何费用的一方_____［插入该方名称］。

本保函应受_____［插入管辖担保人的法律名称］法律管辖，并应遵守**国际商会** 2010 年修订出版的第 758 号公布的**即付保函统一规则**的规定。

Signed by: .... Signed by⁽¹⁾: ....
(signature) *(signature)*

....
(name) *(name)*

Date: ....

[1] *Whether one or more signatories for the bank are required will depend on the bank and/or applicable law.*

签字人：_____　　签字人[1]：_____
　　　　　　（签字）　　　　　　　　　　　　　（签字）

_____　　　　　_____
　　　（名称）　　　　　　　　　　　　（名称）

日期：_____

[1] 需要银行一人签字还是多人签字取决于银行和/或适用法律。

Annex G   EXAMPLE FORM OF PAYMENT GUARANTEE BY EMPLOYER

Guarantee No.: .... [insert guarantee reference number]

The Guarantor: ....
.... [insert name and address of place of issue, unless indicated in the letterhead]

Name of Contract/Contract No.: ....
.... [insert reference number or other information identifying the contract between the Applicant and the Beneficiary on which the guarantee is based]

The Beneficiary (the "Contractor"): ....
.... [insert name and address of the Beneficiary]

We have been informed that ....
.... [insert name and address of the Employer] (hereinafter called the "Applicant") is required to obtain a bank guarantee.

At the request of the Applicant, we .... [insert name of Guarantor] hereby irrevocably undertake to pay you, the Beneficiary/Contractor, any sum or sums not exceeding in total the amount of .... [insert in figures and words the maximum amount payable and the currency in which it is payable] upon receipt by us of your demand in writing and your written statement that:

(a)   in respect of a payment due under the Contract, the Applicant has failed to make payment in full by the date fourteen days after the expiry of the period specified in the Contract as that within which such payment should have been made, and

(b)   the amount(s) which the Applicant has failed to pay.

Any demand for payment must be accompanied by a copy of .... [insert list of documents evidencing entitlement to payment and the language in which these documents are to be submitted], in respect of which the Applicant has failed to make payment in full.

Any demand for payment must contain your signature(s) which must be authenticated by your bankers or by a notary public. The authenticated demand and statement must be received by us at the following office [address of office] on or before .... [insert the date six months after the expected expiry of the Defects Notification Period for the Works] when this guarantee shall expire.

The party liable for the payment of any charges: .... [insert the name of the party].

This guarantee shall be governed by the laws of .... [insert the law governing the guarantee] and shall be subject to the Uniform Rules for Demand Guarantees (URDG) 2010 Revision, ICC Publication No. 758.

Signed by: ....   Signed by[(1)]: ....
                (signature)   (signature)

....
                 (name)   (name)

Date: ....

[(1)] *Whether one or more signatories for the bank are required will depend on the bank and/or applicable law.*

附件G　　雇主支付保函范例格式

保函编号：＿＿＿＿＿＿＿＿＿＿＿＿＿＿＿＿＿＿＿＿＿［插入保函编号］

担保人：＿＿＿＿＿＿＿＿＿＿＿＿＿＿＿＿＿＿＿＿＿＿［插入签发地的名称和地址，信头纸有显示的除外］

合同名称／合同编号：＿＿＿＿＿＿＿＿＿＿＿＿＿＿＿＿［插入说明保函依据的**申请人**和**受益人**之间合同编号和其他信息］

受益人（"雇主"）：＿＿＿＿＿＿＿＿＿＿＿＿＿＿＿＿＿［插入**受益人**名称和地址］

我方已获知＿＿＿＿＿＿＿＿＿＿＿＿＿［插入**雇主**名称和地址］（以下称为"**申请人**"）被要求取得银行保函。

应**申请人**请求，我方＿＿＿＿＿＿＿＿＿＿＿［插入**担保人**名称］在此不可撤销地承诺，在我方收到你方书面要求和关于以下情况的书面说明后，向你方，**受益人**／**承包商**，支付全部总额不超过＿＿＿＿＿＿＿［插入最高可付金额的数字和文字以及可付货币］的任何一笔或几笔款额：

(a)　　**申请人**对于**合同**规定的应付的某笔款项，未能在**合同**规定的该笔款项应付清的期限期满后14天内，全部付清；（以及）

(b)　　**申请人**未能支付的款额。

任何关于付款的要求，都必须附一份关于**申请人**未能付清款项的＿＿＿＿＿＿＿＿［插入有权收款的证明文件和这些提交的文件所用语言清单］的抄件。

任何关于付款的要求，都必须有经你方银行或公证人确证的你方签字。经确证的要求和说明必须在＿＿＿＿＿＿＿＿＿＿［插入预定**工程缺陷通知期限**期满后6个月的日期］或其以前，由我方在下述办公地点［插入办公地点］收到，届时本保函将期满。

负责支付任何费用的一方＿＿＿＿＿＿＿＿＿＿＿＿［插入该方名称］。

本保函应受＿＿＿＿＿＿＿＿［插入管辖**担保人**的法律名称］法律管辖，并应遵守**国际商会**2010年修订出版的第758号公布的**即付保函统一规则**的规定。

签字人：＿＿＿＿＿＿＿＿＿＿＿＿＿　　签字人[1]：＿＿＿＿＿＿＿＿＿＿＿＿＿
　　　　　　　　（签字）　　　　　　　　　　　　　　　　　　（签字）

　　　　＿＿＿＿＿＿＿＿＿＿＿＿＿　　　　　　　＿＿＿＿＿＿＿＿＿＿＿＿＿
　　　　　　　　（名称）　　　　　　　　　　　　　　　　　　（名称）

日期：＿＿＿＿＿＿＿＿＿＿＿＿＿＿＿＿＿＿＿＿＿

[1] 需要银行一人签字还是多人签字取决于银行和／或适用法律。

施工合同条件

Conditions of Contract for
CONSTRUCTION

用于由雇主设计的建筑和工程
FOR BUILDING AND ENGINEERING WORKS DESIGNED BY THE EMPLOYER

投标函、中标函、合同协议书
和争端避免/裁决委员会协议书格式
**Forms of Letter of Tender, Letter of Acceptance,
Contract Agreement and DAAB Agreement**

通用条件
GENERAL CONDITIONS

专用条件编写指南和附件：
担保函格式
GUIDANCE FOR
THE PREPARATION OF
PARTICULAR CONDITIONS
AND ANNEXES: FORMS
OF SECURITIES

投标函、中标函、合同协议书
和争端避免/裁决委员会
协议书格式
FORMS OF LETTER OF
TENDER, LETTER OF
ACCEPTANCE, CONTRACT
AGREEMENT AND DAAB
AGREEMENT

LETTER OF TENDER

NAME OF CONTRACT:

TO:

We have examined the Conditions of Contract, Specification, Drawings, Schedules including the Bill of Quantities, the Contract Data and Addenda Nos ....
for the above-named Contract and the words and expressions used herein shall have the meanings assigned to them in the Conditions of Contract. We offer to execute and complete the Works and remedy any defects therein, in conformity with this Tender which includes all these documents, for the sum of

....

[*currency and amount in figures*]

....

[*currency and amount in words*]

or such other amount as may be determined in accordance with the Contract.

We agree to abide by this Tender until .... [*date*] and it shall remain binding upon us and may be accepted at any time before that date.

If this offer is accepted, we will provide the specified Performance Security, commence the Works as soon as is reasonably practicable after the Commencement Date, and complete the Works in accordance with the above-named documents within the Time for Completion.

Unless and until a Contract Agreement is prepared and executed this Letter of Tender, together with your written acceptance thereof, shall constitute a binding contract between us.

We understand that you are not bound to accept the lowest or any tender you may receive.

Signature .... in the capacity of
....

duly authorised to sign tenders for and on behalf of ....

Address: ....

Date: ....

投标函

合同名称：

致：

我方已研究了为实施上述**工程**的**合同条件**、**规范要求**、**图纸**以及包括**工程量清单**、**合同数据**及第_____号**补充文件**在内的**资料表**。对于上述**合同**，本合同中使用的措辞和表述应具有**合同条件**中赋予它们的含义。我方愿按照本**投标书**，包括所有这些文件，实施和完成**工程**并修补其中任何缺陷。以

_____ 的总额
[*货币和用数字表示的金额*]

_____
[*货币和用文字表示的金额*]，

或按照**合同**可能确定的此项其他总额的报价。

我方同意遵守本**投标书**直至_____[*日期*]，在该日期前，本**投标书**对我方一直具有约束力，随时可接受中标。

如果我方中标，我方将提供规定的**履约担保**，将在**开工日期**后，尽早开工，并在**竣工时间**内，按照上述文件完成**工程**。

除非并直到制定和实施**合同协议书**，本**投标函**以及你方书面中标通知，应构成你我双方间具有约束力的合同。

我方理解你方没有必须接受你方可能收到的最低标或任何投标的义务。

签字_____职务_____

正式授权签署投标书代表_____

地址：_____

日期：_____

LETTER OF ACCEPTANCE *

NAME OF CONTRACT: ....

CONTRACT NUMBER: ....

TO: ....

Date: ....

Your Reference: ....

Our Reference: ....

We thank you for your Tender dated .... for the execution and completion of the Works comprising the above-named Contract and remedying of defects therein, all in conformity with the terms and conditions contained in the Contract.

We have pleasure in accepting your Tender for the Accepted Contract Amount of:

....
[currency and amount in figures]

....
[currency and amount in words]

In consideration of you properly and truly performing the Contract, we agree to pay you the Accepted Contract Amount or such other sums to which you may become entitled under the terms of the Contract, at such times and as prescribed by the Contract.

We acknowledge that this Letter of Acceptance creates a binding Contract between us, and we undertake to fulfil all our obligations and duties in accordance with the terms of this Contract.

Signed by: .... (signature)

For and behalf of: ....

Date: ....

- Memoranda (if any) to be annexed [see Sub-Clause 1.1.51]

中标函*

合同名称： ....

合同编号： ....

致： ....

日期： ....

贵方函件编号： ....

我方函件编号： ....

我方感谢贵方于 ....（*日期*）提交的**投标书**，该**投标书**旨在按照**合同**条款和条件实施和完成上述**合同**所包含的**工程**，并修补其中的缺陷。

我方高兴地接受贵方的**投标书**，**中标合同金额**为：

_____
[*货币及用数字表示的金额*]

_____
[*货币及用文字表示的金额*]

考虑到贵方正确而真实地履行**合同**，我方同意按**合同**规定的时间向贵方支付**中标合同金额**，或合同条件规定的贵方可能有权获得的其他此类金额。

我方承认，本**中标函**在你我双方之间建立了具有约束力的**合同**，并且我方承诺根据本合同的条款履行我方的所有义务和责任。

签字人： ....（签字）

授权代表： ....

日期： ....

---

* 附备忘录（如果有）[参见*第 1.1.50 项*]。⊖

⊖ 此处中文按照勘误表改正后的英文翻译。——译者注

CONTRACT AGREEMENT

This Agreement made the _____ day of _____
Between _____ of _____ (hereinafter called "the Employer") of the one part,
and _____ of _____ (hereinafter called "the Contractor") of the other part

Whereas the Employer desires that the Works known as _____ [name and number of the Contract] should be executed by the Contractor, and has accepted a Tender by the Contractor for the execution and completion of these Works and the remedying of any defects therein.

The Employer and the Contractor agree as follows:

1. In this Agreement words and expressions shall have the same meanings as are respectively assigned to them in the Conditions of Contract hereinafter referred to.

2. The following documents shall be deemed to form and be read and construed as part of this Agreement:

    (a) The Letter of Acceptance dated

    (b) The Letter of Tender dated

    (c) The Addenda Nos _____

    (d) The Conditions of Contract

    (e) The Specification

    (f) The Drawings

    (g) The Schedules and

    (h) The JV Undertaking.*

    *[if the Contractor constitutes an unincorporated JV, otherwise delete]

3. In consideration of the payments to be made by the Employer to the Contractor as hereinafter mentioned, the Contractor hereby covenants with the Employer to execute and complete the Works and remedy any defects therein, in conformity with the provisions of the Contract.

4. The Employer hereby covenants to pay the Contractor, in consideration of the execution and completion of the Works and the remedying of defects therein, the Contract Price at the times and in the manner prescribed by the Contract.

In Witness whereof the parties hereto have caused this Agreement to be executed the day and year first before written in accordance with their respective laws.

SIGNED by: _____   SIGNED by: _____
for and on behalf of the Employer in the   for and on behalf of the Contractor in the
presence of   presence of
Witness: _____   Witness: _____
Name: _____   Name: _____
Address: _____   Address: _____
Date: _____   Date: _____

合同协议书

本协议书于＿＿＿＿＿＿＿＿＿＿＿＿＿＿＿＿＿＿＿＿＿＿年＿＿＿月＿＿＿日
由＿＿＿＿＿＿＿＿＿＿的＿＿＿＿＿＿＿＿＿＿＿（以下简称"**雇主**")为一方，
和＿＿＿＿＿＿＿＿＿＿的＿＿＿＿＿＿＿＿＿＿＿（以下简称"**承包商**")为另一方协商签订

鉴于**雇主**愿将名称为＿＿＿＿＿＿＿＿＿［合同规定的名称和数量］的**工程**交由**承包商**实施，并已接受了**承包商**提交的关于实施和完成这些**工程**及修补其中任何缺陷的**投标书**，

雇主和**承包商**达成协议如下：

1. 本**协议书**中的词语和措辞的含义应与下文提到的**合同条件**中分别赋予它们的含义相同。

2. 下列文件应被视为本**协议书**的组成部分，并应作为其一部分阅读和解释。

  （a） 中标函日期

  （b） 投标函日期

  （c） 补充文件编号＿＿＿＿

  （d） 合同条件

  （e） 规范要求

  （f） 图纸

  （g） 资料表；（以及）

  （h） 联营体承诺书 *

    *［除非**承包商**组成非法人**联营体**，否则删除本条］

3. 鉴于**雇主**将按下文所述付给**承包商**各项款项，**承包商**特此与**雇主**签约，保证遵守合同的各项规定，实施和完成本**工程**及修补其任何缺陷。

4. 鉴于**承包商**将承担上述**工程**的实施、竣工及修补其任何缺陷，**雇主**特此立约，保证按合同规定的时间和方式，向**承包商**支付**合同价格**。

本**协议书**由双方根据各自法律签字之日起生效，特立此据。

签字人：＿＿＿＿＿＿＿＿＿＿＿   签字人：＿＿＿＿＿＿＿＿＿＿＿
在下列证人在场下，代表**雇主**签字 在下列证人在场下，代表**承包商**签字

见证人：＿＿＿＿＿＿＿＿＿＿＿   见证人：＿＿＿＿＿＿＿＿＿＿＿
名称：＿＿＿＿＿＿＿＿＿＿＿＿   名称：＿＿＿＿＿＿＿＿＿＿＿＿
地址：＿＿＿＿＿＿＿＿＿＿＿＿   地址：＿＿＿＿＿＿＿＿＿＿＿＿
日期：＿＿＿＿＿＿＿＿＿＿＿＿   日期：＿＿＿＿＿＿＿＿＿＿＿＿

DISPUTE AVOIDANCE/ADJUDICATION BOARD AGREEMENT

[All italicised text and any text within square brackets (except sub-clause headings) in this form of agreement is for use in preparing the form and should be deleted from the final product].

Name and details of the Contract ....

This Agreement made the .... day of .... *[month]*, ... *[year]*, between

Name and contact details of the Employer .... *(name)*
.... *(address)*
.... *(telephone)*
.... *(email / other contact details)*;

Name and contact details of the Contractor .... *(name)*
.... *(address)*
.... *(telephone)*
.... *(email / other contact details)*;

and

Name and contact details of the DAAB Member .... *(name)*
.... *(address)*
.... *(telephone)*
.... *(email / other contact details)*;

(" **DAA Agreement** ")

Whereas:

A. the Employer and the Contractor have entered (or intend to enter) into the Contract;

B. under the Contract, the "**DAAB**" or "**Dispute Avoidance/Adjudication Board**" means the sole member or three members (as stated in the Contract Data of the Contract) so named in the Contract, or appointed under Sub-Clause 21.1 *[Constitution of the DAAB]* or Sub-Clause 21.2 *[Failure to Appoint DAAB Members]* of the Conditions of Contract;

C. the Employer and the Contractor desire jointly to appoint the above-named DAAB Member to act on the DAAB as:

(a) the sole member of the DAAB, and where this is the case, all references to the "Other Members" do not apply; or

(b) one of three members/chairman [delete the one which is not applicable] of the DAAB and, where this is the case, the other two persons are:

.... *(name)* .... *(address)* .... *(telephone)* .... *(email / other contact details)*	.... *(name)* .... *(address)* .... *(telephone)* .... *(email / other contact details)*

the "**Other Members**"; and

争端避免/裁决委员会协议书[一]

[*此协议书格式中的所有斜体文本和方括号内的任何文本（条款的标题除外）均用于编写格式，应从最终文本中删除*]。

合同名称和内容 ....

本协议书于 .... 年 .... 月 .... 日由**雇主**、**承包商**和**争端避免/裁决委员会**成员订立。

雇主名称和联系方式 ....（名称）
....（地址）
....（电话）
....（邮箱/其他联系方式）；

承包商名称和联系方式 ....（名称）
....（地址）
....（电话）
....（邮箱/其他联系方式）；

以及

争端避免/裁决委员会成员名称和联系方式 ....（名称）
....（地址）
....（电话）
....（邮箱/其他联系方式）；

("DAAB 协议书")[一]

鉴于：

A. **雇主**和**承包商**已经（或打算）签订**合同**；

B. 按照**合同**规定，"DAAB"或"**争端避免/裁决委员会**"系指**合同**中这样命名的，或按照合同条件第 21.1 款 [*争端避免/裁决委员会的组成*] 或第 21.2 款 [*未能任命争端避免/裁决委员会成员*] 的规定任命的（如合同的**合同数据**中所述的）唯一成员或三名成员；

C. **雇主**和**承包商**希望共同任命上述**争端避免/裁决委员会**成员作为**争端避免/裁决委员会**代表其行事：

(a) **争端避免/裁决委员会**的唯一成员，在这种情况下，对"其他成员"的所有引用均不适用；（或）

(b) **争端避免/裁决委员会**三名成员/主席中的一位 [*删除不适用的成员*]，在这种情况下，另外两名成员是：[一]

....（名称） ....（地址） ....（电话） ....（邮箱/ 其他联系方式）	....（名称） ....（地址） ....（电话） ....（邮箱/ 其他联系方式）

"其他成员"；（以及）

[一] 此处中文按照勘误表改正后的英文翻译。——译者注

D.   the DAAB Member accepts this appointment.

The Employer, Contractor and DAA Member jointly agree as follows:

1.   The conditions of this DAA Agreement comprise:

   (a)   Clause 21 [*Disputes and Arbitration*] of the Conditions of Contract, and any other provisions of the Contract that are applicable to the DAAB's Activities; and

   (b)   the "General Conditions of Dispute Avoidance/Adjudication Board Agreement", which is appended to the General Conditions of the "Conditions of Contract for Construction" Second Edition 2017 published by FIDIC ("GCs"), as amended and/or added to by the following provisions.

2.   [Details of amendments to the GCs, if any. For example:
     In the procedural rules annexed to the GCs, Rule _ is deleted and replaced by: " ... "]

3.   The DAAB Member shall be paid in accordance with Clause 9 of the GCs. The currency of payment shall be ....

In respect of Sub-Clauses 9.1 and 9.2 of the GCs, the amounts of the DAAB Member's monthly fee and daily fee shall be:
monthly fee .... per month, and
daily fee of .... per day
(or as otherwise set under Sub-Clause 9.3 of the GCs).

4.   In consideration of the above fees, and other payments to be made to the DAAB Member in accordance with the GCs, the DAAB Member undertakes to act as DAAB Member in accordance with the terms of this DAA Agreement.

5.   The Employer and the Contractor shall be jointly and severally liable for the DAAB Member's fees and other payments to be made to the DAAB Member in accordance with the GCs.

6.   This DAA Agreement shall be governed by the law of .... (if not stated, the law that governs the Contract under Sub-Clause 1.4 of the Conditions of Contract).

SIGNED by:	SIGNED by:	SIGNED by: ....
Print name:	Print name:	the DAAB Member
Title:	Title:	Title: ....
for and on behalf of the Employer	for and on behalf of the Contractor	
in the presence of	in the presence of	in the presence of
Witness:	Witness:	Witness: ....
Name:	Name:	Name: ....
Address:	Address:	Address: ....
Date:	Date:	Date: ....

D. 争端避免/裁决委员会成员接受该任命。

雇主、承包商和争端避免/裁决委员会成员共同达成协议如下：

1. 本争端避免/裁决委员会协议书的条件包括：⊖

    （a） 合同条件第 21 条 [*争端和仲裁*]，以及适用于**争端避免/裁决委员会**活动的任何其他合同条款；（以及）

    （b） 菲迪克（FIDIC）发布的 2017 年第 2 版《施工合同条件》通用条件（GCS）所附的《**争端避免/裁决委员会**协议书一般条件》"一般条件"，以及对下列条款的修订和/或增补。

2. [对**通用条件**的修订的细节（如果有）。例如：
    从附在**通用条件**后的程序规则中，删去**规则** _____，并代之以："…"]

3. **争端避免/裁决委员会成员**应根据**通用条件**的第 9 条规定获得付款。支付货币为 _____。

根据**通用条件**第 9.1 款和第 9.2 款，**争端避免/裁决委员会成员**的月酬金和日酬金应为：

月酬金 _____ 每月；（以及）
日酬金 _____ 每天
（或根据**通用条件**第 9.3 款设定）。

4. 基于上述酬金，并根据**通用条件**应向**争端避免/裁决委员会成员**支付的其他款项，**争端避免/裁决委员会成员**承诺，根据本**争端避免/裁决委员会**协议书的条款担任**争端避免/裁决委员会成员**职务。⊖

5. **雇主和承包商**应根据**通用条件**的规定，共同并各自负责向**争端避免/裁决委员会成员**支付**争端避免/裁决委员会成员**酬金和其他款项。

6. 本**争端避免/裁决委员会协议书**应受 _____ 法律管辖（如未规定，则为**合同条件**第 1.4 款规定的管辖**合同**的法律）。⊖

| 签字人：_____ | 签字人：_____ | 签字人：_____ |

打印名称：_____   打印名称：_____   **争端避免/裁决委员会成员**
职务：_____   职务：_____   职务：_____
代表雇主签字   代表承包商签字

在场人员   在场人员   在场人员
见证人：_____   见证人：_____   见证人：_____
名称：_____   名称：_____   名称：_____
地址：_____   地址：_____   地址：_____

日期：_____   日期：_____   日期：_____

⊖ 此处中文按照勘误表改正后的英文翻译。——译者注

ERRATA to the FIDIC Conditions of Contract for Construction Second Edition 2017

The following significant errata **are not** included in the content of the Second Edition of the Conditions of Contract for Construction. Several minor typographical errors and layout irregularities **have been corrected and applied** to the content.

GENERAL CONDITIONS

Page 12	Sub-Clause 1.1.10:	On the fourth line, delete "the Contractor's Proposal, ".
Page 24	Sub-Claue 1.1.77:	On the second line after "Payment Certificate under", add "Sub-Clause 14.2.1 [*Advance Payment Guarantee*] (if applicable), ".
Page 24	Sub-Clause 1.1.81:	On the first line, delete "the Contractor's Proposal,".
Page 64	Sub-Clause 4.6:	On the second-last line of the first paragraph before "Contractor's", add "of the".
Page 66	Sub-Clause 4.7.3:	In the second bullet-point of sub-paragraph (b) - before "if the items of reference", add "when examining the items of reference within the period stated in sub-paragraph (a) of Sub-Clause 4.7.2," - on the second and third lines, delete "and the Contractor's Notice is given after the period stated in sub-paragraph (a) of Sub-Clause 4.7.2".
Page 80	Sub-Clause 4.22	On the third line of the second paragraph before "4.17", add "Sub-Clause".
Page 84	Sub-Clause 5.2.2:	In sub-paragraph (a) on the first line before "Subcontractor", add "nominated".
Page 144	Sub-Clause 14.2.1:	On the fifth and sixth lines of the first paragraph, replace "based on the sample form included in the Tender documents" with "in the form annexed to the Particular Conditions".
Page 160	Sub-Clause 14.12:	On the seventh line of the first paragraph, replace "Sub-Clause 21.6 [*Arbitration*]" with "Clause 21 [*Disputes and Arbitration*]".
Page 176	Sub-Clause 17.1:	On the fourth and fifth lines of the first paragraph, replace "Date of Completion of the Works" with "issue of the Taking-Over Certificate for the Works".
Page 180	Sub-Clause 17.3:	On the first line of the second paragraph, replace "notice" with "a Notice".
Page 190	Sub-Clause 19.2.1:	On the last line of the second paragraph, delete "Clause 12 [*Tests after Completion*]."

APPENDIX – GENERAL CONDITIONS OF DISPUTE AVOIDANCE/ADJUDICATION AGREEMENT

Page 214	Title	Replace "General Conditions of Dispute Avoidance/Adjudication Agreement" with "General Conditions of DAAB Agreement".

《施工合同条件》2017 年第 2 版勘误表[⊖]

第 2 版《施工合同条件》的内容中不包括以下重要修改。一些细微的排版错误和不规范版式已经纠正并应用于内容。

通用条件

第 12 页	第 1.1.10 项：	在第四行，删除"the Contractor's Proposal,"。
第 24 页	第 1.1.77 项：	在第二行"Payment Certificate under"后，添加"Sub-Clause 14.2.1 [*Advance Payment Guarantee*] (if applicable),"。
第 24 页	第 1.1.81 项：	在第一行，删除"the Contractor's Proposal,"。
第 64 页	第 4.6 款：	在第一段倒数第二行的"Contractor's"前，添加"of the"。
第 66 页	第 4.7.3 项：	在（b）段的第二点

- 在"if the items of reference"前，添加"when examining the items of reference within the period stated in sub-paragraph (a) of Sub-Clause 4.7.2,"。

- 在第二行和第三行，删除"and the Contractor's Notice is given after the period stated in sub-paragraph (a) of Sub-Clause 4.7.2"。

第 80 页	第 4.22 款：	在第二段第三行的"4.17"前，添加"Sub-Clause"。
第 84 页	第 5.2.2 项：	在第一行（a）段的"Subcontractor"前，添加"nominated"。
第 144 页	第 14.2.1 项：	在第一段第五行和第六行，以"in the form annexed to the Particular Conditions"替换"based on the sample form included in the Tender documents"。
第 160 页	第 14.12 款：	在第一段第七行，以"Clause 21 [*Disputes and Arbitration*]"替换"Sub-Clause 21.6 [*Arbitration*]"。
第 176 页	第 17.1 款：	在第一段的第四行和第五行，以"issue of the Taking-Over Certificate for the Works"替换"Date of Completion of the Works"。
第 180 页	第 17.3 款：	在第一行，以"a Notice"替换"notice"。
第 190 页	第 19.2.1 项：	在第二段最后一行，删除"Clause 12 [*Tests after Completion*]"。

附录——争端避免 / 裁决协议书一般条件

第 214 页	标题	以"General Conditions of DAAB Agreement"替换"General Conditions of Dispute Avoidance/Adjudication Agreement"。

[⊖] 中文翻译文本已改正。本勘误表只是应 FIDIC 的要求而翻译，仅供参考。——译者注

Page 214	Sub-Clause 1.2:	On both the first and third lines, replace "DAA Agreement" with "DAAB Agreement".
Page 214	Sub-Clause 1.3:	- on the first line, replace " "Dispute Avoidance/Adjudication Agreement" or "DAA Agreement" means" " with: " "DAAB Agreement" is as defined under the Contract and is". - on the first line of sub-paragraph (c), replace "DAA Agreement" with "DAAB Agreement" - in sub-paragraph (c)(ii), replace "chairman" with "chairperson".
Pages 214 to 230	Sub-Clause 1.7 to Clause 12	Replace all instances of "DAA Agreement" with "DAAB Agreement"
Page 228	Sub-Clause 11.1	On the second line, delete the text: ", or in the case of a three-member DAAB the Other Members jointly, ".

ANNEX – DAAB PROCEDURAL RULES

Page 234	Rule 4.2	On the fourth line, replace "chairman" with "chairperson".
Page 240	Rule 8.3	On the sixth line, replace "chairman" with "chairperson".

GUIDANCE FOR THE PREPARATION OF PARTICULAR CONDITIONS

Page 270	INTRODUCTION	Insert in the beginning of GP5: "Unless there is a conflict with the governing law of the Contract,…" allowing GP5 to read: "**GP5**: Unless there is a conflict with the governing law of the Contract, all formal disputes must be referred to a Dispute Avoidance/Adjudication Board (or a Dispute Adjudication Board, if applicable) for a provisionally binding decision as a condition precedent to arbitration."
Pages 284 to 350	Guidance for Sub-Clauses 1.6, 1.14, 4.2, 14.2, 14.9, 14.15 and 21.1	Replace all instances of "Sample Forms" with "Forms".
Page 324	Guidance for Sub-Clause 9.1	On the second line of the first paragraph, replace "Contractor's Proposal" with "Tender".

Form of LETTER OF ACCEPTANCE

Page 394	Footnote	Replace "1.1.51" with "1.1.50".

Form of DISPUTE AVOIDANCE/ADJUDICATION AGREEMENT

Pages 398 and 400		Replace all instances of "DAA Agreement" with "DAAB Agreement"
Page 398	Recital C	On the first line of sub-paragraph (b), replace "chairman" with "chairperson".

| 第 214 页 | 第 1.2 款： | 在第一行和第三行，以"DAAB Agreement"替换"DAA Agreement"。 |

第 214 页	第 1.3 款：	- 在第一行，以"'DAAB Agreement' is as defined under the Contract and is"替换"'Dispute Avoidance/Adjudication Agreement' or 'DAA Agreement' means"。
		- 在（c）段的第一行，以"DAAB Agreement"替换"DAA Agreement"。
		- 在（c）(ii) 段，以"chairperson"替换"chairman"。

| 第 214 页
至 230 页 | 第 1.7 款
至第 12 条 | 以"DAAB Agreement"替换所有情况的"DAA Agreement"。 |

| 第 228 页 | 第 11.1 款 | 在第二行，删除：", or in the case of a three-member DAAB the Other Members jointly,"。 |

附件——争端避免/裁决委员会程序规则

| 第 234 页 | 规则 4.2 | 在第四行，以"chairperson"替换"chairman"。 |
| 第 240 页 | 规则 8.3 | 在第六行，以"chairperson"替换"chairman"。 |

专用条件编写指南

| 第 270 页 | 引言 | 在 GP5 开头插入："Unless there is a conflict with the governing law of the Contract, ……"
将 GP5 改为： |

"**GP5**: Unless there is a conflict with the governing law of the Contract, all formal disputes must be referred to a Dispute Avoidance/Adjudication Board (or a Dispute Adjudication Board, if applicable) for a provisionally binding decision as a condition precedent to arbitration."

以"Forms"替换所有情况的"Sample Forms"。

| 第 284 页
至第 350 页 | 第 1.6 款、第 1.14 款、
第 4.2 款、第 14.2 款、
第 14.9 款、第 14.15 款
和第 21.1 款 | |

| 第 324 页 | 第 9.1 款指南 | 在第一段的第二行，以"Tender"替换"Contractor's Proposal"。 |

中标函格式

| 第 394 页 | 脚注 | 以"1.1.50"替换"1.1.51"。 |

争端避免/裁决协议书格式

| 第 398 页和第 400 页 | | 以"DAAB Agreement"替换所有情况的"DAA Agreement"。 |
| 第 398 页 | 序言 C | 在第（b）段的第一行，以"chairperson"替换"chairman"。 |